Chemical Alternatives Assessments

ISSUES IN ENVIRONMENTAL SCIENCE AND TECHNOLOGY

How to obtain future titles on publication:

A subscription is available for this series. This will bring delivery of each new volume immediately on publication and also provide you with online access to each title via the Internet. For further information visit http://www.rsc.org/issues or write to the address below.

For further information please contact:
Sales and Customer Care, Royal Society of Chemistry, Thomas Graham House, Science Park, Milton Road, Cambridge, CB4 0WF, UK
Telephone: +44 (0)1223 432360, Fax: +44 (0)1223 426017, Email: booksales@rsc.org
Visit our website at www.rsc.org/books

ISSUES IN ENVIRONMENTAL SCIENCE AND TECHNOLOGY

EDITORS: R.E. HESTER AND R.M. HARRISON

36
Chemical Alternatives Assessments

RSC Publishing

Issues in Environmental Science and Technology No 36

ISBN: 978-1-84973-605-3
ISSN: 1350-7583

A catalogue record for this book is available from the British Library

© The Royal Society of Chemistry 2013

Published by The Royal Society of Chemistry,
Thomas Graham House, Science Park, Milton Road,
Cambridge CB4 0WF, UK

Registered Charity Number 207890

For further information see our web site at www.rsc.org

Printed in the United Kingdom by CPI Group (UK) Ltd, Croydon, CR0 4YY, UK

Preface

Chemicals are an essential part of everyday living which the general public, to a large degree, takes for granted. Public attitudes, heavily influenced by the media, show little appreciation of the beneficial uses of chemicals, but much concern over those uses which may be deleterious to public health or the environment. Such concern has quite rightly led governments in the developed world to establish programmes such as REACH in Europe and the HPV Challenge Program in the United States which provide a hazard screening of chemicals and identify those representing the highest level of potential risk to health and the environment. Such screening is typically based on the three criteria of persistence, bio-accumulation and toxicity (PBT). Although even a relatively simplistic assessment of this kind requires a great deal of test data and is effort intensive, it is primarily designed to identify substances in the high hazard category and it does not provide a mechanism for ranking chemicals possessing some but not all of the adverse characteristics. This leads to a risk that in eliminating the use of a substance of concern, an alternative compound may be increased in use despite the fact that it may present high, albeit different, risks to health and the environment. Clearly, what is needed is a means of assessing and ranking chemicals, and that is the topic of this volume.

The first chapter by Margaret Whittaker and Lauren Heine is an overview of chemical alternatives assessment and the tools available to conduct that assessment. The chapter includes a very useful summary table of many of the more important tools available for comparative chemical hazard assessment. In the following chapter, Adrian Beard addresses the very topical area of flame retardants, which are widely used in consumer electrical goods and other products and many of which present a serious hazard to the environment. It describes the formation of an industry group to develop and promote more environmentally compatible products, mainly non-halogenated alternatives.

The following two chapters are more philosophical in nature. Jay Bolus, Rachel Platin and Christoph Semisch describe the concept of cradle to cradle product evaluation and certification. For those familiar with the cradle to grave concept, cradle to cradle may seem rather strange but the chapter explains how

Issues in Environmental Science and Technology, 36
Chemical Alternatives Assessments
Edited by R.E. Hester and R.M. Harrison
© The Royal Society of Chemistry 2013
Published by the Royal Society of Chemistry, www.rsc.org

this title embodies the concept of how waste products should always be regarded as starting materials for another product or process, leading to maximal resource efficiency and eliminating waste. The following chapter deals with alternative assessments in the building industry. Brandon Zang, Raefer Wallis and Ryan Dick describe the operation of a system within China for assessment of building products which is proving highly successful in influencing the use of more appropriate substances. In the following chapter, Gregory Morose and Monica Becker, using plasticizers for wire and cable as a case study, demonstrate how collaboration between companies and universities to evaluate safer alternatives can provide valuable insights.

One of the more important protocols for chemical hazard assessment in the context of development of alternatives is the GreenScreen for safer chemicals. The working of the protocol is described in some depth by Lauren Heine and Shari Franjevic and in the following chapter, Helen Holder and colleagues describe the application of the GreenScreen for safer chemicals by the Hewlett-Packard Company. Another company using GreenScreen and other approaches is DSM, and Thomas Wegman and co-authors provide a picture of how the ethos of their company has developed towards creating products with an improved environmental performance and how tools such as GreenScreen have proved valuable in that transition.

Clive Davies and colleagues from the US Environmental Protection Agency explain how the Agency's Design for the Environment chemical alternatives assessments are conducted and how they are working towards harmonised methodologies initially for use in the United States and subsequently in other countries. Operating with Europe, ChemSec is an environmental NGO which has been influencing the phase-out of hazardous chemicals with key stakeholders such as policy-makers, progressive companies and financial investors. Jerker Ligthart shows how chemicals management has been promoted through dialogue and also through specific tools designed to make the necessary assessments. Finally, Joel Tickner and colleagues of the University of Massachusetts, Lowell, explore the justification and rationale for requiring alternatives assessments that provide informed substitution and they provide an historical and current overview of chemical restriction and alternatives assessment policies. They set out valuable guidelines for the future development of chemical alternatives assessment.

The editors are grateful to Margaret Whittaker who provided invaluable assistance in identifying authors for this volume. We are delighted that, with her advice, we were able to attract authors from industry, government and non-governmental organizations so as to provide a balanced picture of this highly topical subject which we believe will be of value to scientists involved in chemicals manufacture and assessment, to policy-makers and to students in a wide range of courses concerned with chemicals and/or the environment.

Ronald E. Hester
Roy M. Harrison

Contents

Issues in Environmental Science and Technology, 36
Chemical Alternatives Assessments
Edited by R.E. Hester and R.M. Harrison
© The Royal Society of Chemistry 2013
Published by the Royal Society of Chemistry, www.rsc.org

China's Implementation of Alternatives Assessment in the Building Industry: GIGA
Brandon Zang, Raefer K. Wallis and Ryan D. Dick

A Collaborative Industry and University Alternative Assessment of Plasticizers for Wire and Cable
Gregory Morose and Monica Becker

Chemical Hazard Assessment and the GreenScreen™
for Safer Chemicals   **129**
Lauren G. Heine and Shari A. Franjevic

Hewlett-Packard's Use of the GreenScreen™ for Safer Chemicals   **157**
H. A. Holder, P. H. Mazurkiewicz, C. D. Robertson and C. A. Wray

DSM's Sustainability Journey Towards a Proactive Ingredient Policy for Gaining Effectiveness in the Design of Better Products   177

Thomas A. J. Wegman, Fredric Petit, Annette Wilschut, Theo Jongeling and Gaelle M. Nicolle

US Environmental Protection Agency's Design for the Environment (DfE) Alternatives Assessment Program   **198**

Clive Davies, Melanie Adams, Emily Connor, Elizabeth Sommer,
Caroline Baier-Anderson, Emma Lavoie, Laura Romano and
David DiFiore

Editors

Ronald E. Hester, BSc, DSc (London), PhD (Cornell), FRSC, CChem

Ronald E. Hester is now Emeritus Professor of Chemistry in the University of York. He was for short periods a research fellow in Cambridge and an assistant professor at Cornell before being appointed to a lectureship in chemistry in York in 1965. He was a full professor in York from 1983 to 2001. His more than 300 publications are mainly in the area of vibrational spectroscopy, latterly focusing on time-resolved studies of photoreaction intermediates and on biomolecular systems in solution. He is active in environmental chemistry and is a founder member and former chairman of the Environment Group of the Royal Society of Chemistry and editor of 'Industry and the Environment in Perspective' (RSC, 1983) and 'Understanding Our Environment' (RSC, 1986). As a member of the Council of the UK Science and Engineering Research Council and several of its sub-committees, panels and boards, he has been heavily involved in national science policy and administration. He was, from 1991 to 1993, a member of the UK Department of the Environment Advisory Committee on Hazardous Substances and from 1995 to 2000 was a member of the Publications and Information Board of the Royal Society of Chemistry.

Roy M. Harrison, BSc, PhD, DSc (Birmingham), FRSC, CChem, FRMetS, Hon MFPH, Hon FFOM, Hon MCIEH

Roy M. Harrison is Queen Elizabeth II Birmingham Centenary Professor of Environmental Health in the University of Birmingham. He was previously Lecturer in Environmental Sciences at the University of Lancaster and Reader and Director of the Institute of Aerosol Science at the University of Essex. His more than 400 publications are mainly in the field of environmental chemistry, although his current work includes studies of human health impacts of atmospheric pollutants as well as research into the chemistry of pollution phenomena. He is a past Chairman of the Environment Group of the Royal Society of Chemistry for whom he has edited 'Pollution: Causes, Effects and Control' (RSC, 1983;

Fourth Edition, 2001) and 'Understanding our Environment: An Introduction to Environmental Chemistry and Pollution' (RSC, Third Edition, 1999). He has a close interest in scientific and policy aspects of air pollution, having been Chairman of the Department of Environment Quality of Urban Air Review Group and the DETR Atmospheric Particles Expert Group. He is currently a member of the DEFRA Air Quality Expert Group, the Department of Health Committee on the Medical Effects of Air Pollutants, and Committee on Toxicity.

List of Contributors

1. *Chemicals Alternatives Assessment (CAA): Tools for Selecting Less Hazardous Chemicals*

Margaret H. Whittaker (corresponding author), ToxServices LLC, 1367 Connecticut Avenue, N.W., Suite 300, Washington, DC 20036, USA, E-mail: mwhittaker@toxservices.com

Lauren G. Heine, Lauren Heine Group LLC and Clean Production Action, 2986 Foster Avenue, Juneau, AK 99801, USA, E-mail: lauren@lheinegroup.com

2. *European Initiatives for Selecting Sustainable Flame Retardants*

Adrian Beard (corresponding author), Clariant Corporation, Chemiepark, D0354 Hueth-Knapsack, Germany, Email: adrian.beard@clariant.com

3. *MBDC Cradle to Cradle® Product Evaluation and Certification Program*

Jay J. Bolus (corresponding author), McDonough Braungart Design Chemistry (MBDC), 1001 E. Market Street, Suite 201, Charlottesville, VA 22901, USA, Email: jay.bolus@mbdc.com

Rachel Platin, EPEA Internationale Umweltforschung GmbH, Trostbrücke 4, D20457 Hamburg, Germany

Christoph Semisch, EPEA Internationale Umweltforschung GmbH, Trostbrücke 4, D20457 Hamburg, Germany

4. *China's Implementation of Alternatives Assessment in the Building Industry: GIGA*

Raefer K. Wallis (corresponding author), GIGA 循, 135 Guangyuan Lu, 2F, Shanghai 200030, China, Email: rkw@gigabase.org

Brandon Zang, GIGA 循, 135 Guangyuan Lu, 2F, Shanghai 200030, China, Email: brandon@gigabase.org

Ryan D. Dick, GIGA 循, 135 Guangyuan Lu, 2F, Shanghai 200030, China, Email: rdd@gigabase.org

5. *A Collaborative Industry and University Alternative Assessment of Plasticizers for Wire and Cable*

Gregory Morose (corresponding author), Toxics Use Reduction Institute, University of Massachusetts Lowell, 600 Suffolk Street, Lowell, MA 01854, USA, Email: gregory_morose@uml.edu

Monica Becker, Monica Becker & Associates Sustainability Consulting, 278 Warrington Drive, Suite 2, Rochester, NY 14618, USA

6. *Chemical Hazard Assessment and the GreenScreen™ for Safer Chemicals*

Lauren G. Heine (corresponding author), Lauren Heine Group LLC and Clean Production Action, 2986 Foster Avenue, Juneau, AK 99801, USA, E-mail: lauren@lheinegroup.com
Shari A. Franjevic, Transform to Green, P.O. Box 1355, Duvall, WA 98019, USA

7. *Hewlett-Packard's Use of the GreenScreen™ for Safer Chemicals*

H. A. Holder (corresponding author), Hewlett-Packard Development Company, 1501 Page Mill Road, M/S 1222, Palo Alto, CA 94304, USA, Email: helen. holder@hp.com
P. H. Mazurkiewicz, Hewlett-Packard Development Company, 3404 East Harmony Road., Mail Code 34, Fort Collins, CO 80528, USA
C. D. Robertson, Hewlett-Packard Development Company, 11311 Chinden BlvdBoulevard., Boise, ID 83714, USA
C. A. Wray, Hewlett-Packard Development Company, 3404 East Harmony Road., Mail Code 34, Fort Collins, CO 80528, USA

8. *DSM's Sustainability Journey Towards a Proactive Ingredient Policy for Gaining Effectiveness in the Design of Better Products*

Gaelle M. Nicolle (corresponding author), DSM Head Office, Koninklijke DSM N.V., Het Overloon 1, 6401 JH Heerlen, The Netherlands, Email: Gaelle. Nicolle@dsm.com
Thomas Wegman, DSM Composite Resins, Ceintuurbaan 5, 8022 AW Zwolle, The Netherlands
Fredric Petit, DSM Engineering Plastics, Poststraat 1, 6135 KR Sittard, The Netherlands
Annette Wilschut, DSM Head Office, Koninklijke DSM N.V., Het Overloon 1, 6401 JH Heerlen, The Netherlands
Theo Jongeling, DSM Innovation Center, Mauritslaan 49, 6129 EL Urmond, The Netherlands

9. *US Environmental Protection Agency's Design for the Environment (DfE) Alternatives Assessment Program*

Clive Davies (corresponding author), Design for the Environment Branch, Economics, Exposure and Technology Division, Office of Pollution Prevention and Toxics, United States Environmental Protection Agency, 1200 Pennsylvania, Avenue N.W., (7406M) Washington, DC, 20460–0001, USA, Email: davies.clive@epamail.epa.gov

Melanie Adams, Design for the Environment Branch, Economics, Exposure and Technology Division, Office of Pollution Prevention and Toxics, United States Environmental Protection Agency, 1200 Pennsylvania, Avenue N.W., (7406M) Washington, DC, 20460–0001, USA

Emily Connor, Abt Associates Inc., 4550 Montgomery Avenue, Suite 800 North, Bethesda, Maryland MD, 20814, USA

Elizabeth Sommer, Design for the Environment Branch, Economics, Exposure and Technology Division, Office of Pollution Prevention and Toxics, United States Environmental Protection Agency, 1200 Pennsylvania, Avenue N.W., (7406M) Washington, DC, 20460–0001, USA

Caroline Baier-Anderson, Design for the Environment Branch, Economics, Exposure and Technology Division, Office of Pollution Prevention and Toxics, United States Environmental Protection Agency, 1200 Pennsylvania, Avenue N.W., (7406M) Washington, DC, 20460–0001, USA

Emma Lavoie, Design for the Environment Branch, Economics, Exposure and Technology Division, Office of Pollution Prevention and Toxics, United States Environmental Protection Agency, 1200 Pennsylvania, Avenue N.W., (7406M) Washington, DC, 20460–0001, USA

Laura Romano, Abt Associates Inc., 4550 Montgomery Avenue, Suite 800 North, Bethesda, Maryland MD, 20814, USA

David DiFiore, Design for the Environment Branch, Economics, Exposure and Technology Division, Office of Pollution Prevention and Toxics, United States Environmental Protection Agency, 1200 Pennsylvania, Avenue N.W., (7406M) Washington, DC, 20460–0001, USA

10. *NGO Initiatives in the EU – Identifying Substances of Very High Concern (SVHCs) and Driving Safer Chemical Substitutes in Response to REACH*

Jerker J. Ligthart, ChemSec, International Chemical Secretariat, Norra allégatan 5, 41301 Gothenburg, Sweden, Email: jerker.ligthart@chemsec.org

11. *Alternatives Assessment in Regulatory Policy: History and Future Directions*

J. A. Tickner (corresponding author), University of Massachusetts Lowell, 1 University Avenue, Lowell, MA 01854, USA, Email: joel_tickner@ uml.edu

K. Geiser, University of Massachusetts Lowell, 1 University Avenue, Lowell, MA 01854, USA

C. Rudisill, SRC, Inc., 7502 Round Pond Road, North Syracuse, NY 13212, USA

J. N. Schifano, Occupational Safety and Health Administration, 200 Constitution Avenue N.W., Washington, DC 20210, USA

Chemicals Alternatives Assessment (CAA): Tools for Selecting Less Hazardous Chemicals

MARGARET H. WHITTAKER* AND LAUREN G. HEINE

ABSTRACT

Chemicals alternatives assessment (CAA) is a form of alternatives assessment that focuses on finding alternative chemicals, materials or product designs to substitute for the use of hazardous chemicals. Chemical hazard assessment (CHA) or comparative CHA is a method for comparing chemicals based on their inherent hazard properties. CAA is inclusive of CHA. However, a comprehensive CAA can be much broader and include information such as cost, availability, performance and social and environmental life-cycle attributes. CHA/CAA provides users with hazard-based information to make informed decisions when selecting less hazardous chemical alternatives. There are multiple CAA methods in use around the world and these methods share a common goal, namely, to support the intelligent design, use and substitution of chemicals to benefit humankind in a manner that will not harm our environment and its inhabitants. Ideally, a CAA/CHA will completely characterize a chemical's intrinsic human health and environmental hazards, in the process promoting the selection of less hazardous chemical ingredients, in addition to avoiding unintended consequences of switching to a poorly characterized chemical substitute.

CHA methods typically share common hazard endpoints related to human toxicity, environmental toxicity and environmental fate. The

---

*Corresponding author

Issues in Environmental Science and Technology, 36
Chemical Alternatives Assessments
Edited by R.E. Hester and R.M. Harrison
© The Royal Society of Chemistry 2013
Published by the Royal Society of Chemistry, www.rsc.org

endpoints are evaluated based on criteria that allow for the use of measured or predicted data. Human health criteria in CHA evaluate endpoints such as potential carcinogenicity, mutagenicity, reproductive and developmental toxicity, endocrine disruption, acute and chronic or repeat dose toxicity, dermal and eye irritation and dermal and respiratory sensitization. Acute and chronic aquatic toxicity, terrestrial toxicity, persistence and bioaccumulation are commonly evaluated to predict a chemical's environmental toxicity and fate. Finally, some CHAs (such as Green-Screen™) also evaluate a chemical's physical characteristics such as flammability and reactivity.

Of the CAA methods listed, only the US Environmental Protection Agency (EPA)'s DfE program, CPA's GreenScreen™ and MBDC's Cradle to Cradle® paradigms are fully transparent and publicly available methods of assessment. Most other CAAs in use around the world do not fully disclose all of their reasoning or resources used for establishing threshold values for hazard criteria, prioritization of hazard endpoints and life-cycle concerns. Some CAA methods are limited to a focus on CHA whereas others such as MBDC's Cradle to Cradle® expand the focus to consider some life-cycle attributes. Whether the CAA method includes additional attributes or not, CHA can be used in a modular way, combining with other needed information to inform decision-making.

CAA provides a powerful means to improve upon the *status quo* by establishing methods to inform chemical substitution in a scientifically rigorous and defensible manner. Recognizing the value of CAA and fostering greater adoption of CAA methods provide stakeholders with much-needed tools to address a serious deficiency in the way in which chemicals are used in society, as maintaining the *status quo* is analogous to giving up. As humankind's understanding of the full costs and benefits of chemicals matures, it is critical that we cease using those chemicals that can permanently impair human health or the environment.

1.1   Introduction to Chemicals Alternatives Assessments

Chemicals alternatives assessment (CAA) is a form of alternatives assessment that focuses on finding alternative chemicals, materials or product designs to substitute for the use of hazardous chemicals in products. Chemical hazard assessment (CHA) or comparative CHA is a method for comparing chemicals based on their inherent hazard properties. CAA is inclusive of CHA. CHA/CAA provides users with hazard-based information to make informed decisions when selecting less hazardous chemical alternatives. The approach is used to assess a chemical's impact on human health and the environment. Hazard can be defined as the way in which a chemical, object or situation may cause harm. The degree of a chemical's capacity to harm depends on its intrinsic properties, such as its capacity to interfere with normal biological processes and its capacity to burn, explode or corrode (*e.g.*, non-life-threatening allergic skin reaction to nickel jewelry, lethal egg-shell thinning in avian species attributed to exposure to the organochlorine pesticide DDT).[1] The goal of a CAA is to find a science-based solution that identifies and completely characterizes chemical hazards, promoting the selection of less hazardous chemical ingredients, in addition avoiding unintended consequences of switching to a poorly characterized chemical substitute.

Hazard assessment is a systematic process of assessing and classifying hazards across an entire spectrum of endpoints and levels of severity It involves a characterization of the nature and strength of the evidence of causation.[2] A comparative CHA is a type of hazard assessment that evaluates hazards from two or more agents, with the intent to guide decision-making toward the use of the least hazardous options via a process of informed substitution, as illustrated in Figure 1.1.

In practical terms, comparative CHA is a term that describes the practice of assessing hazards for specific items (such as chemicals, materials, products or technologies) and then comparing these hazards following a structured approach. Ideally, CHA minimizes subjectivity in hazard classification since a

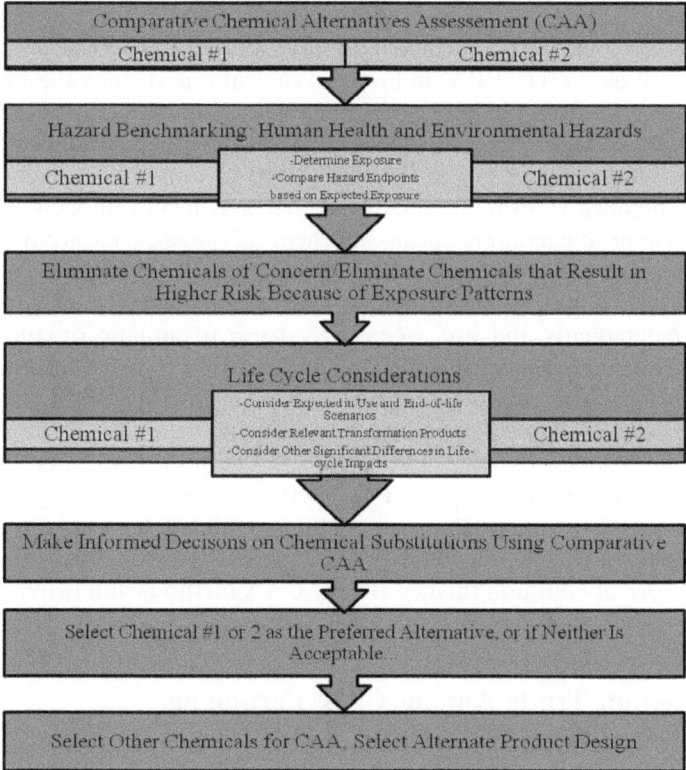

Figure 1.1 Comparative chemical alternatives assessments.

structured approach is used to assign hazards. Over the past 10 years, the number of comparative hazard tools has continued to increase. The Toxics Use Reduction Institute (TURI) at the University of Massachusetts presented a collection of over 100 tools for comparing hazard characteristics of different chemicals.[3] CAAs have numerous applications, including the following:

- Enabling the prioritization of chemicals for reduction or phase-out:
  ○ from any phase in product life-cycle (*e.g.*, manufacturing, product design);
  ○ from the whole supply chain.
- Assisting in the selection of alternatives for the following:
  ○ banned or restricted chemicals or materials;
  ○ chemicals that are perceived as hazardous by the public;
  ○ identifying and classifying Restricted Use Materials (RUMs);
  ○ developing environmentally preferred products.

Some CHA methods, such as GreenScreen™, focus solely on individual chemicals or materials and their subsequent health or environmental impacts while other CAA methods such as Cradle to Cradle® incorporate CHA in addition to certain life-cycle-based considerations such as energy use, water quality and efficiency, social responsibility and potential for material reuse.

To date, various CAA partnerships have brought together environmental agencies, such as the US EPA, industry organizations such as the Phosphorus, Inorganic and Nitrogen Flame Retardants Association (pinfa), academia (such as the University of Massachusetts at Lowell) and non-governmental organizations such as the United States Green Chemistry and Commerce Council (GC3) and Europe's ChemSec to evaluate environmental and health impacts of potential alternatives to problematic chemicals and chemical classes, such as phthalate esters (ubiquitously used in flexible plastics), flame retardants in furniture and printed circuit boards and nonylphenol ethoxylate surfactants (which are commonly used in laundry detergents and are exceedingly toxic to aquatic organisms). Such partnerships demonstrate that CAA can be employed to benefit both producers and users of the chemical to the improvement of ecological and human health.

The purpose of this collection of chapters in the *Issues in Environmental Science and Technology* series is to describe and exemplify several existing CAA methodologies currently being used in North America, Europe and China and to make suggestions on how to improve the overall CAA process, fostering the greater adoption of CAA around the world. This introductory chapter identifies a number of common themes among CAA methods and provides a broad overview of such methods.

1.2 Common Traits Among CAA Paradigms

CAA paradigms share a number of similarities that are implemented as part of a CAA and all have the common goal of identifying less hazardous chemicals. CAAs use standardized procedures to assess whether alternatives have the potential for an improved health and environmental profile. CAAs assess whether the adoption of an alternative chemical is likely to result in lasting environmental or public health improvement. Ideally, a CAA will also address whether chemical alternatives are commercially available, perform well and are cost-effective.

1.2.1 Step One: Hazard Assessment Through Literature Search and Data Identification

As a first step towards characterizing the human health and/or environmental hazards of a chemical, a CHA is performed. The practitioner assesses hazards for each chemical alternative across a range of health effects and environmental endpoints. Such endpoints generally include the following: acute and repeat dose toxicity, endocrine activity, carcinogenicity and mutagenicity, reproductive and developmental toxicity, neurotoxicity, respiratory and dermal sensitization, skin and eye irritation, acute and chronic aquatic toxicity, terrestrial toxicity and persistence and bioaccumulation. When measured data are not available or adequate for an endpoint, a hazard concern level can be assigned based on quantitative structure–activity relationships (QSARs or SARs) and expert judgment. This practice ensures that all endpoints are considered as part of the hazard assessment and that alternatives are evaluated based on a complete understanding of their potential human health and environmental hazards. A level of confidence associated with studies is often assigned.

Sources of information to evaluate and characterize human health and environmental hazards in a CAA include one of more of the following:

- Publicly available experimental data obtained from a literature review
  ○ Sources of such toxicological and environmental fate and effects data include online databases indexing scientific literature, such as:
    - ChemIDplus: http://www.cleanproduction.org/library/greenscreen-translator-benchmark1-possible%20benchmark1.pdf
    - EPA High Production Volume Information System (HPVIS): http://www.epa.gov/hpvis/index.html
    - UNEP OECD (Organisation for Economic Co-operation and Development) Screening Information Datasets (SIDS): http://www.chem.unep.ch/irptc/sids/OECDSIDS/sidspub.html
    - European Chemical Substances Information System IUCLID Chemical Data Sheets (ESIS): http://esis.jrc.ec.europa.eu/index.php?PGM = dat
    - United States National Toxicology Program (NTP): http://ntp.niehs.nih.gov/
    - International Agency for the Research on Cancer (IARC): http://www.inchem.org/pages/iarc.html
    - Human and Environmental Risk Assessment on ingredients of household cleaning products (HERA): http://www.heraproject.com/RiskAssessment.cfm

  – European Chemicals Agency (ECHA): http://echa.europa.eu/
  – ExPub (Expert Publishing): http://www.expub.com
  ○ Experimental data that are not publicly available (such as industry- or
    trade association-sponsored studies)
  ○ SAR-based estimations from predictive methods such as US EPA
    models (*e.g.*, EpiSuite, Ecosar), European Union (*e.g.*, VEGA, ToxTree)
    or OECD (*e.g.*, OECD Toolbox), Derek, Topkat, among other
    predictive software algorithms.

1.2.2 Step Two: Hazard Classification and Benchmarking of Relevant Data

Once the literature search has been performed, relevant studies have been
retrieved and data collected, the second step of a CHA generally entails
assigning hazard scores for the criteria evaluated. For example, a Green-
Screen™ will assign hazard scores of low, moderate, high (and, for some endpoints, very low or very high) for 18 health and environmental fate and toxicity
endpoints, as illustrated for an example chemical in Figure 1.2.[4] Criteria for
assigning hazard scores in a CAA are often based on the Global Harmonized
System of Classification and Labeling of Chemicals (GHS) criteria,[1] in addition
to criteria from other authoritative lists. As an example, a complete version of
GreenScreen™'s hazard criteria for each endpoint can be found at the CPA's
website.[4]

  For several endpoints, such as acute mammalian toxicity, systemic toxicity,
acute and chronic aquatic toxicity, persistence and bioaccumulation, hazard
scores are often assigned based on specific dose thresholds and/or ranges. For
example, the US EPA's Design for the Environment (DfE) Alternatives
Assessment Criteria for Hazard Evaluation will assign a chemical a hazard

DEHT was assigned a GreenScreen™ Benchmark Score of 3_{DG} as it does not meet the data gap requirements for a
GreenScreen™ Benchmark score of 4. Data gaps (dg) exist for Neurotoxicity (N) and Respiratory Sensitization (SnR).[16]

Green Screen Hazard Ratings: Di(2-ethylhexyl) terephthalate (DEHT)																			
Group I Human						Group II and II* Human								Ecotox		Fate		Physical	
C	M	R	D	E	AT	ST		N		SnS*	SnR*	IrS	IrE	AA	CA	P	B	Rx	F
						single	repeat*	single	repeat*										
L	L	L	L	L	L	dg		L	dg	dg	L	dg	L	L	L	L	vL	L	L	L

Note: Hazard levels [Very High (vH), High (H), Moderate (M), Low (L), Very Low (vL)] in *italics* reflect estimated values and
lower confidence. Hazard levels in **Bold** font reflect values based on test data.[16]

AA	Acute Aquatic Toxicity	E	Endocrine Activity	P	Persistence
AT	Acute Mammalian Toxicity	F	Flammability	R	Reproductive Toxicity
B	Bioaccumulation	IrE	Eye Irritation/Corrosivity	Rx	Reactivity
C	Carcinogenicity	IrS	Skin Irritation/Corrosivity	SnS	Sensitization-Skin
CA	Chronic Aquatic Toxicity	M	Mutagenicity and Genotoxicity	SnR	Sensitization-Respiratory
D	Developmental Toxicity	N	Neurotoxicity	ST	Systemic/Organ Toxicity

Figure 1.2   Example GreenScreen™ hazard ratings: di(2-ethylhexyl) terephthalate
(DEHT).

score of Low for acute mammalian toxicity based on an oral LD_{50} of 2000 mg kg^{-1} or greater and a hazard score of Moderate for persistence based on a half-life in water that falls between 16 and 60 days.

For other endpoints, such as carcinogenicity, mutagenicity and reproductive and developmental toxicity, professional judgment and a clear understanding of the available data are necessary to draw a conclusion and assign a hazard classification. If a 2 year carcinogenicity study performed by the US National Toxicology Program (NTP) shows a statistically significant increase of hepatic tumors in rats, US EPA DfE and GreenScreen™ alternatives assessment criteria would assign a hazard score of High for carcinogenicity. In contrast, if a reproductive toxicity study reports a Lowest Observed Adverse Effect Level (LOAEL) of 250 mg $kg\text{-}bw^{-1}$ d^{-1} in mice based on an effect such as decreased weight gain, US EPA and GreenScreen™ alternatives assessment criteria may or may not consider this effect relevant for purposes of assigning a hazard for reproductive toxicity depending on the chemical's mechanism of action. Ultimately, the hazard classification may come down to the professional opinion of the scientist performing the CAA. It is imperative to the integrity of a CAA that all hazard scores are based on sound scientific knowledge and can be properly supported and defended, if necessary.

Points to consider when determining the validity of available data include whether a study was performed following Good Laboratory Practices (GLP) or whether a study was conducted following a specific test guideline (*e.g.*, OECD test guidelines). The level of detail reported in a study is also important, as is the source of the study. Primary sources such as peer-reviewed studies are preferred; however, high-quality secondary sources are acceptable, particularly when supported by a Klimisch score of 1 or 2 (1 = reliable without restrictions; 2 = reliable with restrictions), which provides an indication as to the reliability of the actual data.[5]

In the case of conflicting data, weight of evidence should be used to assign the ultimate hazard score for a specific health effect/environmental endpoint. All professional judgments must be fully justified within the section of that endpoint. The justification for a final hazard score must be transparent and easily understood by all who may read the CAA.

1.2.3 Step Three: CAA Report Preparation

Once the hazard assessment and classification portion of the CAA has been completed, a CAA report is written to provide contextual and supplemental information designed to aid in decision-making and may include descriptions of manufacturing processes, use patterns and life-cycle stages that may pose special exposure concerns. The CAA report may contain a description of the cost of use and the potential economic impacts associated with the selection of alternatives and may also contain information on alternative technologies that might result in safer chemicals, manufacturing processes and practices.

Examples of alternatives assessments can be found for numerous types and classes of chemicals, including nonylphenol ethoxylate surfactants, flame retardants in furniture and printed circuit boards, diethylhexyl phthalate and perchloroethylene.[6,7]

1.3   Life-cycle Assessment and Chemicals Alternatives Assessment

In comparing alternatives, it is important to consider life-cycle impacts in order to avoid shifting burdens between stages of a material's life-cycle. Life-cycle assessment (LCA) is a standardized and quantifiable approach to assessing life-cycle impacts, such as ISO Standard 14040 (Environmental Management: Life Cycle Assessment-Principles and Framework).[8,9] Production of all chemicals or substances requires the extraction of resources from the earth, including water, energy and other raw materials. Energy and other resources are used for manufacture, transportation and during the use phase. At the end of a product's useful life, the product may be released into the environment. This can be problematic if the product contains hazardous materials. As climate change becomes more of a global concern, manufacturers are beginning to assess how their processes affect the environment. Many businesses have responded by identifying 'greener' raw materials and using 'greener' processes to manufacture their products. They are able to reduce the environmental impact their actions have by utilizing LCA. There are four main phases of an LCA:

1. *Defining the goals and scope of the LCA* – Define and describe the product, process or activity.
2. *Performing a life-cycle inventory* – Identify and quantify energy, water and materials usage and environmental releases (*e.g.*, air emissions, solid waste disposal, waste water discharges);
3. *Performing a life-cycle impact assessment* – Assess the potential human and ecological effects of energy, water and material usage and the environmental releases identified in the inventory analysis;
4. *Interpretation* – Evaluate the results of the inventory analysis and impact assessment to select the preferred product, process or service with a clear understanding of the uncertainty and the assumptions used to generate the results.

LCA helps to avoid shifting environmental burdens from one step in a product's life-cycle to another. For example, when choosing between two materials, Option 1 may appear environmentally beneficial because it is more easily biodegradable. However, after conducting an LCA, it may become obvious that Option 2 will produce less of an environmental impact because it uses less energy to manufacture and its breakdown products may pose less of a risk to aquatic ecosystems. An ideal CAA paradigm will take into account life-cycle thinking when evaluating a chemical or material and seek to reduce hazard across the life-cycle. Commercially available LCA software packages such as GaBi and SimaPro are commonly used for LCA, with growing interest in the freeware OpenLCA, which requires inputting of life-cycle inventory data.

1.4 Chemical Alternatives Assessment Paradigms in Use: a Critical Evaluation

Currently, there are a number of CAA paradigms in use around the world to evaluate environmental and human health impacts of potential alternatives to problematic chemicals, as presented in Table 1.1. Often, CAAs employ life-cycle considerations to predict and understand specific phases, from development to manufacture, where industry can make changes to realize environmental and health benefits.[6,10]

Table 1.1 North American and European Alternatives Assessment Paradigms.

Name of Alternative Assessment	Organization	Endpoints
DfE Alternatives Assessment Criteria	Design for the Environment (DfE)	Acute mammalian toxicity, carcinogenicity, mutagenicity/genotoxicity, reproductive/developmental toxicity, neurotoxicity, repeat dose toxicity, sensitization, eye/skin irritation, endocrine activity, aquatic toxicity, persistence, bioaccumulation
GreenScreen™	Clean Production Action (CPA)	Carcinogenicity, mutagenicity/genotoxicity, reproductive/developmental toxicity, endocrine disruption, neurotoxicity, acute toxicity (inhalation, dermal, oral toxicity), corrosion/irritation of the skin or eye, sensitization of the skin or respiratory system, immune system effects, systemic toxicity/organ effects, acute and chronic aquatic toxicity, persistence, bioaccumulation, explosivity, flammability
Cradle to Cradle® (C2C)	McDonough Braungart Design Chemistry, LLC (MBDC)	Material health [*i.e.* carcinogenicity, endocrine disruption, mutagenicity, reproductive toxicity, teratogenicity, acute and chronic toxicity, irritation of the skin or mucous membranes, sensitization, other (carrier function, skin penetration potential, *etc.*), vertebrate toxicity (fish), invertebrate toxicity (daphnia), aquatic plant toxicity (algae), persistence/biodegradation, bioaccumulation, contents of halogenated organics, metal content, climate relevance/ozone depletion] Material reutilization/design for environment; energy; water; social responsibility

Table 1.1   Continued.

Name of Alternative Assessment	Organization	Endpoints
Pollution Prevention Options Analysis System (P2OASys)	Massachusetts Toxics Use Reduction Institute (TURI)	Acute human effects (*i.e.* dermal/ocular/respiratory irritation, skin absorption, inhalation LC_{50}, oral/dermal LD_{50}); Chronic human effects (*i.e.* carcinogenicity, mutagenicity, reproductive/developmental toxicity, neurotoxicity, respiratory sensitivity/disease, other organ effects); Aquatic hazards (*i.e.* water quality criteria, aquatic LC_{50}, plant EC_{50}, fish NOAEC, observed ecological effects); persistence, bioaccumulation; atmospheric hazards (*i.e.*, ozone depletor, greenhouse gas, acid rain formation); chemical hazards (*i.e.*, volatile organic compounds, flammability, reactivity/instability, pH); disposal hazard; energy/resource use; product hazard; exposure potential
Chemical Scoring and Ranking Assessment Model (SCRAM)	Michigan Department of Environmental Quality, Surface Water Quality Division (SWQD) and Michigan State University, National Food Safety and Toxicology Center	Subchronic/chronic toxicity (based on repeat dose toxicity, general organ toxicity, carcinogenicity, mutagenicity, neurotoxicity, immunotoxicity, reproductive and developmental toxicity, endocrine disruption), acute aquatic toxicity, acute terrestrial toxicity, persistence, bioaccumulation
Chemicals Assessment and Ranking System (CARS)	Zero Waste Alliance	Proprietary, but may include carcinogenicity, teratogenicity, endocrine disruption, persistence, bioaccumulation (PBT status), aquatic toxicity, impact on climate, ozone; considers frequency of use, human health, safety impacts, ecological and global impacts, costs
GreenList™	SC Johnson	Proprietary, but may include acute human toxicity, carcinogenicity, mutagenicity and reproductive toxicity, persistence, bioaccumulation (PBT status), aquatic toxicity, biodegradability
PRIO	Swedish Chemicals Agency	Phase-out substances: carcinogenicity, mutagenicity, reproductive toxicity, endocrine disruption, hazardous metals, ozone depletion, persistence, bioaccumulation. Priority risk-reduction substances: acute and chronic toxicity (inhalation, dermal, oral toxicity, sensitization, mutagenicity) persistence, bioaccumulation

Table 1.1 Continued.

Name of Alternative Assessment	Organization	Endpoints
Quick Scan	Dutch Ministry of Housing, Spatial Planning and the Environment	Persistence, bioaccumulation, ecotoxicity, carcinogenicity, mutagenicity, reproductive toxicity, inhalation or dermal toxicity, hormone disruption
Column Model and GHS Column Model	German Institute for Occupational Safety (BIA)	Acute and chronic health hazards (including inhalation, dermal, oral toxicity, sensitization; carcinogenicity and mutagenicity, reproductive toxicity, bioaccumulation), environmental hazards, explosivity, flammability, exposure potential, hazards caused by procedures
Evaluation Matrix	German Federal Environmental Agency	Inclusion in lists of problematic substances, physico-chemical properties, human toxicity, problematic properties related to the environment, mobility within the environment, origin of raw materials, greenhouse gas emission, resource consumption, persistence, bioaccumulation

As mentioned above, CAAs are inclusive of CHA and share common endpoints. CAA methods evaluate chemicals based on their measured or predicted human and ecological hazards, in addition to their environmental fate. Human health criteria in a CAA evaluate endpoints such as potential carcinogenicity, mutagenicity, reproductive and developmental toxicity, endocrine disruption, acute and chronic or repeat dose toxicity, irritation and sensitization. Acute and chronic aquatic toxicity, terrestrial toxicity, persistence and bioaccumulation are commonly evaluated to predict a chemical's environmental toxicity and fate. Finally, some CAAs (such as GreenScreen™) also evaluate a chemical's physical characteristics. such as flammability and reactivity.

Of the CAA methods listed, only the US EPA's DfE program and CPA's GreenScreen™ are fully transparent and publicly available methods of assessment. A number of the other CAA methods identified in Table 1.1 do not fully disclose all of their reasoning or resources used for establishing threshold values for hazard criteria, prioritization of hazard endpoints and life-cycle concerns.

This section describes a number of CAA methods. Although each method uses its own set of criteria when evaluating the hazards of a chemical, the most commonly used benchmarks used when assigning a hazard for a specific endpoint mirror those developed by the US EPA, the OECD, and/ or the Globally Harmonized System of Classification and Labeling of Chemicals (GHS). Tables 1.2 and 1.3 provide an overview of how each

Table 1.2 Comparison of human health hazard evaluation criteria among CAA methods.

Method	Human health endpoint criteria								
	Acute mammalian toxicity	Carcinogenicity	Mutagenicity/genotoxicity	Reproductive/developmental toxicity	Repeat dose toxicity (also referred to as systemic/organ effects)	Irritation/corrosion	Sensitization	Neurotoxicity	Endocrine disruption
DfE Alternatives Assessment Criteria[12]	**Low:** • LD_{50} >2000 mg kg⁻¹ (oral) • LD_{50} >2000 mg kg⁻¹ (dermal) • LC_{50} >20 mg L⁻¹ (gas/vapor) • LC_{50} >5 mg L⁻¹ d⁻¹ (dust/mist/fume) **Moderate:** • LD_{50} ≥ 300–2000 mg kg⁻¹ (oral) • LD_{50} ≥ 1000–2000 mg kg⁻¹ (dermal) • LC_{50} ≥ 10–20 mg L⁻¹ (gas/vapor) • LC_{50} 1–5 mg L⁻¹ d⁻¹ (dust/mist/fume) **High:** • LD_{50} >50–300 mg kg⁻¹ (oral) • LD_{50} ≥ 200–1000 mg kg⁻¹ (dermal) • LC_{50} ≥ 2–10 mg L⁻¹ (gas/vapor) • LC_{50} ≥ 0.5–1 mg L⁻¹ d⁻¹ (dust/mist/fume) **Very High:** • LD_{50} ≤50 mg kg⁻¹ (oral) • LD_{50} ≤2000 mg kg⁻¹ (dermal)	**Low:** • Negative studies • Negative robust mechanism-based SAR **Moderate:** • Limited or marginal evidence of carcinogenicity in animals • Inadequate evidence in humans **High:** • Suspected human carcinogen (GHS Cat. 2) **Very High:** • Known or presumed human carcinogen (GHS Cat. 1A and 1B)	**Low:** • Negative for chromosomal aberrations and gene mutations • No structural alerts **Moderate:** • Evidence of mutagenicity supported by positive results in in vitro or in vivo somatic cells of humans or animals **High:** • Substances causing concern for humans due to possible induction of heritable mutations in the germ cells of humans • OR Evidence of mutagenicity supported by positive results in in vitro AND in vivo somatic cells and/or germ cells of humans or animals (GHS Cat. 2) **Very High:** • Substances known to	**Very Low:** • NOAEL >1000 mg kg-bw⁻¹ d⁻¹ (oral) • NOAEL >2000 mg kg-bw⁻¹ d⁻¹ (dermal) • NOAEC >20 mg L⁻¹ d⁻¹ (gas/vapor) • NOAEC >5 mg L⁻¹ d⁻¹ (dust/mist/fume) **Low:** • NOAEL >250–1000 mg kg-bw⁻¹ d⁻¹ (oral) • NOAEL >500–2000 mg kg-bw⁻¹ d⁻¹ (dermal) • NOAEC >2.5–20 mg L⁻¹ d⁻¹ (gas/vapor) • NOAEC >0.5–5 mg L⁻¹ d⁻¹ (dust/mist/fume) **Moderate:** • NOAEL 50–250 mg kg-bw⁻¹ d⁻¹ (oral) • NOAEL 100–500 mg kg-bw⁻¹ d⁻¹ (dermal) • NOAEC 1–2.5 mg L⁻¹ d⁻¹ (gas/vapor) • NOAEC 0.1–0.5 mg L⁻¹ d⁻¹ (dust/mist/fume) **High:** • NOAEL <50 mg kg-bw⁻¹ d⁻¹ (oral) • NOAEL <100 mg kg-bw⁻¹ d⁻¹ (dermal) • NOAEC <1 mg L⁻¹ d⁻¹ (gas/vapor) • NOAEC <0.1 mg L⁻¹ d⁻¹ (dust/mist/fume)	• All values based on 90 day study, triple for 28 day study: **Low:** • NOAEL >100 mg kg-bw⁻¹ d⁻¹ (oral) • NOAEL >200 mg kg-bw⁻¹ d⁻¹ (dermal) • NOAEC >1 mg L⁻¹ (6 h)⁻¹ d⁻¹ (gas/vapor) • NOAEC >0.2 mg L⁻¹ (6 h)⁻¹ d⁻¹ (dust/mist/fume) **Moderate:** • NOAEL 10–100 mg kg-bw⁻¹ d⁻¹ (oral) • NOAEL 20–200 mg kg-bw⁻¹ d⁻¹ (dermal) • NOAEC 0.2–1 mg L⁻¹ (6 h)⁻¹ d⁻¹ (gas/vapor) • NOAEC 0.02–0.2 mg L⁻¹ (6 h)⁻¹ d⁻¹ (dust/mist/fume) **High:** • NOAEL <10 mg kg-bw⁻¹ d⁻¹ (oral) • NOAEL <20 mg kg-bw⁻¹ d⁻¹ (dermal) • NOAEC <0.2 mg L⁻¹ (6 h)⁻¹ d⁻¹ (gas/vapor)	**Very Low:** • Not irritating (eye and skin) **Low:** • Clearing in less than 24 h, mildly irritating (eye) • Mild or slight irritation at 72 h (skin) **Moderate:** • Clearing in 7 d or less, moderately irritating (eye) • Moderate irritation at 72 h (skin) **High:** • Clearing in 8–21 d, severely irritating (eye) • Severe irritation at 72 h (skin) **Very High:** • Irritation persists for >21 d or corrosive (eye) • Corrosive (skin)	**Low:** • Adequate data available and not GHS Cat. 1A or 1B (skin) • Adequate data available indicating lack of respiratory sensitization **Moderate:** • Low to moderate frequency of sensitization in humans and/or low to moderate potency in animals (GHS Cat. 1B) (skin) • EC3 >2% (local lymph node assay) • ≥ 30–60% responding at ≤0.1–≤1% intradermal dose or ≥ 30% at >1% dermal dose (guinea pig maximization test) • ≥ 15–<60% responding at >0.2–≤20% topical dose or ≥ 15% at >20% topical dose (Buehler assay) • Limited evidence, including presence of structural alerts (respiratory) **High:** • High frequency of sensitization in humans and/or high potency in animals (GHS Cat. 1A) (skin) • EC3 ≤2% (local lymph node assay)	• All values based on 90 day study, triple for 28 day study: **Low:** • NOAEL >100 mg kg-bw⁻¹ d⁻¹ (oral) • NOAEL >200 mg kg-bw⁻¹ d⁻¹ (dermal) • NOAEC >1 mg L⁻¹ (6 h)⁻¹ d⁻¹ (gas/vapor) • NOAEC >0.2 mg L⁻¹ (6 h)⁻¹ d⁻¹ (dust/mist/fume) **Moderate:** • NOAEL 10–100 mg kg-bw⁻¹ d⁻¹ (oral) • NOAEL 20–200 mg kg-bw⁻¹ d⁻¹ (dermal) • NOAEC 0.2–1 mg L⁻¹ (6 h)⁻¹ d⁻¹ (gas/vapor) • NOAEC 0.02–0.2 mg L⁻¹ (6 h)⁻¹ d⁻¹ (dust/mist/fume) **High:** • NOAEL <10 mg kg-bw⁻¹ d⁻¹ (oral) • NOAEL <20 mg kg-bw⁻¹ d⁻¹ (dermal) • NOAEC <0.2 mg L⁻¹ (6 h)⁻¹ d⁻¹ (gas/vapor) • NOAEC <0.02 mg L⁻¹ (6 h)⁻¹ d⁻¹ (dust/mist/fume)	• Evidence of a chemical having endocrine activity will be summarized in a narrative

- $LC_{50} \leq 2$ mg L^{-1} (gas/vapor)
- $LC_{50} \leq 0.5$ mg L^{-1} d^{-1} (dust/mist/fume)

induce heritable mutations or to be regarded as if they induce heritable mutations in the germ cells of humans (GHS Cat. 1A or 1B)

- NOAEC <0.02 mg L^{-1} (6 h)$^{-1}$ d^{-1} (dust/mist/fume)

- ≥ 30% responding at ≤0.1% intradermal dose or ≥ 60% at >0.1–≤1% intradermal dose (guinea pig maximization test)
- ≥ 15% responding at ≤0.2% topical dose or ≥ 60% at >0.2–≤20% topical dose (Buehler assay)
- Occurrence in humans or evidence of sensitization in humans based on animal or other tests (GHS Cat. 1A and 1B) (respiratory)

GreenScreen version 1.2[4]					
Low: • No basis for concern identified • LD_{50} >2000 mg kg^{-1} (oral) • LD_{50} >2000 mg kg^{-1} (dermal) • LC_{50} >20 mg L^{-1} (gas/vapor) • LC_{50} >5 mg L^{-1} (dust/mist/fume) or • GHS Category 5 *Moderate:* • LD_{50} >300–2000 mg kg^{-1} (oral) • LD_{50} >1000–2000 mg kg^{-1} (dermal) • LC_{50} >10–20 mg L^{-1} (gas/vapor) • LC_{50} >1–5 mg L^{-1} (dust/mist/fume) or • GHS Category 4 *High:* • LD_{50} >50–300 mg kg^{-1} (oral)	*Low:* • No basis for concern identified • IARC Group 4 *Moderate:* • Limited or marginal evidence of carcinogenicity in animals • Analog data • Chemical class known to produce toxicity • IARC Group 2B • EU Category 3 or • GHS Category 2 *High:* • Evidence of adverse effects in humans	*Low:* • No basis for concern identified *Moderate:* • Suggestive animal studies • Analog data • Chemical class known to produce toxicity • EU Category 3 or • GHS Category 2 *High:* • Evidence of adverse effects in humans • Weight of evidence demonstrates potential for adverse effects in humans • EU Category 1 or 2 or • GHS Category 1	*Low:* • No basis for concern identified *Moderate:* • Suggestive animal studies • Analog data • Chemical class known to produce toxicity • GU Category 3 or GHS Category 2 *High:* • Evidence of adverse effects in humans • Weight of evidence demonstrates potential for adverse effects in humans • NTP Center for the Evaluation of Risks to Human Reproduction • California Prop 65 • GHS Category 1A or 1B	*Low:* • No basis for concern identified • NOAEL >50 mg $kg^{-1}sd^{-1}$ (oral) • NOAEC >200 mg $kg^{-1}sd^{-1}$ (dermal) • NOAEC >1 mg L^{-1} (6 h) day^{-1} (gas/vapor) • NOAEC >0.2 mg L^{-1} (6 h) day^{-1} (dust/mist/fume) *Moderate:* • NOAEL >10–100 mg kg-bw^{-1} d^{-1} (oral) • NOAEL >20–200 mg kg-bw^{-1} d^{-1} (dermal) • NOAEC >0.2–1 mg L^{-1} (6 h)$^{-1}$ d^{-1} (gas/vapor) • NOAEC >0.02–0.2 mg L^{-1} (6 h)$^{-1}$ d^{-1} (dust/mist/fume) • GHS Category 3 single exposure or	*Low:* • No basis for concern identified *Moderate:* • Evidence of reversible effects in humans or animals • GHS Category 3: mild irritant (skin) • GHS Category 2B: mildly irritating (eye) *High:* • GHS Category 2: irritant (skin) • GHS Category 2A: irritating (eye) *Very High:* • Evidence of irreversible effects in studies of human populations • Weight of evidence of irreversible effects in animal studies or • GHS Category 1: corrosive and/or irreversible (skin or eye)

Low:
- No basis for concern identified

Moderate:
- Suggestive animal studies
- Analog data or Chemical class known to produce toxicity
- GHS Category 1B (low to moderate frequency of occurrence)

High:
- Evidence of adverse effects in humans
- Weight of evidence demonstrates potential for adverse effects in humans
- GHS Category 1A (high frequency of occurrence)

Low:
- No basis for concern identified
- EU Category 3a

Moderate:
- Evidence of endocrine activity
- Suggestive animal studies
- Analog data
- Chemical class known to produce toxicity
- EU Draft List – Category 1 or 2

High:
- Evidence of endocrine activity and adverse effects in humans or

Table 1.2 Continued.

Human health endpoint criteria

Method	Acute mammalian toxicity	Carcinogenicity	Mutagenicity/genotoxicity	Reproductive/developmental toxicity	Repeat dose toxicity (also referred to as systemic/organ effects)	Irritation/corrosion	Sensitization	Neurotoxicity	Endocrine disruption
	• LD$_{50}$ >200–1000 mg kg^{-1} (dermal) • LC$_{50}$ >2.0–10 mg L^{-1} (gas/vapor) • LC$_{50}$ >0.5–1 mg L^{-1} (dust/mist/fume) • US EPA Extremely Hazardous Substance List or GHS Category 3 *Very High:* • LD$_{50}$ ≤50 mg kg^{-1} (oral) • LD$_{50}$ ≤200 mg kg^{-1} (dermal) • LC$_{50}$ ≤2 mg L^{-1} (gas/vapor) • LC$_{50}$ ≤0.5 mg L^{-1} (dust/mist/fume) • US EPA Extremely Hazardous Substance List or GHS Category 1 or 2	• Weight of evidence demonstrates potential for adverse effects in humans • NTP known or reasonably anticipated to be human carcinogen • OSHA carcinogen • California Prop 65 • IARC Group 1 or 2A • EU Category 1 or 2 or GHS Category 1A or 1B	• GHS Category 1A or 1B		• Category 2 repeated exposure *High:* • NOAEL ≤10 mg kg-bw^{-1} d^{-1} (oral) • NOAEL ≤20 mg kg-bw^{-1} d^{-1} (dermal) • NOAEC ≤0.2 mg L^{-1} (6 h)$^{-1}$ d^{-1} (gas/vapor) • NOAEC ≤0.02 mg L^{-1} (6 h)$^{-1}$ d^{-1} (dust/mist/fume) • GHS Category 2 single exposure or Category 1 repeated exposure *Very High:* • GHS Category 1 single exposure			*Very High:* • GHS Category 1 single exposure	• Weight of evidence demonstrates that mechanisms of action lead to adverse effects • EU Draft List – Category 1 or 2
C2C version 3.0 (Includes only Ingredient Characterization Criteria)[19]	*Green:* • LD$_{50}$ >2000 mg (oral/dermal) • LC$_{50}$ >2000 ppmV (gas) • LC$_{50}$ >20mg/L (4 h vaper) • LC$_{50}$ >5mg/L (4 h dust/mist) *Yellow:* • LD$_{50}$ 300–2000 mg/kg (oral) • LD$_{50}$ 1000–2000 mg/kg (dermal)	*Green:* • Not known or suspected carcinogen *Yellow:* • Not classifiable as a human carcinogen *Red:* • Known or suspected carcinogen	*Green:* • Product has been tested and is not mutagenic or clastogenic invitro at concentration up to 100mg/L *Yellow:* • Negative mutagenicity at concentration up to 100mg/L	*Green:* • Not known or suspected of being a reproductive toxin or teratogen • NOAEL >500 mg/kg bw/day (oral) • NOAEL >2.5 mg/l 6–8 hr/day (inhalative) *Yellow:* • Evidence of being a secondary non-specific reproductive toxin or teratogen	Chronic toxicity not listed as an explicit hazard in v 3.0.	*Green:* • Single Exposure Organ, Sub-Chronic, and Chronic Toxicity Endpoints for Green Rating *Yellow:* • Single Exposure Organ, Sub-Chronic, and Chronic Toxicity Endpoints for Yellow Rating	*Green:* • Not sensitizing to skin or airways via experience or test *Yellow:* • Equivocal sensitization data *Red:* • Listed as a skin or airway sensitizer or has tested positive in sensitization test	*Green:* • Single Exposure Organ, Sub-Chronic, and Chronic Toxicity Endpoints for Green Rating *Yellow:* • Single Exposure Organ, Sub-Chronic, and Chronic Toxicity Endpoints for Yellow Rating	*Green:* • Not known or suspected of being an endocrine disruptor *Yellow:* • Insufficient evidence of being an endocrine disruptor

P2OASys (levels of concern are characterized as numerical hazard scores the smaller the number, the less hazardous)[20]							

P2OASys (levels of concern are characterized as numerical hazard scores the smaller the number, the less hazardous)[20]

Column — Acute toxicity (oral/dermal/inhalation)

2.00:
- LD_{50} >5000 mg kg⁻¹ (oral/dermal)
- LC_{50} >10 000 ppm (inhalation)

4.00:
- LD_{50} 500–5000 mg kg⁻¹ (oral/dermal)
- LC_{50} 1000–10 000 ppm (inhalation)

6.00:
- LD_{50} 50–500 mg kg⁻¹ (oral/dermal)
- LC_{50} 150–1000 ppm (inhalation)

8.00:
- LD_{50} 5–50 mg kg⁻¹ (oral/dermal)
- LC_{50} 15–150 ppm (inhalation)

10.00:
- LD_{50} <5 mg kg⁻¹ (oral/dermal)
- LC_{50} <15 ppm (inhalation)

Column — IARC/EPA

2.00:
- IARC/EPA Class 4,E

4.00:
- IARC/EPA Class 3,D

6.00:
- IARC/EPA Class 2B,C

8.00:
- IARC/EPA Class 2A,B

10.00:
- IARC/EPA Class 1,A

Column — Mutagenicity

Red:
- Positive mutagenicity or clastogenic invitro at concentration up to 100mg/L or GHS 1A, 1B, or 2

Column — Reproductive toxicity / NOAEL

- NOAEL = 50–500 mg/kg bw/day (oral)
- NOAEL = 0.25 mg–2.5 mg/l 6–8 hr/day (inhalative)

Red:
- Known as a reproductive toxin or teratogen as GHS 1A, 1B, or 2
- NOAEL <50 mg/kg bw/day (oral)
- NOAEL <0.25 mg/l 6–8 hr/day (inhalative)

Column — rating

2.00:
- L

4.00:
- L/M

6.00:
- M

8.00:
- M/H

10.00:
- H

Column — Acute LC_{50} (gas/vapor/dust)

- LC_{50} 2500–20000 ppmV (gas)
- LC_{50} 10–20 mg/m³ (4 hr vapor)
- LC_{50} 1–5 mg/m³ (4 hr dust/mist)

Red:
- LD_{50} <300 mg/kg (oral)
- LD_{50} <1000 mg/kg (dermal)
- LC_{50} <2500 ppmV (gas)
- LC_{50} <10 mg/m³ (4 hr vapor)
- LC_{50} <1 mg/m³ (4 hr dust/mist)

Column — For skin, eye, respiratory tract:

2.00:
- L

4.00:
- L/M

6.00:
- M

8.00:
- M/H

10.00:
- H

Column — For respiratory sensitivity/disease:

2.00:
- L

4.00:
- L/M

6.00:
- M

8.00:
- M/H

10.00:
- H

Column — Single Exposure Organ / Neurotoxicity

Red:
- Single Exposure Organ, Sub-Chronic, and Chronic Toxicity Endpoints for Red Rating or Listed in Grandjean et al. text for neurotox effects

Column — rating

2.00:
- L

4.00:
- L/M

6.00:
- M

8.00:
- M/H

10.00:
- H

Column — Endocrine disruption

Red:
- Listed as known or suspected endocrine disruptor supported by peer reviewed science

N/A

Table 1.2 Continued.

Human health endpoint criteria

Method	Acute mammalian toxicity	Carcinogenicity	Mutagenicity/genotoxicity	Reproductive/developmental toxicity	Repeat dose toxicity (also referred to as systemic/organ effects)	Irritation/corrosion	Sensitization	Neurotoxicity	Endocrine disruption
SCRAM[a21] (levels of concern are characterized as numerical hazard scores the smaller the number, the less hazardous)	*1.0:* • E/LD$_{50}$ >5000 mg kg⁻¹ (oral/mammal) • E/LD$_{50}$ >100 lb acre⁻¹ (plant) *2.0:* • E/LD$_{50}$ 500–5000 mg kg⁻¹ (oral/mammal) • E/LD$_{50}$ 10–100 lb acre⁻¹ (plant) *3.0:* • E/LD$_{50}$ 50–500 mg kg⁻¹ (oral/mammal) • E/LD$_{50}$ 1–10 lb acre⁻¹ (plant) *4.0:* • E/LD$_{50}$ 5–50 mg kg⁻¹ (oral/mammal) • E/LD$_{50}$ 0.1–1 lb acre⁻¹ (plant) *5.0:* • E/LD$_{50}$ ≤5 mg kg⁻¹ (oral/mammal) • E/LD$_{50}$ ≤0.1 lb acre⁻¹ (plant)	*1.0:* • 1/ED$_{10}$ ≤1.5^b mg kg⁻¹ d⁻¹ *2.0:* • 1/ED$_{10}$ 1.5–5^b mg kg⁻¹ d⁻¹ *3.0:* • 1/ED$_{10}$ 5–15^b mg kg⁻¹ d⁻¹ *4.0:* • 1/ED$_{10}$ 15–45^b mg kg⁻¹ d⁻¹ *5.0:* • 1/ED$_{10}$ >45^b mg kg⁻¹ d⁻¹	*1.0:* • No known mutagenic effects *2.0:* • Possible somatic line *3.0:* • Positive somatic line *4.0:* • Possible germ line *5.0:* • Positive germ line	*1.0:* • NOAEL > 1000 mg kg⁻¹ d⁻¹ (≥ 90 d/human) *2.0:* • NOAEL 100–1000 mg kg⁻¹ d⁻¹ (≥ 90 d/human) *3.0:* • NOAEL 10–100 mg kg⁻¹ d⁻¹ (≥ 90 d/human) *4.0:* • NOAEL 1–100 mg kg⁻¹ d⁻¹ (≥ 90 d/human) *5.0:* • NOAEL ≤1 mg kg⁻¹ d⁻¹ (≥ 90 d/human)	*1.0:* • NOAEL >1000 mg kg⁻¹ d⁻¹ (≥ 90 d/mammal) • L/NOAEL >5000 mg kg⁻¹ d⁻¹ (invertebrate) • L/NOAEL >100 lb acre⁻¹ (plant) *2.0:* • NOAEL 100–1000 mg kg⁻¹ d⁻¹ (≥ 90 d/mammal) • L/NOAEL 1000–5000 mg kg⁻¹ d⁻¹ (invertebrate) • L/NOAEL 10–100 lb acre⁻¹ (plant) *3.0:* • NOAEL 10–100 mg kg⁻¹ d⁻¹ (≥ 90 d/mammal) • L/NOAEL 100–1000 mg kg⁻¹ d⁻¹ (invertebrate) • L/NOAEL 1–10 lb acre⁻¹ (plant) *4.0:* • NOAEL 1–100 mg kg⁻¹ d⁻¹ (≥ 90 d/mammal) • L/NOAEL 10–100 mg kg⁻¹ d⁻¹ (invertebrate) • L/NOAEL 0.1–1 lb acre⁻¹ (plant)	N/A	N/A	*1.0:* • NOAEL >1000 mg kg⁻¹ d⁻¹ (≥ 90 d/human) *2.0:* • NOAEL 100–1000 mg kg⁻¹ d⁻¹ (≥ 90 d/human) *3.0:* • NOAEL 10–100 mg kg⁻¹ d⁻¹ (≥ 90 d/human) *4.0:* • NOAEL 1–100 mg kg⁻¹ d⁻¹ (≥ 90 d/human) *5.0:* • NOAEL ≤1 mg kg⁻¹ d⁻¹ (≥ 90 d/human)	*1.0:* • Low potential *2.0:* Moderate potential *3.0:* High potential

CARS GreenList™	Proprietary	Proprietary								
PRIO[24] (chemicals are divided into two categories: phase-out and risk-reduction substances)	*Risk-reduction:* • LD_{50} <200 mg kg^{-1} • EU R25, R26, R27, R28, R39/26, R39/27 and R39/28	*Phase-out:* • CMR Cat. 1 and 2 • EU R45 or R49 *Risk-reduction:* • CMR Cat. 3 • EU R68	*Phase-out:* • CMR Cat. 1 and 2 • EU R46 *Risk-reduction:* • CMR Cat. 3	*Phase-out:* • CMR Cat.1 and 2 • EU R60 and R61	*Phase-out:* • NOEC <30 mg kg^{-1} *Risk-reduction:* • EU R48/23, 48/24 and R48/25	N/A	*Risk-reduction:* • EU R42 and R43	N/A	*5.0:* • NOAEL ≤1 mg kg^{-1} d^{-1} (≥90 d/mammal) • L/NOAEL ≤10 mg kg^{-1} d^{-1} (invertebrate) • L/NOAEL ≤0.1 lb acre^{-1} (plant)	There are no generally accepted criteria for endocrine-disruptive substances. An assessment is made on a case-by-case basis
Quick Scan[25] (the lower the hazard number, the higher the hazard)	*G4:* • Not classified *G3:* • Harmful or possible risk of irreversible effects through inhalation (EU R20, R68/20), in contact with skin (EU R21, R68/21) or if swallowed (EU R22, R68/22) • Harmful: may cause lung damage if swallowed (EU R65) *G2:* • Toxic or danger of very serious irreversible effects through inhalation (EU R23, R39/23), in contact with skin (EU R24, R39/24) or if swallowed (EU R25, R39/25)	*C4:* • Not classified *C2:* • Limited evidence of carcinogenicity (EU R40 CMR Cat. 3) *C1:* • May cause cancer (EU R45, R49 CMR Cat. 1 and 2)	*M4:* • Not classified *M1:* • May cause heritable genetic damage (EU R46)	*R4:* • Not classified *R2:* • Possible risk of impaired fertility (EU R62) • Possible risk of harm to the unborn child (EU R63) • May cause harm to breastfed babies (EU R63) *R1:* • May impair fertility (EU R60) • May cause harm to the unborn child (EU R61)	*G4:* • Not classified • Repeat exposure may cause skin dryness or cracking (EU R66) *G3:* • Harmful: danger of serious damage to health by prolonged exposure (EU R48/20, R48/21, R48/22) *G1:* • Toxic: danger of serious damage to health by prolonged exposure (EU R48/23, R48/24, R48/25)	*G4:* • Not classified • Irritating to eyes (EU R36), respiratory system (EU R37) or skin (EU R38) *G3:* • Risk of serious damage to eyes (EU R41) *G2:* • Causes severe burns (EU R34, R35)	*G4:* • Not classified *G2:* • May cause sensitization by skin contact (EU R43) • May cause sensitization by inhalation (EU R42)	*H4:* • Not classified		*H2:* • TBD

Table 1.2 Continued.

	Human health endpoint criteria								
Method	Acute mammalian toxicity	Carcinogenicity	Mutagenicity/ genotoxicity	Reproductive/developmental toxicity	Repeat dose toxicity (also referred to as systemic/organ effects)	Irritation/corrosion	Sensitization	Neurotoxicity	Endocrine disruption
	GI: • Very toxic: danger of very serious irreversible effects through inhalation (EU R26, R39/26), in contact with skin (EU R27, R39/27) or if swallowed (EU R28, R39/28)								
Column Model[26]	*Low Risk:* • Substances/preparation which may cause lung damage if swallowed (EU R65) • Vapors causing drowsiness and dizziness *Medium Risk:* • Substances/preparation harmful to health (EU R20, R21, R22) • Non-toxic gases may cause suffocation by air displacement (e.g. nitrogen) *High Risk:* • Toxic substances/preparations (EU R23, R24, R25) • Substances/preparation which may liberate toxic gases when in contact with water or acids (EU R29, R31)	*High Risk:* • Carcinogenic substances of category 3 (Carc. Cat. 3, K3, EU R40) • Preparations containing carcinogenic or mutagenic substances of category 3 in concentrations ≥ 1% *Very High Risk:* • Carcinogenic substances of category 1 or 2 (Carc. Cat. 1, K1, Carc. Cat. 2, K2, EU R45, R49) • Preparations containing carcinogenic substances of category 1 or 2 in concentrations ≥ 0.1%	*High Risk:* • Mutagenic substances of category 3 (Mut. Cat. 3, M3, EU R68) • Preparations containing carcinogenic or mutagenic substances of category 3 in concentrations ≥ 1% *Very High Risk:* • Mutagenic substances of category 1 or 2 (Mut. Cat. 1, M1, Mut. Cat. 2, M2, EU R46) • Preparations containing mutagenic substances of category 1 or 2 in concentrations ≥ 0.1%	*Medium Risk:* • Substances which may accumulate in breast milk (EU R64) • Substances toxic to reproduction of category 3 (Repr. Cat. 3, R_E3, R_F3, EU R62, R63) • Preparations containing substances of category 3 toxic to reproduction in concentrations ≥ 5% (gases ≥ 1%) *High Risk:* • Substances toxic to reproduction of category 1 or 2 (Repr. Cat.1, R_E1, R_F1, Repr. Cat.2, R_E2, R_F2, EU R60, R61) • Preparations containing substances toxic to reproduction of category 1 or 3 in concentrations ≥ 0.5% (gases ≥ 0.2%)	*Low Risk:* • Otherwise chronically affecting substances (no R-phrase, but nonetheless a hazardous substance) *High Risk:* • Substances which can accumulate in the human body (EU R33)	*Low Risk:* • Irritant substances/preparations (EU R36, R37, R38) • Skin-affecting substances/preparations (EU R66) • Skin effects when working in wet environment *Medium Risk:* • Substances/preparations causing burns (corrosive) (EU R34, pH ≤2 or ≥ 11.5) • Substances harmful to eyesight (EU R41) *High Risk:* • Substances/preparations causing severe burns (highly corrosive) (EU R35)	*High Risk:* • Skin-sensitizing substances (EU R42, Sh) • Substances sensitizing the respiratory tract (EU R42, Sa) • Preparations containing skin or respiratory tract-sensitizing substances in a concentration ≥ 1% (gases ≥ 0.2%)	N/A	N/A

GHS Column Model[27]							
Very High Risk: • Very toxic substances/preparation (EU R26, R27, R28) • Substances/preparation which may liberate very toxic gases when in contact with acids (EU R32) *Low Risk:* • Substances/preparation which may cause lung damage if aspirated (H304) • Substances/mixtures with specific target organ toxicity (single exposure), Cat. 3: drowsiness, dizziness (H336) *Medium Risk:* • Substances/mixtures, Cat. 4 (H302, H312, H332) • Non-toxic gases may cause suffocation by air displacement (e.g. nitrogen) *High Risk:* • Acutely toxic substances/preparations, Cat. 3 (H301, H311, H331) • Substances/mixtures that may liberate toxic gases when in contact with water or acids (EUH029, EU H031)	*High Risk:* • Substances/mixtures mutagenic to germ cells, Cat. 2 (AGS: M3, H341) *Very High Risk:* • Substances/mixtures mutagenic to germ cells, Cat 1A or 1B (AGS: M1, M2, H340)	*High Risk:* • Carcinogenic substances of category 2 (SGS: K3, H351) *High Risk:* • Carcinogenic substances of categories A1.1B (AGS: K1, K2, H350, H350i) • Carcinogenic activities or processes according to TRGS 906	*Medium Risk:* • Substances/mixtures that can harm babies via their mothers' milk (H362) • Substances/mixtures toxic to reproduction of category 2 (AGS: $R_{E}3$, $R_{F}3$, H361, H361f, H361d, H361fd) *High Risk:* • Substances toxic to reproduction, Cat. 1A or 1B (AGS' RE1, $R_{F}1$, $R_{E}2$, $R_{F}2$, H360, H360D, H360F, H360Fd, H360Df)	*Low Risk:* • Substances/mixtures with specific target organ toxicity (single exposure), Cat. 3: irritation of the respiratory organs (H335) • Substances chronically harmful in other ways (no H-phrase, but still a hazardous substance) *Medium Risk:* • Substances/mixtures with specific target organ toxicity (single exposure), Cat. 2: organ damage (H371) • Substances/mixtures with specific target organ toxicity (repeated exposure), Cat. 2: organ damage (H373) *High Risk:* • Substances/mixtures with specific target organ toxicity (single exposure), Cat. 1: organ damage (H370)	*Low Risk:* • Irritant substances/preparations: skin or eye (H315, H319) • Skin-affecting substances/preparations (EU R66) • Skin effects when working in wet environment • Skin-damaging substances/mixtures (EUH066) *Medium Risk:* • Substances/mixtures with corrosive effect on respiratory organs (EUH071) • Substances corrosive to the skin (H314, pH ≤2 or ≥ 11.5) *High Risk:* • Substances/mixtures toxic in contact with eyes (EUH070) • Eye-damaging substances/mixtures (H318)	*High Risk:* • Skin-sensitizing substances/mixtures (H317, Sh) • Substances sensitizing the respiratory tract (H334, Sa)	N/A
					N/A		
							N/A

Table 1.2 Continued.

	Human health endpoint criteria								
Method	Acute mammalian toxicity	Carcinogenicity	Mutagenicity/genotoxicity	Reproductive/developmental toxicity	Repeat dose toxicity (also referred to as systemic/organ effects)	Irritation/corrosion	Sensitization	Neurotoxicity	Endocrine disruption
	Very High Risk: • Acutely toxic substances/mixtures, Cat. 1 and 2 (H300, H310, H330) • Substances/preparations which may liberate very toxic gases when in contact with acids (EUH032)				• Substances/mixtures with specific target organ toxicity (repeated exposure), Cat. 1: organ damage (H372)				
Evaluation Matrix[28]	*Green:* • Substance is not dangerous to human health • Substance has EU R65, R67 *Yellow:* • Substance may damage health • Substance has EU R20, R22, R23, R25, R29, R31, R39/23, R39/25 *Red:* • Substance may cause severe health damage • Substance has EU R26, R28, R32, R39/26, R39/28	*Yellow:* • Substance has EU R40 *Red:* • Carcinogen Category 1 or 2 • Substance has EU R45, R49	*Red:* • Mutagen Category 1 or 2 • Substance has EU R46	*Yellow:* • Substance has EU R62, R63 *Red:* • Reproductive Toxicant Category 1, 2 or 3 • Substance has EU R60, R61, R64	*Green:* • Substance has EU R66 *Yellow:* • Substance has EU R48/20, R48/22, R68/20, R68/22 *Red:* • Substance has EU R48/23, R48/25	*Green:* • Substance has only light skin effects • Substance has EU R36, R37 *Yellow:* • Substance damages skin • Substance has EU R21, R24, R34, R38, R39/24, R41, R48/21, R48/24, R68/21 *Red:* • Substance may cause health damage if taken up via the skin • Substance has EU R35, R24, R27, R34, R39/27	*Yellow:* • Substance has EU R42 *Red:* • Substance has EU R43	N/A	*Green:* • There is evidence/tests show that the substance is not endocrine disrupting *Yellow:* • Substance is a suspected endocrine disruptor • Test results are ambiguous *Red:* • Substance is on the list of endocrine-disrupting substances

[a] These values are multiplied by an uncertainty score before a final composite score is calculated.

[b] For carcinogenicity, multiply the $1/ED_{10}$ value by a weight of evidence factor:
- 'known human carcinogen' = 3
- 'likely human carcinogen' = 2
- 'suggestive evidence of carcinogenicity' or 'conflicting data' = 1.

Use the corrected value to score the chemical.

Table 1.3 Comparison of ecotoxicity, environmental fate and physical properties criteria among CAA methods.

Method	Ecotoxicity and environmental fate endpoint criteria				Eutrophication	Explosivity	Flammability
	Acute aquatic toxicity	*Chronic aquatic toxicity*	*Persistence: half-life (d)*	*Bioaccumulation*			
DfE Alternatives Assessment Criteria[12]	*Low:* • LC$_{50}$/EC$_{50}$ >100 mg L^{-1} *Moderate:* • LC$_{50}$/EC$_{50}$ 10–100 mg L^{-1} *High:* • LC$_{50}$/EC$_{50}$ 1–10 mg L^{-1} *Very High:* • LC$_{50}$/EC$_{50}$ <1 mg L^{-1}	*Low:* • LOEC >10 mg L^{-1} *Moderate:* • LOEC 1–10 mg L^{-1} *High:* • LOEC 0.1–1 mg L^{-1} *Very High:* • LOEC <0.1 mg L^{-1}	*Very Low:* • Passes Ready Biodegradability test with 10 day window[b] *Low:* • Half-life in water, soil or sediment <16 d OR passes Ready Biodegradability test not including the 10 day window[b] *Moderate:* • Half-life in water, soil or sediment 16–60 d *High:* • Half-life in water, soil or sediment 60–180 d *Very High:* • Half-life in water, soil or sediment >180 d or recalcitrant	*Very Low:* • BCF/BAF <100 *Low:* • BCF/BAF 100–1000 *High:* • BCF/BAF 1000–100000 *Very High:* • BCF/BAF >100000	*Low:* • Total phosphorus in cleaning product limited to 0.5 wt% • Inorganic phosphate is not present	N/A	N/A
GreenScreen version 1.2[4]	*Low:* • LC$_{50}$/EC$_{50}$/IC$_{50}$ >100 mg L^{-1} *Moderate:* • LC$_{50}$/EC$_{50}$/IC$_{50}$ 1–100 mg L^{-1} or	*Low:* • NOEC >10 mg L^{-1} *Moderate:* • NOEC 0.1–10 mg L^{-1} or • GHS Category 2, 3 or 4	*Low:* • Half-life in soil, sediment <30 d or • Half-life in water <7 d or • Ready biodegradability	*Low:* • BCF/BAF <500 or • Absent such data, log K$_{ow}$ <4 *Moderate:* • BCF/BAF 500 to 1000 or	N/A	*Low:* • No basis for concern identified *Moderate:* • GHS Category: Divisions 1.4, 1.5	*Low:* • No basis for concern identified *Moderate:* • GHS Category 2 – Flammable Gases or

Table 1.3 Continued.

Method	Ecotoxicity and environmental fate endpoint criteria						
	Acute aquatic toxicity	Chronic aquatic toxicity	Persistence: half-life (d)	Bioaccumulation	Eutrophication	Explosivity	Flammability
	• GHS Category 2 or 3 *High:* • $LC_{50}/EC_{50}/IC_{50}$ <1 mg L^{-1} or • GHS Category 1	*High:* • NOEC <0.1 mg L^{-1} or • GHS Category 1	*Moderate:* • Half-life in soil, sediment 30-60 d or • Half-life in water 7-40 d *High:* • Half-life in soil, sediment >60-180 d or • Half-life in water >40-60 d or • Potential for long-range environmental transport *Very High:* • Half-life in soil or sediment >180 d or • Half-life in water >60 d	• Absent such data, log K_{ow} >4-4.5 or • Suggestive evidence of bioaccumulation in humans or wildlife *High:* • BCF/BAF >1000-5000 or • Absent such data, log K_{ow} >4.5-5 or • Weight of evidence demonstrates bioaccumulation in humans or wildlife *Very High:* • BCF/BAF >5000 or • Absent such data, log K_{ow} >5		*High:* • GHS Category: Unstable Explosives or Division 1.1, 1.2 or 1.3	*High:* • GHS Category 2 – Flammable Aerosols or • GHS Category 3 or 4 – Flammable Liquids *High:* • GHS Category 1 – Flammable Gases or • GHS Category 1 – Flammable Aerosols or • GHS Category1 or 2 – Flammable Liquids
C2C version 3.0[19] (Includes only Ingredient Characterization Criteria)	*Green:* • LC_{50} >100 mg/L (96 hr) (fish) • $L(E)C_{50}$ >100 mg/L (48 hr) (daphnia) • $L(E)C_{50}$ >100 mg/L (72/96 hr) (algae) *Yellow:* • LC_{50} = 10-100 mg/L (96 hr) (fish) • $L(E)C_{50}$ = 10-100 mg/L (48 hr) (daphnia) • $L(E)C_{50}$ = 10-100 mg/L (72/96 hr) (algae) *Red:* • LC_{50} <10 mg/L (96 hr) (fish)	*Green:* • NOEC >10 mg/L *Yellow:* • NOEC = 1-10 mg/L *Red:* • NOEC <1 mg/L	*Green:* • Half-life in water, soil, sediment <30/90 days; • Readily biodegradable (based on OECD tests) *Yellow:* • Half-life in air, soil, sediment = 30/90-60/180 days; • Ultimately biodegradable *Red:* • Half-life in water, soil >60/180 days; • Not readily biodegradable	*Green:* • BCF <100 *Yellow:* • BCF = 100-500 *Red:* • BCF >500	N/A	N/A	N/A

• L(E)C$_{50}$ <10 mg/L (48 hr) (daphnia) • L(E)C$_{50}$ <10 mg/L (72/96 hr) (algae)							
P2OASys[20] (levels of concern are characterized as numerical hazard scores; the smaller the number, the less hazardous)	2.00: • LC$_{50}$ >1000 mg L^{-1} (aquatic) • EC$_{50}$ >100 mg L^{-1} (plant) 4.00: • LC$_{50}$ 50–1000 mg L^{-1} (aquatic) • EC$_{50}$ 10–100 mg L^{-1} (plant) 6.00: • LC$_{50}$ 1–50 mg L^{-1} (aquatic) • EC$_{50}$ 1–10 mg L^{-1} (plant) 8.00: • LC$_{50}$ 0.1–1 mg L^{-1} (aquatic) • EC$_{50}$ 0.1–1 mg L^{-1} (plant) 10.00: • LC$_{50}$ <0.1 mg L^{-1} (aquatic) • EC$_{50}$ <0.1 mg L^{-1} (plant)	2.00: • NOAEC >0.2 mg L^{-1} (fish) 4.00: • NOAEC 0.02 mg L^{-1} (fish) 6.00: • NOAEC 0.002 mg L^{-1} (fish) 8.00: • NOAEC 0.0002 mg L^{-1} (fish) 10.00: • NOAEC <0.00002 mg L^{-1} (fish)	2.00: • L • BOD/hydrolysis half-life 4 d 4.00: • L/M • BOD/hydrolysis half-life 10 d 6.00: • M • BOD/hydrolysis half-life 100 d 8.00: • M/H • BOD/hydrolysis half-life 500 d 10.00: • H • BOD/hydrolysis half-life >500 d	2.00: • BCF <10 • Log K_{ow} <1 4.00: • BCF 100 • Log K_{ow} 2 6.00: • BCF 200 • Log K_{ow} 4 8.00: • BCF 1000 • Log K_{ow} 6 10.00: • BCF >1000 • Log K_{ow} >6	N/A	2.00: • 0.0 4.00: • 1.0 6.00: • 2.0 8.00: • 3.0 10.00: • 4.0	2.00: • 0.0 4.00: • 1.0 6.00: • 2.0 8.00: • 3.0 10.00: • 4.0
SCRAM[a21] (levels of concern are characterized as numerical hazard scores; the smaller the number, the less hazardous)	1.0: • E/LC$_{50}$ >1000 mg L^{-1} (aquatic animals) • E/LC$_{50}$ >1000 mg L^{-1} (plant) 2.0: • E/LC$_{50}$ 100–1000 mg L^{-1} (aquatic animals)	1.0: • NOEC >100 mg L^{-1} (fish/amph.) • NOEC >5000 mg L^{-1} (invertebrate) • NOEC >100 mg L^{-1} (plant) 2.0: • NOEC 10–100 mg L^{-1} (fish/amph.)	1.0: • Half-life <4 d (biota, air, water, soil, sediment) 2.0: • Half-life 4–20 d (biota, air, water, soil, sediment)	1.0: • BCF/BAF ≤100 2.0: • BCF/BAF 100–1000 3.0: • BCF/BAF 1000–10 000	N/A	N/A	N/A

Table 1.3 Continued.

	Ecotoxicity and environmental fate endpoint criteria						
Method	Acute aquatic toxicity	Chronic aquatic toxicity	Persistence: half-life (d)	Bioaccumulation	Eutrophication	Explosivity	Flammability
	• E/LC$_{50}$ 100–1000 mg L^{-1} (plant) *3.0:* • E/LC$_{50}$ 10–100 mg L^{-1} (aquatic animals) • E/LC$_{50}$ 10–100 mg L^{-1} (plant) *4.0:* • E/LC$_{50}$ 1–10 mg L^{-1} (aquatic animals) • E/LC$_{50}$ 1–10 mg L^{-1} (plant) *5.0:* • E/LC$_{50}$ ≤1 mg L^{-1} (aquatic animals) • E/LC$_{50}$ ≤1 mg L^{-1} (plant)	• NOEC 1000–5000 mg L^{-1} (invertebrate) • NOEC 10–100 mg L^{-1} (plant) *3.0:* • NOEC 1–10 mg L^{-1} (fish/amph.) • NOEC 100–1000 mg L^{-1} (invertebrate) • NOEC 1–10 mg L^{-1} (plant) *4.0:* • NOEC 0.1–1 mg L^{-1} (fish/amph.) • NOEC 10–100 mg L^{-1} (invertebrate) • NOEC 0.1–1 mg L^{-1} (plant) *5.0:* • NOEC ≤0.1 mg L^{-1} (fish/amph.) • NOEC ≤10 mg L^{-1} (invertebrate) • NOEC ≤0.1 mg L^{-1} (plant)	*3.0:* • Half-life 20–50 d (biota, air, water, soil, sediment) *4.0:* • Half-life 50–100 d (biota, air, water, soil, sediment) *5.0:* • Half-life >100 d (biota, air, water, soil, sediment)	*4.0:* • BCF/BAF 10 000–100 00 *5.0:* • BCF/BAF >100 000			
CARS GreenList™	Proprietary Proprietary						
PRIO (chemicals are divided into two categories: phase-out and risk-reduction substances)[24]	*Risk-reduction:* • E/LC$_{50}$ <0.1 mg L^{-1}	*Phase-out:* • NOEC <0.01 mg L^{-1}	*Phase-out:* • Half-life in seawater >60 d • Half-life in fresh-water >40 d • Half-life in marine sediment >180 d	*Phase-out:* • BCF >2000 *Risk-reduction:* • Log K_{ow} ≥ 4.5	N/A	N/A	N/A

Method						
Quick Scan[25] (the lower the hazard number, the higher the hazard)	*T4:* • LC_{50} >10 mg L^{-1} *T3:* • LC_{50} ≤10 mg L^{-1} *T2:* • LC_{50} ≤1 mg L^{-1} *T1:* • LC_{50} ≤0.1 mg L^{-1}	*T4:* • NOEC >1 mg L^{-1} *T3:* • NOEC ≤1 mg L^{-1} *T2:* • NOEC ≤0.1 mg L^{-1} *T1:* • NOEC 0.01 mg L^{-1}	• Half-life in fresh-water sediment >120d • Half-life in soil >120 d *Risk-reduction:* • Substances that are not readily or inherently biodegradable *P4:* • Readily biodegradable *P3:* • Inherently biodegradable; adaptive or incomplete *P2:* • Inherently biodegradable; slow *P1:* • Not inherently biodegradable; no fast abiotic degradation	*B4:* • BCF <100 *B3:* • BCF ≥ 100 *B2:* • BCF ≥ 500 *B1b:* • BCF ≥ 2000 *B1a:* • BCF ≥ 5000	N/A	N/A
Column Model[26]	*Low Risk:* • Not water-polluting substances/preparations (NWG, formerly WGK 0) *Moderate Risk:* • Substances/preparations of the German water pollution class WGK 1	*Low Risk:* • Not water-polluting substances/preparations (NWG, formerly WGK 0) *Moderate Risk:* • Substances/preparations of the German water pollution class WGK 1	N/A	N/A	*Very High Risk:* • Explosive substances/preparations (EU R2, R3) • Oxidizing substances/preparations (EU R7, R8, R9) • Substances/preparations with specific properties (EU R1, R4, R5, R6, R7, R14, R16, R18, R19, R30, R44)	*Low Risk:* • Inflammable or hardly flammable substances/preparations (for liquids flashpoint >100 °C) *Moderate Risk:* • Hardly flammable substance/preparations (flashpoint 55–100 °C)

Table 1.3 Continued.

	Ecotoxicity and environmental fate endpoint criteria						
Method	Acute aquatic toxicity	Chronic aquatic toxicity	Persistence: half-life (d)	Bioaccumulation	Eutrophication	Explosivity	Flammability
	High Risk: • Substances/preparations without warning symbols N, but with hazards indications EU R52, R3 • Substances/preparations of the German water pollution class WGK 2 *Very High Risk:* • Substances/preparations with the warning symbol N and hazards indications EU R50, R51, R53, R54, R55, R56, R57, R58, R59 • Substances/preparations of the German water pollution class WGK 3	*High Risk:* • Substances/preparations without warning symbols N, but with hazards indications EU R52, R3 • Substances/preparations of the German water pollution class WGK 2 *Very High Risk:* • Substances/preparations with the warning symbol N and hazards indications EU R50, R51, R53, R54, R55, R56, R57, R58, R59 • Substances/preparations of the German water pollution class WGK 3					*High Risk:* • Flammable substances/preparations (EU R10) *Very High Risk:* • Extremely flammable gases and liquids (EU R12) • Spontaneously flammable substances/preparations (EU R17) • Highly flammable substances/preparations (EU R11) Substances/preparations liberating extremely flammable gases when in contact with water (EU R15)
GHS Column Model[27]	*Low Risk:* • Substances/mixtures of the German water pollution class WGK 1 • Substances/mixtures chronically hazardous to the aquatic environment, Cat. 4 (H413)	*Low Risk:* • Substances/mixtures of the German water pollution class WGK 1 • Substances/mixtures chronically hazardous to the aquatic environment, Cat. 4 (H413)	N/A	N/A	N/A	*Low Risk:* • Self-reactive substances/mixtures, Type G (no H-phrase) • Organic peroxides, Type G (no H-phrase) *Moderate Risk:* • Self-reactive substances/mixtures,	*Low Risk:* • Not readily flammable substances/mixtures/flashpoint >60–100 °C, no H-phrase *Moderate Risk:* • Flammable aerosols, Cat. 2 (H223) • Flammable liquids, Cat. 3 (H226)

Moderate Risk:
- Substances/preparations of the German water pollution class WGK 2
- Substances/mixtures chronically hazardous to the aquatic environment, Cat. 3 (H412)

High Risk:
- Substances/mixtures chronically hazardous to the aquatic environment, Cat. 1 (H410) and Cat. 2 (H411)
- Substances hazardous to the ozone layer (H420)

Very High Risk:
- Substances/mixtures acutely hazardous to the aquatic environment, Cat. 1 (H400)
- Substances/preparations of the German water pollution class WGK 3

Moderate Risk:
- Substances/preparations of the German water pollution class WGK 2
- Substances/mixtures chronically hazardous to the aquatic environment, Cat. 3 (H412)

High Risk:
- Substances/mixtures chronically hazardous to the aquatic environment, Cat. 1 (H410) and Cat. 2 (H411)

Very High Risk:
- Substances/mixtures acutely hazardous to the aquatic environment, Cat. 1 (H400)
- Substances/preparations of the German water pollution class WGK 3

Types E and F (H242)
- Organic peroxides, Types E and F (H242)
- Self-heating substances/mixtures, Cat. 2 (H252)
- Oxidizing liquids or solids, Cat. 3 (H272)
- Gases under pressure (H280, H281)
- Substances/mixtures corrosive to metals (H290)

High Risk:
- Self-reactive substances/mixtures, Types C and D (H242)
- Organic peroxides Types C and D (H242)
- Self-heating substances/mixtures Cat. 1 (H251)
- Oxidizing gases, Cat. 1 (H270)
- Oxidizing liquids or solids, Cat. 2 (H272)
- Substances/mixtures with certain properties (EUH001, EUH006, EUH014, EUH018, EUH019, EUH044)

Very High Risk:
- Unstable explosive substances/mixtures (H200)
- Explosive substances/mixtures/products, divisions 1.1 (H201), 1.2 (H202), 1.3

Flammable solids, Cat. 2 (H228) Substances/mixtures which in contact with water emit flammable gases, Cat. 3 (H261)

High Risk:
- Flammable aerosols, Cat. 1 (H222)
- Flammable liquids, Cat. 2 (H225)
- Flammable solids, Cat. 1 (H228)
- Substances/mixtures which in contact with water emit flammable gases, Cat. 2 (H261)

Very High Risk:
- Flammable gases, Cat. 1 (H220) and Cat. 2 (H221)
- Flammable liquids, Cat. 1 (H224)
- Substances/mixtures which in contact with water emit flammable gases, Cat. 1 (H260)

Table 1.3 Continued.

Method	Ecotoxicity and environmental fate endpoint criteria						
	Acute aquatic toxicity	Chronic aquatic toxicity	Persistence: half-life (d)	Bioaccumulation	Eutrophication	Explosivity	Flammability
						(H203), 1.4 (H204), 1.5 (H205) and 1.6 (without H-phrase) • Self-reactive substances/mixtures, Types A (H240) and B (H241) • Organic peroxides, Types A (H240) and B (H241) • Pyrophoric liquids or solids, Cat. 1 (H250) • Oxidizing liquids or solids, Cat. 1 (H271)	
Evaluation Matrix[28]	*Green:* • Low aquatic toxicity *Yellow:* • LC_{50} <0.1 mg L^{-1} • Substance has EU R50, R51, R52	*Green:* • Low aquatic toxicity *Yellow:* • Substance has EU R50, R51, R52 *Red:* • NOEC <0.01 mg L^{-1}	*Green:* • There is evidence that the substance is not a PBT or vPvB • Substance is not classified as dangerous to the environment *Yellow:* • Based on available information, it cannot be excluded that the substance is a PBT/vPvB	*Green:* • There is evidence that the substance is not a PBT or vPvB • Substance is not classified as dangerous to the environment *Yellow:* • Based on available information, it cannot be excluded that the substance is a PBT/vPvB	N/A	*Green:* • Substance has no EU R-phrase *Red:* • Substance is explosive, oxidizing	*Green:* • Substance has no EU R-phrase *Yellow:* • Substance has EU R10, R11, R15 *Red:* • Substance is very flammable or pyrophoric • Substance has EU R7, R8, R9

Red:

- Substance fulfills the **REACH PBT** criteria: half-life in seawater >40 d half-life in freshwater >40 d half-life in marine sediment >180 d half-life in freshwater sediment >120 d half-life in soil >120 d
- Substance fulfills the **REACH vPvB** criteria: half-life in seawater >60 d half-life in freshwater >60 d half-life in marine sediment >180 d half-life in freshwater sediment >180 d half-life in soil >180 d

Red:

- Substance fulfills the **REACH PBT** criteria: BCF >2000
- Substance fulfills the **REACH vPvB** criteria: BCF >5000

---

[a]These values are multiplied by an uncertainty score before a final composite score is calculated.

[b]No degradation products of concern.

CAA paradigm evaluates human health and environmental hazard criteria, and it is important to remember that these criteria are dynamic and change over time.

1.4.1   US EPA's Design for the Environment (DfE)

The US EPA's DfE program works with industry, environmental groups and academia to reduce risk to human health and the environment.[6] The US EPA DfE Alternative Assessments program employs a variety of design approaches to reduce the overall human health and environmental impact of a product or process. DfE's Alternatives Assessment framework is a hazard-based assessment tool for evaluating and differentiating among chemicals based on their concern for human health and environmental hazard, in the process promoting informed substitution.[10,11] DfE's Alternatives Assessments follow six broad steps, as illustrated in Figure 1.3.[10]

As part of a DfE Alternatives Assessment, chemicals are evaluated for numerous health effect and environmental endpoints, including carcinogenicity, mutagenicity, reproductive and developmental toxicity, acute and repeat dose toxicity, toxicity to aquatic organisms and environmental fate.[11] For most hazard endpoints, DfE criteria define 'High,' 'Moderate,' and 'Low' concern. Authoritative sources [the United Nation's Globally Harmonized System (GHS) for the Classification and Labeling of Hazard Substances and US EPA programs] are the basis for these distinctions. Both experimental and modeled data are used in assigning a hazard designation of High, Moderate or Low. In the absence of experimental data, measured data from a suitable analog are preferred over estimated data. Approved modeling tools include EPI Suite, ECOSAR, OncoLogic and the Endocrine Disruptor Screening Program to predict possible hazards.

DfE CAAs do not specify a favored alternative, but do promote informed substitution when combined with cost, performance and national and international regulatory initiatives and requirements. DfE's Alternatives Assessments are used by the Office of Pollution Prevention and Toxics (OPPT) in EPA's Office of Chemical Safety and Pollution Prevention (OCSPP) to seek safer alternatives. Currently, Chemical Action Plans are the primary source for identifying chemical candidates for risk management and specifying actions. EPA proposes to further evaluate the chemicals and address risks. DfE has applied their alternatives assessment paradigm to flame retardants in furniture and printed circuit boards, along with identifying alternatives to chemicals for which there are Agency Action Plans, including nonylphenol ethoxylate surfactants, bisphenol A alternatives in thermal and paper and alternatives to the flame retardants decabromodiphenyl ether (decaBDE) and hexabromocyclododecane.[12]

1.4.2   CPA's GreenScreen™

The GreenScreen™ for Safer Chemicals (version 1.2) is a quantitative chemical screening method designed to help manufacturers identify chemicals of

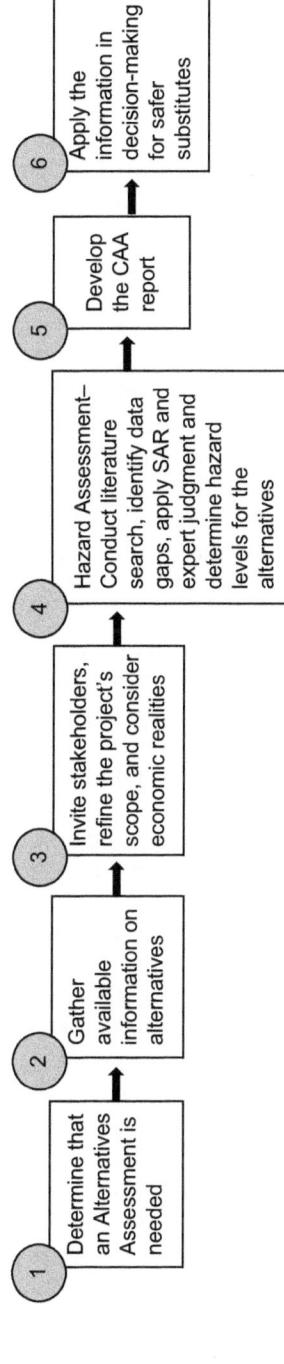

Figure 1.3 US EPA's six broad steps in conducting a DfE alternatives assessment.[10]

concern and inherently less hazardous alternatives to benefit humans and the environment.[13] The structure of the GreenScreen™ method builds from the alternatives assessment approach developed by the US EPA's DfE CAA program.[11] Similarly to a DfE CAA, the GreenScreen™ was developed as a comparative hazard assessment method that assists in selecting chemicals that are inherently safer for humans and the environment. Both Green-Screen™ DfEs are based on the principles of Green Chemistry[14] and focus on managing chemical risk by reducing hazard rather than controlling exposure to potentially toxic chemicals. As part of a GreenScreen™ evaluation process, each ingredient or chemical is assigned a hazard concern level. Individual hazards are evaluated for 18 hazard endpoints (such as carcinogenicity, reproductive toxicity, neurotoxicity aquatic toxicity, persistence and bioaccumulation) and then a level of concern of High, Moderate or Low is assigned for each endpoint for each chemical. Two hazards, persistence and bioaccumulation, have an additional level of concern of Very High to reflect the growing international consensus in defining very persistent and very bioaccumulative (vPvB) chemicals. The threshold values for each hazard endpoint are based on those established by the EPA's DfE program and also the GHS where available. After a chemical has been evaluated against criteria for each hazard endpoint, the results are collectively scored to one of the following GreenScreen™ Benchmark scores:[16,17]

- Benchmark 1: Avoid (Chemical of High Concern).
- Benchmark 2: Use (But Search for Safer Substitutes).
- Benchmark 3: Use (But Still Opportunity for Improvement).
- Benchmark 4: Prefer (Safer Chemical).
- Benchmark 'U': Undetermined (Insufficient Data).

To progress from a lower GreenScreen™ Benchmark score (more hazardous) to a higher (less hazardous) GreenScreen™ Benchmark score, a chemical and its feasible and relevant transformation products must pass all criteria specified under the lower Benchmark. The criteria become increasingly more stringent for environmental and human health and safety and also for data completeness for each Benchmark. Depending on the type and number of data gaps, a chemical may either receive either a downgraded Benchmark score or be assigned Benchmark 'U' to indicate insufficient data. The criteria for Benchmark One align with those that governments in Europe, Canada and the USA use to characterize substances of very high concern.

Figure 1.2 identifies the hazard-based scores for the plasticizer di(2-ethylhexyl) terephthalate (DEHT), which result in a GreenScreen™ Benchmark score of 3: 'Use (But Still Opportunity for Improvement).'

CPA's GreenScreen™ (version 1.0) was initially developed to assess only organic chemicals.[18] Because some inorganic chemicals are recalcitrant and others are persistent as stable moieties, hazard criteria for persistence are not always relevant to inorganics. Persistence alone does not indicate that a chemical is hazardous. Chemicals that are persistent as well as bioaccumulative and toxic are of high concern, as their concentration in the environment increases over time, allowing for more opportunity to exert a toxic effect on human health or aquatic or terrestrial organisms. In 2011, GreenScreen™ (version 1.2) was expanded to address hazards from inorganics such as mineral oxides to allow for comparison of inorganic chemicals used as flame retardants.[17] Under GreenScreen™ version 1.2 criteria, a persistent inorganic chemical with a Low hazard rating for human- and eco-toxicity across all hazard endpoints and a Low hazard rating for bioaccumulation will not be considered problematic. Inorganic chemicals that are only persistent may reach Benchmark 4 [the best score: Prefer (Safer Chemical)]. The GreenScreen™ is not intended to address all critical elements of sustainability. It does not consider social equity or important life-cycle impacts such as energy quantity and quality like other alternatives assessments. A GreenScreen™ assessment does not necessarily include reagents used to synthesize a chemical. Rather, the GreenScreen™ focuses on hazard assessment of chemicals from their point of generation. The GreenScreen™ is a CHA that can be used in a modular way with any other desired components of a full CAA.

1.4.3 Cradle to Cradle® (C2C)

The Cradle to Cradle® method (C2C) began as a proprietary product certification program that incorporates CHA and can be used for CAA. It is currently being maintained and administered by the Cradle to Cradle Products Innovation Institute (C2CPII)[19] and is moving toward greater transparency. C2C uses the metabolism of Nature as a model for human industry – products or materials that cannot be metabolized by the natural world should never enter it. Rather, they should be considered as 'technical nutrients' and designed to flow in technical cycles. Evaluation criteria are grouped into the following categories: 'Material Health,' 'Material Reutilization,' 'Renewable Energy and Carbon Management,' 'Water' and 'Social Fairness.' Materials and products are certified as Basic, Bronze, Silver, Gold or Platinum, based on how the criteria are met. For the Material Health evaluation, all ingredients are 'scored' based on their impact on human and environmental health. Ingredients are evaluated against all common endpoints mentioned above and scored as Green (Little to No Risk), Yellow (Low to Moderate Risk), Red (High Risk) or Grey (Incomplete Data). The C2C paradigm goes further to evaluate a product based on Material Reutilization. The percentage of the product that is considered 'Recyclable or Compostable' is combined with the percentage of the product that is manufactured from 'Recycled or Renewable Content' to calculate a Nutrient Reutilization Score.

In addition to the materials used, the amount of energy and water (both quantity and quality) required for product manufacture and assembly is evaluated for certification. The ultimate goal of C2C design is to have all energy inputs come from 'current solar income' (*i.e.* geothermal, wind, biomass, hydro and photovoltaic energy sources). For Gold certification, at least 50% of purchased electricity is renewably sourced or offset with renewable energy projects, and 50% of direct onsite emissions are offset. For Silver, Gold and Platinum certification levels, the applicant must create or adopt a set of principles or guidelines to illustrate the manufacturing facility's strategies for protecting and preserving the quality and supply of water resources. An audit of the facility is required for Bronze and higher certification. For Gold certification, the applicant must demonstrate that the facility has optimized all product-related process chemicals in effluent. Finally, the organization must adopt and make public one or more statements regarding their social and ethical performance goals such as fair labor practices, corporate and personal ethics, customer service and local community outreach. As detailed in the Cradle to Cradle Certified[CM] Products Standard, requirements for the Social Fairness category increase relative to the specific certification level.[19]

1.4.4   TURI's Pollution Prevention Options Analysis System (P2OASys)

In 1997, the Toxics Use Reduction Institute (TURI) at the University of Massachusetts developed the Pollution Prevention Options Assessment System (P2OASys) tool to help companies determine the potential environmental, worker and public health impacts of alternative technologies aimed at reducing toxics use. The P2OASys tool is a downloadable software package that assists industry in two ways: it systematically examines the potential environmental and worker impacts of toxic use reduction options in a comprehensive manner, examining the total impacts of process changes, rather than simply those of chemical change, and compares toxic use reduction options with the company's current process based on quantitative and qualitative factors.[20] The user inputs both quantitative and qualitative data on a chemical's toxicity, ecological effects and physical properties. Data needed for P2OASys are available through vendors, existing databases or the Toxics Use Reduction Institute. In addition, data for up to three alternatives can be entered as a comparison with the original material. Embedded formulae in the P2OASys program can provide a numerical hazard score for each endpoint for the ingredient(s) in question. For example, if a chemical has an oral LD_{50} of $5000\,\text{mg kg}^{-1}$, it may be assigned a standardized hazard score of 2; the higher the score, the greater is the hazard. Endpoints include human effects, aquatic hazards, persistence/bioaccumulation, atmospheric hazard, disposal hazard, energy/resource use and exposure potential.

1.4.5 The Chemical Scoring and Ranking Assessment Model (SCRAM)

The Chemical Scoring and Ranking Assessment Model (SCRAM) was developed to rank or score chemicals based on persistence, bioaccumulation and toxicity.[21] The program was initially developed for use in the Great Lakes area but is not site specific. This program consists of a simple spreadsheet system that allows the assessor to calculate an index (scores range from 1 to 5) based on the potential exposure and toxicity of chemicals. In addition, unlike other assessment tools, SCRAM addresses the uncertainty of these rankings due to a lack of data. Instead of ignoring a chemical with a large data gap, SCRAM will assign an uncertainty score that, along with its persistence, bioaccumulation and toxicity scores, is used to rank the chemical relative to others in question. In-house libraries and on-line databases are searched for data describing persistence, bioaccumulation and toxicity of a chemical in question. A minimum of one data point is required for bioaccumulation and is scored on the basis of bioaccumulation factors (BAF), bioconcentration factors (BCF) or the octanol/water partition coefficients (K_{ow}). A minimum of one data point is required for scoring a chemical in the persistence category. Persistence is scored based on half-lives in five environmental compartments: biota, air, soil, sediment and water. Measured data take priority over estimated data; however, multi-media models such as the US EPA's EpiSuite and ECOSAR are acceptable when measured data are not available. Uncertainty points are assigned based on the source of the data, whether they are measured or estimated or surrogate data. The final bioaccumulation chemical score is multiplied by the final persistence score and the result is then multiplied by a weighing factor of 1.4. The environmental fate of a chemical is emphasized in SCRAM because of the potential for a chemical deemed not toxic during laboratory studies later potentially to be found to cause toxicity through other mechanisms. A minimum of one data point in at least one toxicity category is required. Acute toxicity scores are composed of two components, acute aquatic and acute terrestrial toxicity, and are based on E/LC_{50} values for aquatic organisms and E/LD_{50} values for terrestrial organisms. Subchronic/chronic scores are based on LO(A)ELs and NO(A)ELs for aquatic and terrestrial organisms in addition to human toxicity values. The subchronic/chronic scores represent repeat dose toxicity, general organ toxicity and also carcinogenicity, mutagenicity, neurotoxicity, immunotoxicity, reproductive and developmental toxicity and endocrine disruption. The most conservative (or lowest) acute and subchronic/chronic data are used when scoring for toxicity. Both the acute toxicity score and the subchronic/chronic toxicity scores are added to uncertainty scores before being summed with the bioaccumulation and persistence scores for a final chemical score. The final chemical score is summed together with a final uncertainty score to give a final composite score that can be compared with those of other chemicals. The relative rankings of chemicals can aid scientists in determining which chemicals need more regulation and/or additional research.

1.4.6 Chemicals Assessment and Ranking System (CARS)

The Chemicals Assessment and Ranking System (CARS) is a decision support tool developed by the Zero Waste Alliance (ZWA) that provides a process for inventorying, assessing and ranking chemicals.[22] Similarly to the SCRAM paradigm, chemicals are ranked according to their potential impacts on human health and safety, ecological health and ecosystem-wide impacts. CARS has been used in support of environmental management systems such as those defined by ISO 14000. The first step is an inventory of chemicals used within an organization based on Material Safety Data Sheets (MSDS). The next step is to screen those chemicals against the CARS database to identify any suspected or potential carcinogens, teratogens, persistence bioaccumulative toxins, global warming gases, ozone-depleting chemicals and more. The CARS database utilizes hazard lists from sources such as the US EPA, the American Conference of Governmental Industrial Hygienists (ACGIH), the National Toxicology Program (NTP) and the International Agency for Research on Cancer (IARC). A Prioritization Criteria Worksheet summarizes the results of the screen and provides a summary of chemicals of concern and the products that contain them. Products are then ranked based on their frequency of use, potential human health and safety impacts, ecological and global impacts and life-cycle costs associated with storage, disposal, training and management. CARS is used to prioritize chemical products for replacement. Specific methods and criteria used in CARS to rank chemicals are not publicly accessible.

1.4.7 SC Johnson & Son's Greenlist™

In 2001, the US company SC Johnson & Son (SCJ) developed a system known as Greenlist™ to classify ingredients found in SCJ products based on each chemical's impact on the environment and human health.[23] So far, SCJ has used Greenlist™ to rate over 95% of their products, and although the Greenlist™ criteria can be obtained from SCJ, the list of chemicals evaluated under Greenlist™ is not made public. The Greenlist™ process has been validated by the UK's Forum for the Future and the US EPA. Each potential ingredient receives a rating from 0 (Restricted Use) to 3 (Best). The company strives to continually improve its overall Greenlist™ ratings. According to SCJ, 27% of ingredients used in its products are classified as 'best' ingredients.[23] Greenlist™ chemical ratings are confidential, so individual chemical and product scores are not available. SCJ licenses the overall GreenList™ process to other companies free of charge,[23] but the total number of companies who have adopted GreenList™ and tailored this process to their own needs is unknown.

1.4.8 PRIO

PRIO is an automated, web-based tool intended to reduce the risks to human health and the environment from chemicals.[24] Developed by the Swedish Chemicals Agency, this model recommends phasing out high-priority chemicals to achieve a non-toxic environment. The tool is not intended to rate or score chemicals based on their human health and environmental hazards, rather it is used to identify the hazardous properties of a chemical. PRIO applies only to chemicals of high concern and categorizes them into two groups: phase-out substances and priority risk-reduction substances. Phase-out substances are those that are of such high concern that they should not be used, such as PBT (Persistent, Bioaccumulative and Toxic) chemicals. Priority risk-reduction substances are those to which special attention should be paid. The criteria used by PRIO have been based on REACH legislation and EU risk phrases. The only criteria against which phase-out substances are evaluated are carcinogenicity, mutagenicity, reproductive toxicity and endocrine disruption. In addition to these, compounds are also assessed as to whether or not they contain any hazardous metals (mercury, cadmium and lead). Priority risk-reduction substances are evaluated for acute and chronic toxicity, sensitization and mutagenicity. Both categories are evaluated for their environmental hazards: persistence and bioaccumulation. Phase-out substances are also assessed for their ozone depletion properties.

1.4.9 The Quick Scan

The Quick Scan method was developed by the Dutch Ministry of Housing, Spatial Planning and the Environment to implement a chemicals substitution policy for chemicals with high hazards.[25] The Quick Scan attempts to develop chemical profiles based on hazard data, classify chemicals into categories of concern and assist industry in taking action for chemicals of high concern. The endpoints assessed during a Quick Scan include persistence (P), bioaccumulation (B), (eco)toxicity (T), health damage in humans (He), carcinogenicity (C), mutagenicity (M), reprotoxicity (R) and hormone disruptive effects (Ho). Hazard data on the chemical in question are gathered and then assessed in order to assign the chemical the appropriate hazard level. Chemicals are also categorized based on the level of concern: 'Very High Concern,' 'High Concern,' 'Concern,' 'Low Concern,' 'No Data, Very High Concern.' Concern categories are adjusted based on the likelihood of exposure. For example, the Quick Scan has four categories of exposure potential based on chemical use: intermediates, industrial applications, professional use and consumer use. Chemicals of Very High Concern are not to be used whereas substances of High Concern are not permitted for use in consumer products or in open profession use. Substances of Concern are permitted with limitations. The Quick Scan does not lead the user to a final selection, but only screens chemicals into broad categories based on a level of concern.

1.4.10   The Column Model and GHS Column Model

The Column Model was developed by the German Institute for Occupational Safety to comply with the German Hazardous Substances Ordinance that requires companies to replace hazardous substances with substances with a lower health risk. Since being introduced in 2001, the Column Model[26] has been updated to reflect GHS hazard classifications.[27] Similarly to the Dutch Quick Scan, chemical hazard data are presented in tabular form including six endpoints: acute health hazards, chronic health hazards, environmental hazards, fire and explosion hazards, exposure potential and hazards caused by procedures. Each chemical or ingredient is evaluated against each end-point and assigned one of five hazard levels: very high, high, medium, low and negligible. Unlike the Quick Scan, the Column Method does not cate-gorize chemicals based on levels of concern. The criteria used to assign a hazard level are determined primarily by risk phrases (R-phrases) (for the original Column Model) or hazard phrases (H-phrases) (for the newer GHS Column Model). Assessors are responsible for identifying and analyzing data. The Column Model and GHS Column Model are not comparative methods; however, the models can be used in two different ways: dominance analysis and positional analysis. In dominance analysis, two alternatives are compared and if one is 'dominated' by the other (*i.e.* poses a greater risk), the dominated alternative is discarded and another selected for comparison until one, non-dominated alternative exceeds all those analyzed. In positional analysis, the decision is made based on the criteria considered most impor-tant to the user.

1.4.11   Evaluation Matrix

Developed for the German Federal Environmental Agency, the Evaluation Matrix is an aggregated data method, similar to the Column Method, which defines risk levels for specific endpoints and uses.[28] The Evaluation Matrix is a template that allows the assessor to perform an evaluation based on sus-tainability using substance-specific criteria. Chemicals are evaluated using tables with specific indicators and the colors green, yellow, red and white are used to indicate the result. Eight substance-specific criteria are evaluated and involve checking to see if that chemical is present on several lists of problematic substances, the possible dangerous physico-chemical properties of the substance, human toxicity, problematic properties related to the environment, mobility within the environment, the origin of the raw materials used, emission of greenhouse gases and resource consumption. Information on the substance's mobility can be found in its MSDS such as EU risk phrases or in other publicly available resources. The criterion used when evaluating whether or not a substance is a PVT or vPvB chemical is that of REACH Annex XIII (see Tables 1.2 and 1.3). Following the evaluation against these endpoints, substances are placed into one of four categories: Green (No action is needed, because available information indicates that the chemical is not problematic), Yellow (No action is needed, because the available information indicates problematic substance properties), Red (There is a high priority to act, because the available information indicates very problematic substance properties), White (There is a need to gather further information, because no or few data are available). A risk index can be created by weighting the endpoints and then summing up the weighted values.

1.5 Challenges Facing Chemicals Alternatives Assessment Methods

Several drawbacks exist when conducting a CAA. The primary challenge is associated with the correct management of hazard and other trade-offs. Switching from a chemical that poses a moderate human health risk to an alternative whose degradation products are aquatically toxic would not be a desirable substitution. This is one example how using CHA and life-cycle thinking would be beneficial when conducting a CAA. It is important to note that although several of the above-mentioned CAA methods use criteria that are more in-depth, not all CAA methods take into account parameters such as resource consumption, energy usage or recyclability. Another primary challenge is complete characterization of the hazard profile for a chemical that has an incomplete data set for human health effects or environmental fate and toxicity.

1.5.1 Chemicals Alternatives Assessments and Data Gaps

In instances where data gaps exist for a chemical (either for a health effects or environmental effects endpoint), it is sometimes possible to characterize the potential hazard of that chemical by evaluating data on chemical surrogates or using software to predict the chemical's potential hazard.

1.5.1.1 Selection of Chemical Surrogates. Hazard characterization data gaps can often be addressed by evaluating hazard data pertaining to one or more structurally similar surrogates. This approach is based on the assumption that a chemical's structure imparts properties that relate to biological activity and that a group of chemicals that produce the same activity have something similar about their chemistry and/or structure. According to the OECD guidelines, an analog selected to fill a data gap must be data rich and share similar physical and chemical properties, including behavior in physical or biological process, with the original compound.[29] Chemicals produced by similar methods by the same company and used for similar purposes make good potential analogs. In addition, degradation products of the parent compound can be used as surrogates, especially if the parent compound is expected to break down readily in the environment.

The US EPA (2010)[30] and OECD (2007)[29] have defined guidelines for identifying similar substances to use analogs based on the following commonalities:

- A common functional group or substance (*e.g.* phenols, aldehydes).
- A common precursor or breakdown product may result in structurally similar chemicals, which can be used to examine related chemicals such as acids/esters/salts (*e.g.* short-chain alkyl methacrylate esters which are metabolized to methacrylic acid).
- An incremental or constant change (*e.g.* increased carbon chain length; typically used for physico-chemical properties such as boiling point).
- Common constituents or chemical class, similar carbon range numbers – used with substances of unknown or variable composition, complex reaction products or biological material.

At all times, the practitioner must include the rationale for his or her choice of analog(s) in the CAA.

1.5.1.2 Software Modeling to Address Data Gaps. If a structurally similar analog is not available, a modeling software program may be suitable to satisfy any data gaps. These computerized systems predict toxicity using structure–activity relationships. Examples of software programs used in CAAs include (but are not limited to) the following:

- OncoLogic (carcinogenicity): http://www.epa.gov/oppt/sf/pubs/oncologic.htm
- Toxicity Estimation Software Tool (TEST): http://www.epa.gov/nrmrl/std/qsar/qsar.html#TEST
- Estimation Program Interface (EPI) Suite (environmental fate): http://www.epa.gov/oppt/exposure/pubs/episuite.htm
- Ecological Structure–Activity Relationships (ECOSAR) (aquatic toxicity): http://www.epa.gov/oppt/newchems/tools/21ecosar.htm
- ToxTree (toxic hazard estimation): http://toxtree.sourceforge.net/
- OECD QSAR Toolbox (www.qsartoolbox.org)
- VEGA (Virtual Models for Evaluating the Properties of Chemicals within a Global Architecture) (www.vega-qsar.eu/download.html).

All output/estimates generated from modeling programs should be appended to a CAA to promote transparency, accuracy and accountability. In some cases, data gaps cannot be filled because viable analogs are not available and the models may not be appropriate. In such situations, data gaps must be clearly presented and weighted in the assessment.

1.6   Conclusion

Finding safer alternatives to problematic chemicals is a growing global concern. The development of frameworks, tools and paradigms has been fueled by regulatory and non-regulatory initiatives at all levels, such as REACH in the European Union and the widespread adoption of the United Nation's GHS around the world. Ideally, a CAA supports the intelligent creation, use and substitution of chemicals to benefit humankind in manners that will not harm the environment or organisms inhabiting the environment.

A number of the CAA methods described in this chapter have advantages over others. Some CAA paradigms place an emphasis on human health hazards, whereas others only address environmental hazards. None of the CAA methods in this chapter are yet automated and require 30–60 h of highly technical work per chemical to characterize its potential hazard. CAA methods require the evaluator to be skilled in toxicology, chemistry, ecotoxicology and environmental science, in addition to having a working knowledge of LCA methods and concepts. Not all CAA paradigms consider LCA attributes. As the complexities of the substance being evaluated (chemical, material, product) increase, so do the complexities of the CAA evaluation. The best CAA paradigms are those that are flexible, adhere strictly to transparency and can be modified in order to meet the specific goals of the evaluator.

Improvement in CAA requires greater transparency in the actual methods employed as part of a CAA. Of the CCA methods discussed in this chapter, only the US EPA's DfE Alternatives Assessment, CPA's GreenScreen™ and Cradle to Cradle® have publicly available criteria and background materials. Most of the CAA methodologies were developed a decade ago, so it is likely that hundreds if not thousands of CAAs have been performed using one of the CAA methods described in this chapter. Despite this, there is no central database that can be accessed to search for completed CAAs. An online database that could index completed CAAs using different CAA paradigms would save time and resources by minimizing duplication of effort among CAA assessors who currently end up performing CAAs that other organizations have likely assessed under one or more of the prevailing CAA methods.

To date, a chemical's human health or environmental footprint has taken a backseat to attributes such as a chemical's impact on a product's performance, reliability or price in the marketplace. This way of thinking is not sustainable or preferable, as only a portion of a chemical's true cost and impact are considered—or paid for—in the marketplace. Similarly, banning or restricting chemicals on an *ad hoc* basis is not a solution to our current system, nor is a system of positive lists allowing the use of only certain chemicals, as that stifles innovation and continuous improvement. Early measures such as the Montreal Protocol and the Basel Convention, and more recently REACH and GHS, demonstrate that faults in the current system are recognized; however, these initiatives do not instill a fundamental change in our way of thinking. CAAs provide a powerful means to improve upon the *status quo* by establishing methods to inform chemical substitution in a scientifically rigorous and defensible manner. Obviously, there is great room for refinement and improvement in CAA methods, as this is a relatively young discipline. It is not the nature of humans—or any living entity—to start out by giving up.[15]

Recognizing the value of CAA and fostering greater adoption of CAA methods provide stakeholders with much-needed tools to address a serious deficiency in the way in which chemicals are used in society, as maintaining the *status quo* is analogous to admitting defeat. As humankind's understanding of the full costs and benefits of chemicals matures, it is critical that we cease using those chemicals that can permanently impair human health or the environment.

References

1. United Nations, *Purpose, Scope and Application of the Globally Harmonized System of Classification and Labelling of Chemicals*, United Nations, Geneva, 4th edn, 2011.
2. D. K. Asante-Duah, *Hazardous Waste Risk Assessment,* Lewis Publishers, Chelsea, MI, 1993, p. 59.
3. S. Edwards, M. Rossi and P. Civie, *Alternatives Assessment for Toxics Use Reduction: a Survey of Methods and Tools*, Technical Report 23, Massachusetts Toxics Use Reduction Institute, Lowell, MA, 2005.
4. Clean Production Action, *GreenScreen™ for Safer Chemicals, Version 1.2. Hazard Criteria*, Clean Production Action, Somerville, Massachusetts, 2012.
5. H. J. Klimisch, M. Andreae and U. Tillmann, *Regul. Toxicol. Pharmacol.*, 1997, **25**, 1.
6. United States Environmental Protection Agency, *Design for the Environment Alternatives Assessments,* US EPA, Washington, DC, 2012.
7. Toxics Use Reduction Institute, *Five Chemicals Alternatives Assessment Study,* Toxics Use Reduction Institute, Lowell, MA, 2006.
8. International Standards Organization, *Environmental Management. Life Cycle Assessment. Principles and Framework (ISO 14040)*, ISO, Geneva, 2006.
9. K. Saur, G. Donato, E. Cobas Flores, P. Frankl, A. Astrup Jensen, E. Kituyi, K. M. Lee, T. Swarr, M. Tawfic and A. Tukker, *Draft Final Report of the LCM Definition Study, Version 3.6,* United Nations Environmental Program/ Society for Environmental Toxicology and Chemistry, Brussels, 2003.
10. E. T. LaVoie, L. G. Heine, M. S. Rossi, R. E. Lee, E. A. Connor, M. A. Vrabel, D. M. DiFiore and C. L. Davies, *Environ. Sci. Technol.*, 2010, **44**, 9.
11. United States Environmental Protection Agency, *Design for the Environment Program Alternatives Assessment Criteria for Hazard Evaluation,* Version 2.0, US EPA, Washington, DC, 2011.
12. United States Environmental Protection Agency, *Existing Chemicals Action Plans,* US EPA, Washington, DC, 2012.
13. Clean Production Action, *GreenScreen™ for Safer Chemicals, Version 1.2,* Clean Production Action, Somerville, Massachusetts, 2011.
14. P. T. Anastas and J. C. Warner, *Green Chemistry: Theory and Practice,* Oxford University Press, Oxford, 1999.

15. A. Rand, *The Fountainhead,* Penguin Books, London, 1968.
16. Clean Production Action, *GreenScreen™ for Safer Chemicals, Version 1.2. Guidance for Hazard Assessment and Benchmarking Chemicals*, Clean Production Action, Somerville, Massachusetts, 2011.
17. Clean Production Action, *GreenScreen™ for Safer Chemicals, Version 1.2. Benchmarks*, Clean Production Action, Somerville, Massachusetts, 2011.
18. Clean Production Action, *GreenScreen™ for Safer Chemicals, Version 1*, Clean Production Action, Somerville, Massachusetts, 2009.
19. B. Kuczenski and R. Geyer, *Chemical Alternatives Analysis: Methods, Models and Tools. Revised Final Report to the Department of Toxic Substances Control*, DTSC Contract 08-T3629, DTSC, Sacramento, CA, 2010.
20. Toxic Use Reduction Institute, *P2OASys Tool to Compare Materials,* University of Massachusetts at Lowell, Lowell, MA, 2012.
21. E. M. Snyder, S. A. Snyder, J. P. Giesy, S. A. Blonde, G. K. Hurlburt, C. L. Summer, R. R. Mitchell and D. M. Bush, *Environ. Sci. Pollut. Res.*, 2000, **7**(1), 51.
22. Zero Waste Alliance, *Chemical Assessment and Ranking System,* Zero Waste Alliance, Portland, OR, 2010.
23. SC Johnson & Son, *Greenlist Process,* SC Johnson & Son, Racine, WI, 2012.
24. Swedish Chemicals Agency, *PRIO – a Tool for Risk Reduction of Chemicals,* Swedish Chemicals Agency, Sundbyberg, 2010.
25. The Netherlands Ministry of Housing, Spatial Planning and the Environment, *Implementation Strategy on Management of Substances,* Ministry of Housing, Spatial Planning and the Environment, The Hague, 2001.
26. Deutsche Gesetzliche Unfallversicherung, *The Column Model. An Aid to Substitute Assessment,* Deutsche Gesetzliche Unfallversicherung, Berlin-Mitte, 2009.
27. Deutsche Gesetzliche Unfallversicherung, *The GHS Column Model. An Aid to Substitute Assessment,* Deutsche Gesetzliche Unfallversicherung, Berlin-Mitte, 2011.
28. A. Reihlen, D. Bunke, R. Grob, D. Jepsen and C. Blum. *Guide on Sustainable Chemicals: a Decision Tool for Substance Manufacturers, Formulators and End Users of Chemicals (Draft),* Umweltbundesamt (German Federal Environmental Agency: International Chemicals Management), Dessau-Rosslau, 2010.
29. Organisation for Economic Co-operation and Development, *Guidance on Grouping of Chemicals*, OECD Environment Health and Safety Publications, Series on Testing and Assessment, Vol. 80, OECD, Paris, 2007.
30. United States Environmental Protection Agency, *Predicting the Toxicities of Chemicals to Aquatic Animals Species,* US EPA, Washington, DC, 2010.

European Initiatives for Selecting Sustainable Flame Retardants

ADRIAN BEARD[†]

ABSTRACT

Flame retardants, also often referred to as fire retardants, are a diverse group of chemical substances or formulations which are added to plastics, wood or textiles to reduce their propensity to ignite. Since the 1990s, flame retardants have started to raise environmental concerns, because some brominated flame retardants [*e.g.* polybrominated diphenyl ethers (PBDEs)] were found to form halogenated dioxins and furans in uncontrolled combustion and because the flame retardants were found in various environmental compartments and biota, including humans. This also led to regulatory restrictions in Europe and elsewhere, *e.g.* the RoHS (European Directive on the Restriction of certain Hazardous Substances in Electric and Electronic Equipment, 2011/65/EU) and WEEE (European Directive on Waste of Electric and Electronic Equipment, 2012/19/EU) legislation. A number of flame retardant producers have responded to these concerns and formed the industry group pinfa (Phosphorus, Inorganic and Nitrogen Flame Retardants Association) in order to develop and promote more environmentally compatible products, focussing on non-halogenated alternatives. This group has successfully engaged in European research initiatives for sustainable flame retardants and also alternatives assessments by the GreenScreen™ methodology or the US Environmental Protection Agency (EPA) design

---

[†]The views and opinions expressed in this chapter are those of the author and do not necessarily reflect the official policy or position of Clariant Corporation.

Issues in Environmental Science and Technology, 36
Chemical Alternatives Assessments
Edited by R.E. Hester and R.M. Harrison
© The Royal Society of Chemistry 2013
Published by the Royal Society of Chemistry, www.rsc.org

for environment projects. Other European drivers and activities towards the sustainability of flame retardants are also presented in this chapter.

2.1 Introduction – What are Flame Retardants?

Flame retardants, also often referred to as fire retardants, are a diverse group of chemical substances or formulations which are added to plastics, wood or textiles to reduce their propensity to ignite. Flame retardants are classified according to their 'active' chemical element – the main groups are products based on bromine, chlorine (grouped together as halogenated flame retardants), phosphorus, nitrogen, aluminium, magnesium and – to a lesser extent – boron, zinc and carbon (expandable graphite). In addition to the different active elements, flame retardants cover a broad range of chemistries, from inorganic small molecules to organic oligomers or substances which can form part of a polymer structure. This great variety stems from the fact that the flame retarding effect can be achieved by several different mechanisms, *e.g.* 'poisoning' of the chemical reactions in the flame zone, formation of char on the surface of the burning item or release of inert gases such as water vapour or nitrogen. Since the mid-1990s, the use of nano-scale materials has been researched, but, for example, nano clay minerals and carbon nanotubes have not yet found significant commercial use. The variety of flame retardant products is necessary, because the materials and products which are to be rendered fire safe are very different in nature and composition. For example, plastics have a wide range of mechanical and chemical properties and differ in combustion behaviour. Therefore, they need to be matched to the appropriate flame retardants in order to retain key material functionalities. Flame retardants are therefore necessary to ensure the fire safety of a wide range of materials including plastics, foam and fibre insulation materials, foams in furniture, mattresses, wood products and natural and man-made textiles. These materials are used in, for example, parts of electrical equipment, cars, aeroplanes and building components.

Flame retardants are designed to minimize the risk of a fire starting in case of contact with a small heat source such as a cigarette or candle or an electrical fault. If the flame retarded material or an adjacent material has ignited, the flame retardant will slow combustion and often prevent the fire from spreading to other items. Many common materials such as paper, wood, textiles and plastics can be easily ignited by smouldering items such as cigarettes or small flames from candles, matches or lighters. The ease of ignition depends not only on the material properties but also very much on the material thickness and orientation (*e.g.* horizontal or vertical ignition). These factors play a crucial role in specific fire tests which define and measure ignitability. Troitzsch compiled a detailed overview of many standards for reaction to fire tests.[1] The amount of flame retardant that one has to add to achieve the desired level of fire safety can range from less than 1% for highly effective flame retardants to more than 50% for inorganic fillers. Typical ranges are 5–20% by

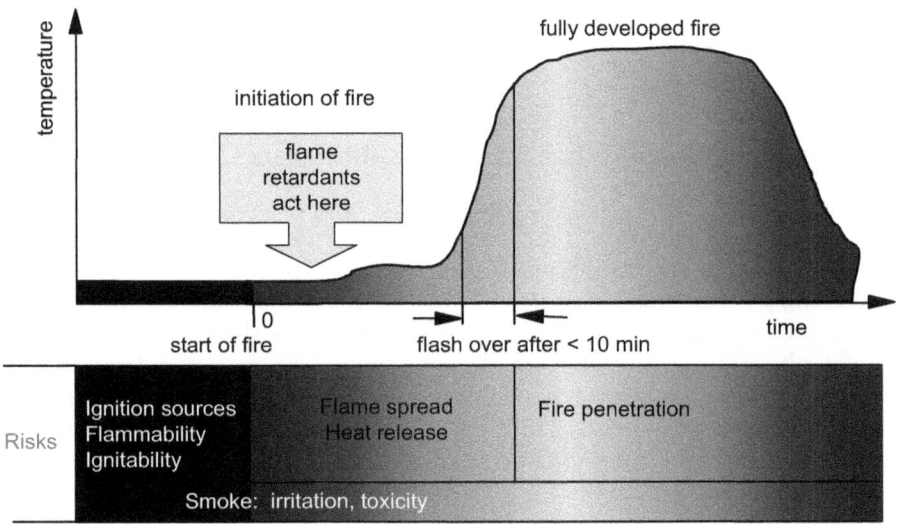

Figure 2.1   A typical fire development curve for an enclosed room with the associated risks along the time line and showing where flame retardants are designed to take effect.

weight. Flame retardants are meant to prevent ignition from small ignition sources and to slow fire spread in the early stages of a developing fire. They are not designed to stop a fully developed fire. However, they may lower the rate of heat release. Figure 2.1 shows where in a fire scenario flame retardants function.

For 2010, the global consumption of flame retardants was estimated at 1.8 million metric tons, which represented a value of US$5600 million.[2] Their regional distribution and division into different chemical groups is shown in Figure 2.2. The consumption in 2010 was almost the same as before the economic crisis in 2008–2009. The overall growth during 2010–2017 is expected to reach almost 7% p.a., but only 3% until 2015.[3] China and Asia-Pacific are main growth drivers, because they are the hub of the electric and electronic (E&E) industry. The transition to halogen-free electronics by leading OEMs (original equipment manufacturers) is expected to shift the market further towards phosphorus, inorganic and nitrogen flame retardants.

2.1.1   *Bromine and Chlorine*

Together, these products are designated halogenated flame retardants (Figure 2.3). There are about 75 different commercial brominated flame retardants (BFRs) but fewer chlorinated products. Their mechanism of action is based on suppression of radical reactions in the flame zone. They come in various forms and can be liquids, powders or pellets. BFRs are commonly used

5622 Mio USD FR Market by Region
Share in percent, 2010

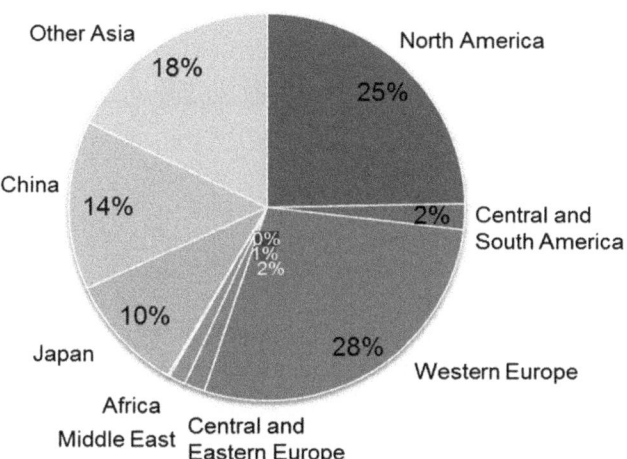

1.78 mio. MT FR Market by Chemistry
Share in percent, 2010

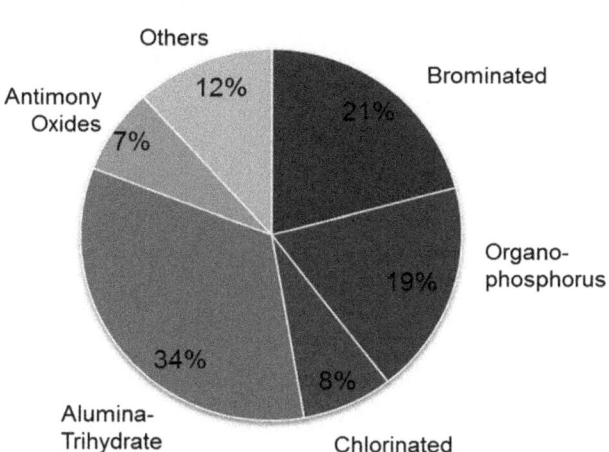

Figure 2.2 Global consumption of flame retardant chemicals, by region (value) and chemistry (tonnage), for 2010.
Source: SRI Consulting.[2]

to prevent fires in electronics and electrical equipment. This area accounts for more than 50% of their applications, for example, in the outer housings of TV sets and computer monitors. The internal circuitry of such devices can heat up and, over time, collect dust, and short-circuits and electrical or electronic malfunctions can occur. Since printed circuit boards are commonly

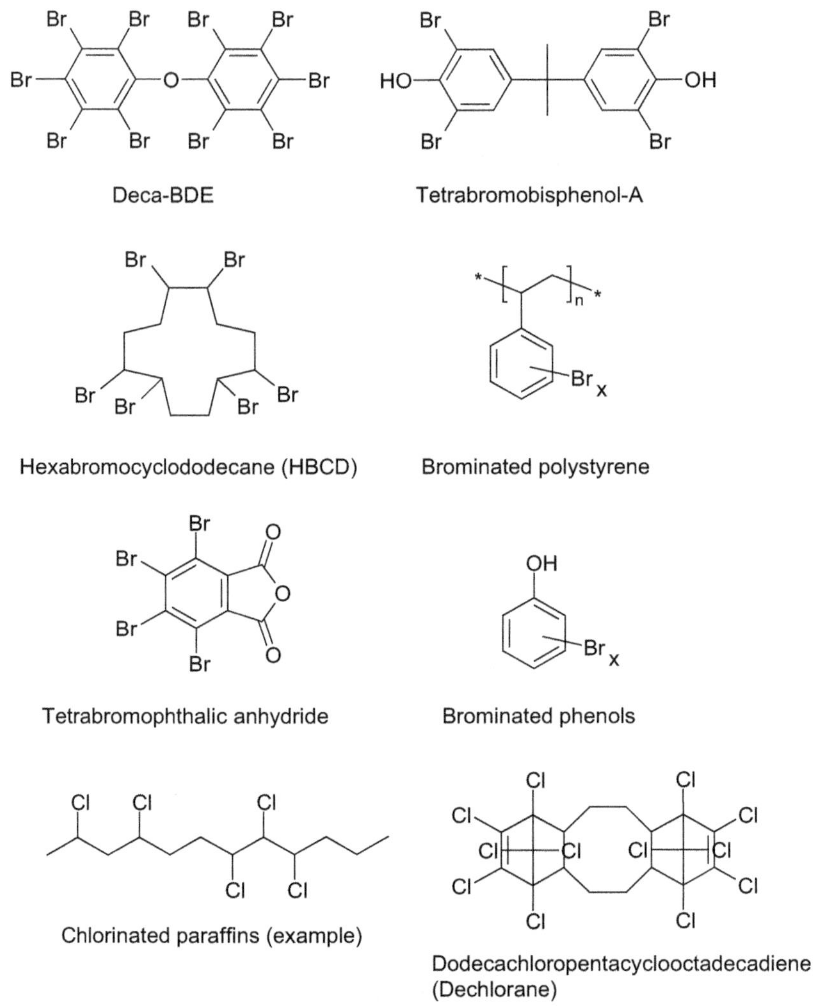

Deca-BDE

Tetrabromobisphenol-A

Hexabromocyclododecane (HBCD)

Brominated polystyrene

Tetrabromophthalic anhydride

Brominated phenols

Chlorinated paraffins (example)

Dodecachloropentacyclooctadecadiene
(Dechlorane)

Figure 2.3 Structures of common brominated and chlorinated flame retardants.

manufactured from flammable epoxy resins, these require fire protection. Here, often a cross-linked brominated epoxy resin polymer manufactured from tetrabromobisphenol-A (TBBPA) is used. In addition, BFRs are applied in wire and cable compounds and also other building materials, such as insulation foams [*e.g.* hexabromocyclododecane (HBCD)]. Furthermore, fabric back-coatings for curtains, seating and furniture in transport and public buildings and also domestic upholstered furniture can be treated with BFRs.

Chlorinated flame retardants are not as common as their brominated counterparts. Chlorinated paraffins represent the largest group. They are straight-chain hydrocarbons ($>C_{10}$) that have been chlorinated up to a chlorine content of 30–70% by weight. Their major applications are in plastics, rubber, coatings, adhesives and textiles. Apart from chlorinated phosphate esters, only a few specialty products have gained commercial importance, *e.g.* Dechlorane Plus® (CAS 13560-89-9), a tricyclic aliphatic compound, for various engineering plastics and thermosets.

Antimony trioxide is commonly used in combination with BFRs or halogenated polymers such as PVC where it produces a synergistic effect. The most important reactions take place in the gas phase and are the result of enhancing the radical chain mechanism of the halogens.

2.1.2 Phosphorus

Phosphorus-containing flame retardants (PFRs) are widely used in standard and engineering plastics, polyurethane (PU) foams, thermosets, coatingsand textiles. The class of phosphorus-containing flame retardants covers a wide range of inorganic and organic compounds and includes both reactive (chemically bound into the material) and additive (integrated into the material by physical mixing only) compounds (Figure 2.4). They have a broad application

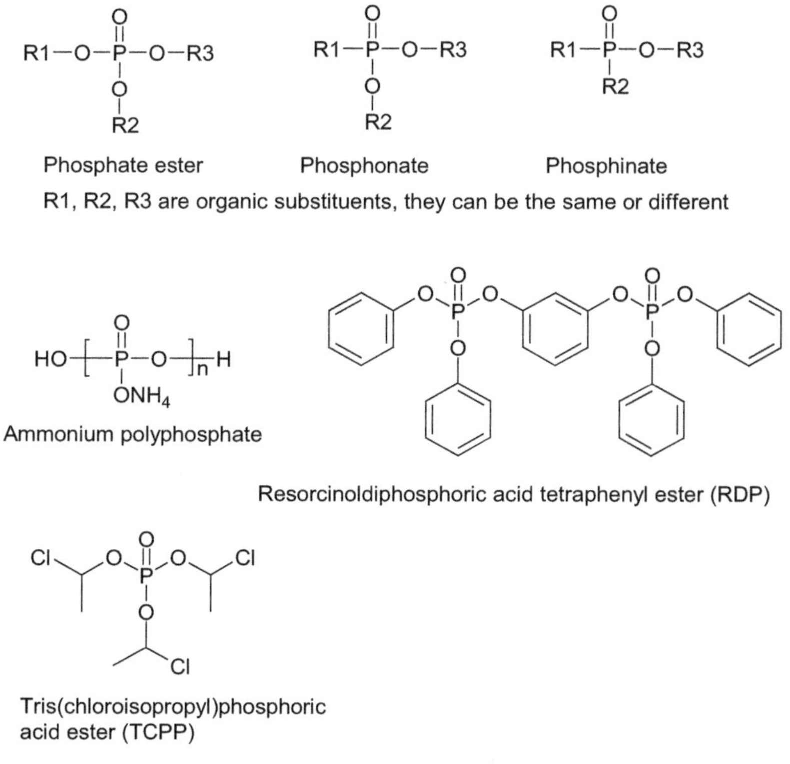

Figure 2.4 Principal structures of phosphorus-based flame retardants and specific examples.

fieldand a good fire safety performance. The most important phosphorus-containing flame retardants are phosphate esters, phosphonates and phosphinates, red phosphorus and ammonium polyphosphate.

Phosphate esters are mainly used as flame retardant plasticizers in poly(vinyl chloride) (PVC) (alkyl/aryl phosphates) and engineering plastics, particularly in styrenic blends for enclosure materials for electrical and electronic equipment. Other applications include phenolic resins and coatings. Additive chlorinated phosphate esters such as tris(2-chloroisopropyl) phosphate (TCPP) and tris(1,3-dichloroisopropyl) phosphate (TDCP) are used in flexible PU foams for upholstered furniture and automotive applications. TCPP is also widely used in rigid PU insulation foams.

Certain phosphates, phosphonates and phosphinates are used as reactive phosphorus-containing flame retardants in flexible PU foams for automotive and building applications. Additive organic phosphinates are a new class of flame retardants for use in engineering plastics, particularly in polyamides and polyesters. Specific reactive phosphorus flame retardants are used in polyester fibres and for wash-resistant flame retardant textile finishes. Other reactive organophosphorus compounds can be used in epoxy resins in printed circuit boards.

Flame retardant grades based on red phosphorus are mainly used in polyamide 6 and 66. Ammonium polyphosphate grades are primarily used in intumescent coatings for steel protection, but also in formulations for rigid and flexible PU foams and polyolefins (injection-moulded types) and in formulations for unsaturated polyesters, phenolics, epoxies and coatings for textiles.

2.1.3 Nitrogen

The largest group of nitrogen-containing flame retardants is based on melamine (Figure 2.5): pure melamine, melamine derivatives, *i.e.* salts with organic or inorganic acids such as boric acid, cyanuric acid, phosphoric acid or pyro-/polyphosphoric acid,and melamine homologues such as melam, melem and

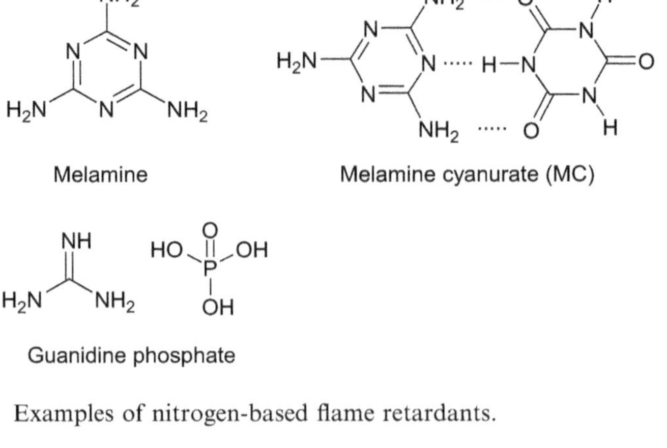

Melamine

Melamine cyanurate (MC)

Guanidine phosphate

Figure 2.5 Examples of nitrogen-based flame retardants.

melon. Nitrogen flame retardants are believed to act by several mechanisms. In the condensed phase, melamine is transformed into cross-linked structures which promote char formation. Ammonia is released in these reactions. A mechanism in the gas phase may involve the release of molecular nitrogen which dilutes the volatile polymer decomposition products. Melamine is mainly used in PU foams, whereas melamine cyanurate is used in nylons or in polypropylene intumescent formulations in conjunction with ammonium polyphosphate. The phosphate and poly- and pyrophosphates of melamine contain both nitrogen and phosphorus and are used in nylons. In some specific formulations, triazines, isocyanurates, urea, guanidine and cyanuric acid derivatives are used as reactive compounds.

2.1.4 Mineral Flame Retardants

With an estimated global consumption of 730 000 metric tons in 2007,[2] aluminium trihydroxide (ATH) is by far the most widely used flame retardant. It is inexpensive, but usually requires high loadings in polymers of up to more than 60%. The flame retardant mechanism is based on the release of water, which cools and dilutes the flame zone (Figure 2.6). Magnesium hydroxide (MDH) is used in polymers which have higher processing temperatures, because it is stable up to temperatures of $\sim 300\,^{\circ}\mathrm{C}$ compared with ATH which decomposes at $\sim 200\,^{\circ}\mathrm{C}$. Fine precipitated ATH and MDH ($<2\,\mu\mathrm{m}$) are used in melt compounding and extrusion of thermoplastics such as PVC or polyolefins for cables. For use in cables, ATH and more often MDH are coated with organic materials to improve their compatibility with the polymer. Coarser ground and air-separated grades can be used in liquid resin compounding of thermosets for electrical applications, seats, panels and vehicle parts. Other inorganic fillers such as talcum and chalk (calcium carbonate) are not flame retardants in the common sense; however, simply by diluting a combustible polymer they reduce its flammability and fire load.

2.1.5 Other Flame Retardants and Synergists: Borates, Zinc Compounds and Expandable Graphite

Since the late 1990s, these 'nano materials' have been intensively studied as potential new flame retardants. However, they have limited effects and require

$$2\,\mathrm{Al(OH)_3} \xrightarrow[+\ 1050\ \mathrm{kJ/kg}]{200\,^{\circ}\mathrm{C}} 3\,\mathrm{H_2O} + \mathrm{Al_2O_3}$$

$$\mathrm{Mg(OH)_2} \xrightarrow[+\ 1300\ \mathrm{kJ/kg}]{300\,^{\circ}\mathrm{C}} \mathrm{H_2O} + \mathrm{MgO}$$

Figure 2.6 Principal reactions of aluminium and magnesium hydroxides.

special processing, so for the time being they are not expected to become viable stand-alone flame retardants. A general update on research up to 2007 was given by Morgan and Wilkie.[4]

A major application of borates is the use of mixtures of boric acids and borax as flame retardants for cellulose (cotton) and of zinc borate for PVC and other plastics such as polyolefins, elastomers, polyamides and epoxy resins. In halogen-containing systems, zinc borate is used in conjunction with antimony oxide, whereas in halogen-free systems, it is normally used in conjunction with aluminium trihydroxide, magnesium hydroxide or red phosphorus. In some particular applications zinc borate can be used alone. Boron-containing compounds act by stepwise release of water and formation of a glassy coating protecting the surface. Zinc compounds were initially developed as smoke suppressants for PVC (zinc hydroxystannate). Later it was found that they also act as flame retardants in certain plastics mainly by promoting char formation. Zinc sulfide shows synergistic effects in PVC and can partly replace antimony trioxide.

Expandable graphite is manufactured from flake graphite by treatment with strong acids such as sulfuric or nitric acid. The acid is trapped in the crystal layers of the graphite ('intercalated'). When it is heated, the graphite starts to expand up to several hundred cubic centimetres per gram, forming a protective layer for the polymer. Expandable graphite is used in plastics, rubbers (elastomers), coatings, textiles and especially in polymeric foams. To achieve perfect flame retardancy, the use of synergists such as ammonium polyphosphate or zinc borate is necessary. The black colour of graphite limits its application in some cases.

2.2 Environmental and Human Health Concerns About Halogenated Fire Retardants

The previous sections have shown the large diversity in flame retardant chemistries. Even within a single retardant category the chemical structure may vary, as the presence of many congeners in, for example, brominated flame retardants demonstrate. These differences in structure may lead to different degrees of environmental impact and it may be difficult to extrapolate the impact derived from one product to another based on these differences. However, on a broad basis, the impact of flame retardants based on organic or inorganic compounds can be well separated. Some reviews provide evidence of the adverse impact of some of these substances in the environment, highlighting the need for future research.[5,6]

Brominated flame retardants (BFRs) are organic substances frequently used in many industries. The high structural diversity of these compounds has been highlighted in the previous sections. From the point of view of possible negative effects on human and animal health, the focus of attention has been devoted mainly to polybrominated diphenyl ethers (PBDEs), which may persist in the environment for extended periodsand which accumulate in living beings.

PBDEs are highly stable lipophilic substances that resemble polychlorinated biphenyls (PCBs) in their structure and physical and chemical characteristics.[7] Nowadays, the most frequently used PBDE-based products are commercially produced decaBDE mixtures. In the past, octaBDE and pentaBDE mixtures were also extensively used. The use of these BFRs has, however, been banned in electrical and electronic equipment in Europe and their use has declined in other regions of the world. In the United States, the three leading producers have declared a phase-out by the end of 2012/2013 after intensive discussions with the US Environmental Protection Agency (EPA).

PBDEs persist in the environment, accumulate in food chains and have toxic effects and they are therefore a potential health risk for both animals and humans. PBDEs have been shown to have a negative effect on processes of hormonal regulation in living organisms and they are therefore regarded as so-called environmental endocrine disruptors.[8,9] Negative effects on the thyroid hormone homeostasis in the serum and plasma of exposed experimental organisms (mainly experimental rodents) have been observed. *In vitro* studies have also demonstrated the ability of these substances to bind to estrogen and androgen receptors, thereby affecting the regulation of sexual steroid hormones. For example, male rats exposed to BFRs show signs of androgenic degeneration, manifest in a delayed onset of puberty, suppressed growth of the ventral prostate gland and seminal vesicles and delayed preputial separation.[10]

As a result of hormonal disruption, experimental animals often show signs of behavioural alterations. Tests on rodents have also demonstrated neurotoxicity of some of the PBDEs, impacting their spontaneous behaviour, learning and memory capacity (*e.g.* Refs 8 and 11). These effects seem to be particularly severe during early life stages and the behavioural alterations become more pronounced as the individuals grow older. Hence a clear correlation exists between the age of individuals and the magnitude of behavioural alteration as a result of exposure to PBDEs.

These findings are of great concern because PBDEs are highly stable substances with only a limited degradation capability. In fact, PBDEs are ubiquitous in the environment and they are even found in remote areas where no heavy industrialization exists (*e.g.* in Greenland). The most frequently detected PBDE congeners in living organisms include congeners BDE-47, BDE-99, BDE-100, BDE-153, BDE-154 and BDE-183, the most important congener in sediments being BDE-209.[5,12]

Sediment is the main source of contaminants that affect biota in aquatic environments. The load of PBDEs in sediments depends chiefly on the amount of industrial pollution in the area. As in the case of sediments, PBDE concentrations detected in fish correlate with the intensity of industrialization in the region. It is interesting that juvenile fish often have lower concentrations of BFRs in their tissues compared with adult or older specimens, which suggests that these substances bioaccumulate during an organism's life span. PBDE concentrations in organisms also depend to a large extent on the trophic position of the organisms in food webs.

BFRs are also readily detected in the terrestrial environment. For example, comparing the concentrations of PBDEs in eggs of wild peregrine falcons (*Falco peregrinus*) from Sweden, Lindberg *et al.*[13] found that the concentration in species living in captivity was substantially lower than that in wild populations. Given the long persistence of BFR in the environment, the results suggest that bioaccumulation of congeners during the birds' lifetime and biomagnification resulting from food web effects explain the high concentration of BFR in these top predators.

Humans are undoubtedly the ultimate top predators on Earth and may therefore inevitably be exposed to BFRs in the food chain (mainly from consumption of fish and other food of animal origin, but also to BFRs from the air).[14–16] While the personnel of industries dedicated to the production and use of BFRs may be particularly at risk of excessive exposure (reviewed in Ref. 17), recent studies have also shown that people who obviously have no direct contact with BFRs may accumulate different levels of these substances in their bodies. Recently, concentrations in breast milk, blood serum and human adipose tissue have been under intensive scrutiny. The review by Mikula and Svoboda[6] suggests that people in the USA suffer from much greater exposure to PBDE s than people in Europe, where the exposure is far lower. However, the authors acknowledged the different sizes of experimental groups and different numbers of PBDE congeners that were monitored, which limits the interpretation of the results.

Higher levels of exposure through recycling operations of electronic waste have also been observed, mainly in industrializing countries in Africa and in China, where these operations often take place under crude conditions. Since the waste is usually thermally treated or even burnt in the open, brominated and chlorinated (and mixed) dibenzofurans and dibenzodioxins are easily formed, with direct emissions to the environment and impact on workers.[12]

Perhaps a very European, at least a typical Swiss, initiative related to flame retardants took place in 2006 and 2007. In a so-called 'consensus platform,' the endocrine effects of brominated flame retardants were evaluated in a 'structured, constructive dialogue between industry, the authorities and scientists; it aims to achieve general agreement on the impact of endocrine disruptors on humans, animals and ecosystems and on action to minimize any detrimental effects. [...]. It involved a total of 19 representatives of the producing and applying industries, the authorities and scientists, in addition to two moderators.'[9] The main conclusions of the group were as follows:

- Tetrabromobisphenol-A (TBBPA) poses no concern as a chemically bound copolymer in the application phase.
- Based on available data, pentabromodiphenyl ether (pentaBDE) is an endocrine active substance, TBBPA and HBCD are 'potentially' endocrine active substances and decaBDE is not.
- Industry should reinforce their efforts to develop and bring to market flame retardants with an improved risk profile.

2.3 European Activities Related to Non-halogenated Flame Retardants

2.3.1 Formation of the Phosphorus, Inorganic and Nitrogen Flame Retardants Association (pinfa)

In response to growing public concern about flame retardants and corresponding legislative activities in Europe and elsewhere, flame retardants based on different chemistries such as phosphorus, nitrogen and inorganic aluminium and magnesium compounds have been developed. The good environmental and health profile of some of these new materials has already been proven in life-cycle studies.[18,19] In 2009, the association of manufacturers for non-halogenated phosphorus, inorganic and nitrogen flame retardants, the Phosphorus, Inorganic and Nitrogen Flame Retardants Association (pinfa), was founded with the goal of generating more information on these products and making it available to the public. In addition, pinfa members wanted to differentiate themselves from the defence-oriented lobby activities of the bromine industry.

Pinfa strives to enter a constructive dialogue with all stakeholders, including environmental non-governmental organizations. The need for (fire) safe products needs to be balanced with environmental concerns and, therefore, flame retardants should be optimized to meet technical, economic and environmental criteria. The pinfa members share the vision of continuously improving the environmental and health profile of their flame retardant products and offering innovative solutions for sustainable fire safety. The following sections highlight some of pinfa's activities and involvements. Table 2.1 shows the current members of pinfa, representing chemical companies of various sizes and also users of flame retardants from the polymer and textile industries.

Pinfa has published two overview brochures on non-halogenated flame retardants in electrical and electronic equipment and also transport applications such as aeroplanes, ships, trains and cars, available on the pinfa website (http://www.pinfa.org/library/brochures.html). The brochures give an overview by application or type of part rather than simply by polymer or flame retardant chemistry, *i.e.* suitable flame retardants for enclosure materials, cables, printed circuit boards, connectors, *etc.*, are listed and described.

Table 2.1 List of pinfa member companies in 2012.

Adeka Palmarole	DSM	Rockwood
BASF	FRX Polymers	Thor
Budenheim	Italmatch Chemicals	Schill & Seilacher
Catena Additives	Krems Chemie	William Blythe
Clariant	Lanxess	Georg H. Luh
Dartex Coatings	Nabaltec	iNEMI (mutual associate
Delamin	Perstorp	membership)
DuPont	Rhodia	

2.3.2   Technology Drivers: Electronics Groups iNEMI and HDPUG

The technical feasibility of alternative flame retardants has been of great concern to the end users such as OEMs. In particular, big brand owners in the electronics sector wanted to switch to halogen-free flame retardants. Therefore, they initiated technical research projects through their associations to compile the state of the art and study technical aspects of a transition.

The High Density Packaging User Group (HDPUG) (www.hdpug.org), with members such as Cisco, IBM, Dell and Intel, was the first to publish an overview of existing halogen-free solutions for the various materials needed in consumer electronics items such as computers and notebooks in 2008.[20] Currently, HDPUG is running technical evaluation projects of halogen free cable materials, *i.e.* alternatives to polyvinyl chloride (PVC) based materials.

The international Electronic Manufacturing Initiative (iNEMI) is another leading group formed to study the technical aspects of halogen-free materials. Some of their key members went forward to publish an 'iNEMI Low-Halogen Position Statement,'[21] which defines the term 'low-halogen' (halogen-free) for these OEMs, because IPC, another larger electronics group, could not come to a conclusion on a standard because of strong political lobbying against it. In addition, iNEMI wants to 'identify technology readiness, supply chain capabilityand reliability characteristics for "HFR-free" alternatives to conventional printed circuit board materials and assemblies' and also to 'define technology limits for HFR-free materials across all market segments'. The challenge for flame retardants in these materials is to fulfil the numerous technical material specifications which are related to harsh processing conditions such as soldering and potential moisture absorption at high temperatures and also electronic insulation or signal damping effects. Pinfa has agreed on a mutual membership with iNEMI to provide input from flame retardant manufacturers in these key technical projects.

2.3.3   European Legislation: RoHS and REACH

After many years of debate, the European Directive on the Restriction of certain Hazardous Substances (RoHS) in Electric and Electronic (E&E) Equipment was published as 2002/95/EC. The goal of this law was to reduce the use of the heavy metals lead, mercury, cadmium and hexavalent chromium in E&E equipment (Figure 2.7). The brominated flame retardant groups of polybrominated biphenyls and diphenyl ethers were also banned. After an initial exemption from the Directive, decaBDE has no longer been permitted in E&E equipment in Europe since July 2008. The legislative process and the debates on the ban on brominated flame retardants created a rising demand for alternative solutions. This was reinforced in the discussions leading to the recast of RoHS which was published as 2011/65/EU. A general ban on brominated

Figure 2.7 The WEEE symbol for electrical and electronic equipment indicating that they must not be discarded as waste but should be recycled.

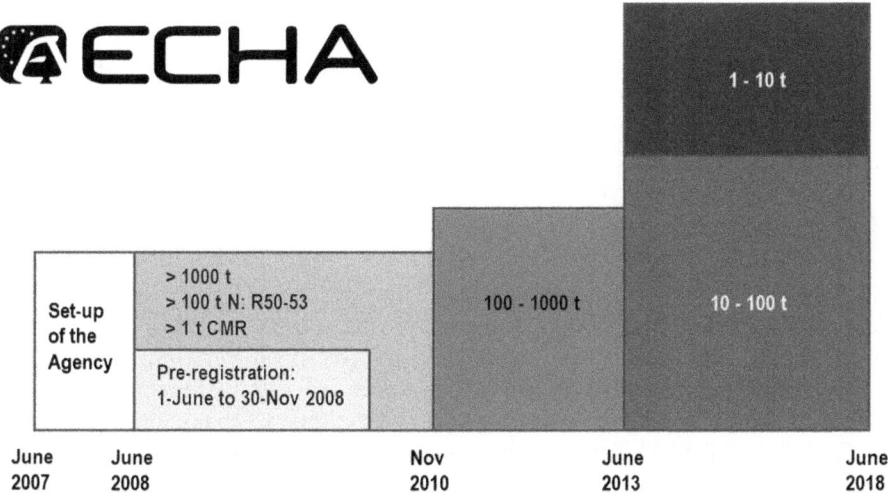

Figure 2.8 Timeline for REACH and substance registrations in Europe.

flame retardants was even proposed. However, in the end no new substance bans were implemented, also with a view to aligning RoHS and REACH more closely. Only hexabromocyclododecane (HBCD) and phthalates are mentioned in the recital as substances which should be evaluated with priority. Nevertheless, from 2008 onwards, a number of leading OEMs made strong commitments with firm deadlines to phase out brominated flame retardants. This 'pull' from the OEMs created a strong demand for alternative solutions together with the readiness of the supply chain to invest in reformulation activities and find technical solutions.

The European chemical legislation on the Registration, Evaluation and Authorization of Chemicals (REACH) came into force in 2007. All chemicals that are marketed in Europe must be registered and information on their environmental and health properties must be submitted. The timeline for registration of different production volume chemicals is shown in Figure 2.8.

Table 2.2   REACH and flame retardants, status May 2012:[22] the Annex 17
restrictions were inherited from the previous chemical legislation
and have been in place since 2004 and before.

Annex 17 Restrictions lists these flame retardants:

- Pentabromodiphenyl ether (pentaBDE, 0.1% w/w)
- Octabromodiphenyl ether (octaBDE, 0.1% w/w)
- Not allowed in articles for skin contact (*e.g.* textiles):
  ○ Tris(aziridinyl)phosphine oxide
  ○ Tris(2,3-dibromopropyl) phosphate (TRIS)
  ○ Polybromobiphenyls (PBBs)

Annex 14 (Candidate) List of Substances of Very High Concern for Authorization:

- Hexabromocyclododecane (HBCD) – persistent, bioaccumulative and toxic (PBT) substance
- Tris(chloroethyl) phosphate (TCEP) – reproductive toxicity category 1b
- Alkanes, C_{10}–C_{13}, chloro (short-chain chlorinated paraffins) – PBT and very persistent and very bioaccumulative (vPvB)
- Boric acid – reproductive toxicity
- Decabromodiphenyl ether (Deca), proposed as SVHC candidate (PBT, vPvB)

REACH has inherited previous chemical restrictions, and those for flame retardants are listed in Table 2.2. However, the emphasis of REACH is clearly to have all chemicals evaluated either by the producers/importers or the authorities. This process has led to the Annex 14 List of Substances of Very High Concern (SVHC) for Authorization – the flame retardants are also shown in Table 2.2. REACH is an elaborate and comprehensive process for compiling data on chemicals. Some chemicals are then further evaluated by authorities, *e.g.* with a view to SVHC criteria. Hence is there any need for further alternatives analysis? The main reasons are as follows:

- REACH provides a plethora of data; however, a comparison of chemicals is difficult even for the expert with the many data points and criteria.
- Data can be contradictory, because many are not consolidated by any authority.[‡]
- Basically, REACH only classifies chemicals into two groups: legal to use (for the registered applications) and restricted by Annexes 14 and 17; in the 'legal to use' group, no further differentiation is made.

[‡]A very revealing example is the database on Classification and Labeling of substances maintained by the European Chemicals Agency (ECHA). It contains entries from any producer or importer with many proven mistakes, some obvious even to the lay person, many others only identifiable for an expert. ECHA has not yet installed a process to appeal data or have erroneous information removed.

2.3.4 GreenScreen™ (see also Chapters 1 and 9)

Pinfa has engaged in a pilot project with Clean Production Action and Hewlett-Packard to have the following substances assessed:

- ammonium polyphosphate, CAS 68333-79-9
- diethylphosphinic acid aluminium salt, CAS 225789-38-8
- aluminium trihydroxide, CAS 21645-51-2
- aluminium oxide hydroxide, CAS 1318-23-6
- melamine polyphosphate, CAS 218768-84-4

Pinfa opted for GreenScreen™, because it promised to be a tool to assess quickly the hazard profile of chemicals and categorize them into an easily understandable grading system. However, the simplified GreenScreen™ approach does not allow for in-depth studies or the inclusion of exposure aspects which is normally done in a risk assessment. Nevertheless, it has proven to be a valuable tool to gain insight quickly into data gaps or ambiguous or contradictory data, often coming from public domain sources. For pinfa, tools such as GreenScreen™ are important to prove and communicate the environmental and health profile of existing or new products. It is also a tool for internally benchmarking products in order continuously to improve them. However, there still seems to be room for improvement within the Green-Screen™ methodology, such as the appraisal of persistence for inorganic materials or the simplification of peer review and criteria review processes. Some of these are being addressed already.

2.3.5 ENFIRO

ENFIRO was a European Commission-funded project which performed a prototypical case study on substitution options for specific brominated flame retardants (BFRs), running from 2009 to 2012[23] (www.enfiro.eu) (Figure 2.9). The project started out from the assumption that not enough environmental and health data were known for alternatives to the established BFRs. The project delivered a comprehensive dataset on viability of production and application, environmental safety and a life-cycle assessment of the alternative flame retardants. Several flame retardant–product combinations were studied in-depth for environmental and toxicological risks, viability of industrial implementation, *i.e.* production of the flame retardant and final products as well as fire safety requirements. The information was used for a risk assessment of the alternative flame retardants. The outcome of that assessment together with socio-economic information was used in a life-cycle assessment. The project followed a pragmatic approach, avoiding final recommendations on environment-compatible substitution options that would not be viable for implementation by industry.

ENFIRO started with a prioritization and selection phase to select the three most promising flame retardant–product combinations for further detailed

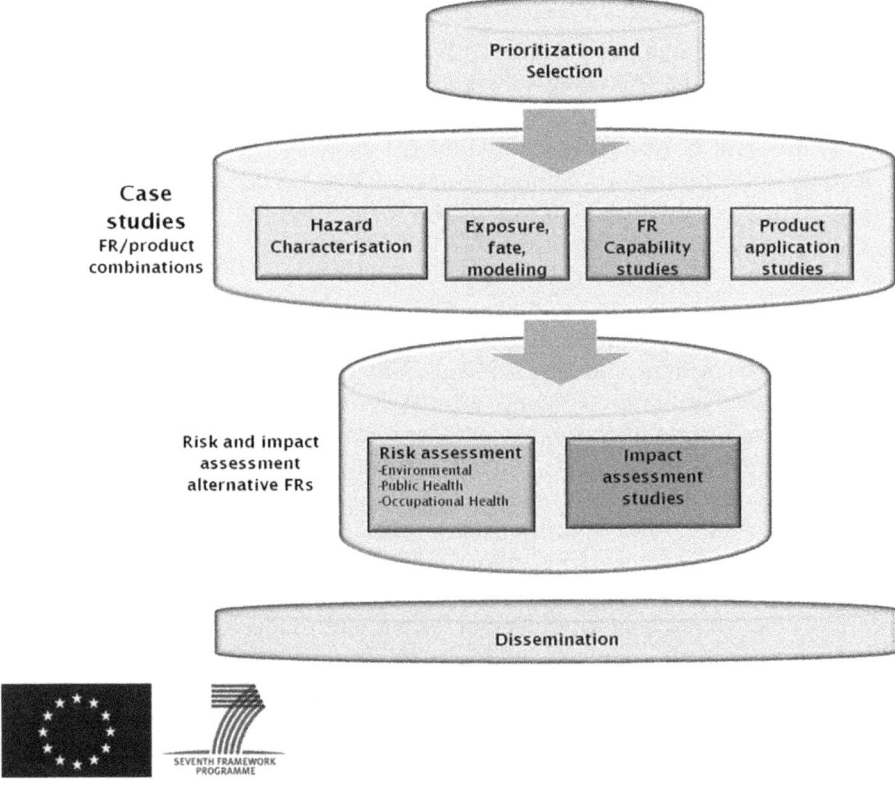

Figure 2.9 ENFIRO approach to study flame retardant substitution options (the European Commission funded this project, No. 226563).

studies. The three selected combinations were studied on hazard characterization, exposure and fate, flame retardants emissions and fire-retarding properties (flame retardant capability studies).

The technical suitability of the flame retardants was assessed when used as such or as mixtures in specific applications such as engineering plastics, printed circuit boards, encapsulants to textiles and intumescent coatings. The collected information was analysed in a risk assessment. After collection of socio-economic information on the flame retardant–product combinations together with the risk assessment, the outcome was digested in a life-cycle assessment, including an analysis of costs and socio-economic aspects. This finally resulted in a recommendation of certain flame retardant–product combinations. A large group of the studied flame retardants were found to have a good environmental and health profile:

- aluminium diethylphosphinate (ALPI)
- aluminium hydroxide (ATH)
- ammonium polyphosphate (APP)

- dihydrooxaphosphaphenanthrene (DOPO)
- magnesium hydroxide (MDH)
- melamine polyphosphate (MPP)
- zinc stannate (ZS)
- zinc hydroxystannate (ZHS).[§]

Overall, they were found to have a much lower tendency to bioaccumulate in fatty tissue than the studied BFRs. The tests on the fire behaviour of materials with different flame retardants revealed that halogen-free flame retardants produce less smoke and toxic fire emissions, with the exception of the aryl phosphates RDP and BDP in styrenic polymers. The leaching experiments showed that the nature of the polymer is a dominant factor and that the leaching behaviour of halogen-free and brominated flame retardants is comparable. The more porous or 'hydrophilic' a polymers is, the more flame retardants can be released. However, moulded plates which represent real-world plastic products showed much lower leaching levels than extruded polymer granules. The impact assessment studies reconfirmed that the improper waste and recycling treatment of electronic products with BFRs can produce dioxins, which is not the case with halogen-free alternatives. The whole project is also summarized in a 20 min documentary film produced by the project partner Callisto – see the ENFIRO website (www.enfiro.eu) for more information.

Pinfa has been actively involved in the 'stakeholder forum' of the project and one member company was even part of the project consortium.

One participating research group has recently published a review of phosphorus flame retardants.[24] It comes to the conclusion that there are some concerns about carcinogenicity for chlorinated alkyl phosphate esters which are commonly found in the environment and also in indoor dust and air, with TCPP [tris(2-chloroisopropyl) phosphate] being the most abundant. While some aryl phosphates show aquatic toxicity, the authors concluded that 'from an environmental perspective, resorcinoldiphosphoric acid tetraphenyl ester (RDP), bisphenol-A diphenyl phosphate (BADP) and melamine polyphosphate may be suitable good substitutes for BFRs.' Other alternatives such as diethyl phosphinates have good (eco-)toxicological properties, but do not degrade easily in the environment. However, this is linked to the required chemical stability in polymers that are processed at elevated temperatures of up to 300 °C.

As part of the project, a review article on the PBT (persistence, bioaccumulation and toxicity) properties of 13 phosphorus, inorganic and nitrogen flame retardants was published[25] based on data that were available before the ENFIRO project. Large data gaps were found for many alternative flame retardants. However, these can be filled with ENFIRO experimental data and

---

[§] ZS and ZHS showed some neurotoxic effects in *in vitro* tests (tests with cells or cell cultures but not whole animals), which were not confirmed in *ex vivo* tests (with organs from animals) and therefore are of limited relevance.

upcoming information from REACH dossiers. So far, a number of non-halogenated flame retardants seem to have acceptable PBT properties.

2.3.6   Ecolabels

Ecolabels have a fairly long history in Europe, with the German Blue Angel being one of the first labels introduced in 1978. There are several other national labels such as the Nordic Swan and the TCO, a private certification organization (Figure 2.10). Within the labels, the environmental criteria for electronic and other products have often banned either specific halogenated flame retardants such as PBDEs or restricted substances based on their hazard properties as defined in risk phrases/hazard statements. The classification and labelling of hazard properties are regulated in chemical legislation, the most recent being the CLP Directive,[26] which implements the Globally Harmonised System of the United Nations.

With its new regulation 66/2010/EC, the European Ecolabel, the 'EU Flower,' has even taken a more rigorous approach by restricting almost all chemicals with environmental and health hazards from ecolabeled products, unless a derogation is granted which is usually substance based. However, this very broad approach has proven unworkable in practice: on the one hand, there are many chemicals (not only flame retardants) with very 'weak' chemical hazards that should not be excluded from products. On the other hand, the

Figure 2.10   Examples of ecolabels in Europe and North America with restrictions on flame retardants.

hazard-based approach was developed for the workplace environment and the direct handling of the chemicals in the first place, not for exposure situations where the chemical is, for example, mixed into a polymer and inside a piece of equipment. Furthermore, the hazard criteria ignore other stages of the life-cycle, *e.g.* recyclability or the formation of dangerous degradation products such as halogenated dioxins and furans. Therefore, a revision process has started that it is hoped will make the criteria more practical and feasible for manufacturers to comply with – pinfa is actively engaged in this process.

The US label EPEAT (www.epeat.net) issued new product requirements for imaging equipment and televisions[27] in late 2012. These have as required criteria that external plastic casings weighing more than 25 g shall contain no more than 0.1 wt% (1000 ppm) of bromine and 0.1 wt% (1000 ppm) of chlorine attributable to brominated flame retardants (BFRs), chlorinated flame retardants (CFRs)and poly(vinyl chloride) (PVC), with some exceptions. There are also optional criteria for eliminating or reducing the BFR/CFR content of printed circuit board laminates and other polymeric materials beyond the enclosures.

2.4 Conclusion

Flame retardants are an essential element of fire safety and will continue to be chosen to protect otherwise flammable materials in applications with fire risks. However, owing to environmental and health concerns regarding some of the chemicals used, the industry is focusing more and more on developing more environmentally benign products. Along with the general trend to replace halogenated flame retardants, reactive systems which become part of the target polymer or textile as well as oligo- or polymeric flame retardants are seen as favourable, because they have no potential to migrate out of finished products. However, monomeric or 'small-molecule' substances can also be viable alternatives, if they have a favourable environmental and health profile. Despite a prolific patent literature, new flame retardant molecules which make it through to commercial success are rare, because the technical hurdles are high: in addition to the compatibility with the target polymer and its processing, many applications demand high mechanical, electrical or endurance criteria. Furthermore, for many applications solutions exist at an attractive price level against which new products have to compete – and switching to a new flame retardant often requires some re-formulation work or re-tooling.

By providing intrinsic fire safety with no maintenance requirements such as with sprinklers and other technical systems, flame retardants are indisputably an important element of fire safety. Flame retardants prevent injuries and fatalities from otherwise highly flammable materials in electronic products, construction materials, furniture and textiles around us. Flame retardants act in the very early stages of a fire: they are meant to prevent ignition from a low-energy source such as an electric spark or a match. In addition, they will slow flame spread in a developing fire. A wide range of inorganic and organic chemical substances will yield a flame retarding effect; most commonly products based on aluminium, bromine, chlorine, phosphorus and nitrogen are used.

However, there are concerns related to release of chemicals into the environment and potential health effects. BFRs have been at the focus of this scrutiny, because environmental organizations brought these to widespread public attention. There have also been findings of phosphate ester flame retardants in indoor air and dust. The European chemical legislation REACH (1907/2006/EC) will further increase the information on hazards and risks of flame retardants in addition to existing scientific literature. In industries such as consumer electronics and computers, there has been a strong trend to replace halogenated flame retardants. Independent studies and research projects have shown that there are a number of flame retardants with good environmental and health profiles. The new industry group for non-halogenated phosphorus, inorganic and nitrogen flame retardants (pinfa) aims to support polymer processors and end users in addition to environmental regulators and groups with necessary product information.

References

1. J. Troitzsch, *Plastics Flammability Handbook*, Hanser Publishers, Munich, 2004, ISBN 3-446-21308-2.
2. H. Janshekar, H. Chinn, W. Yang and Y. Ishikawa, *Flame Retardants Market Report,* SRI Consulting, Englewood, CO, 2011; www.sriconsulting.com (last acceessed 27 December 2012).
3. Industry-Experts, *Flame Retardant Chemicals, a Global Market Overview*, Industry-Experts, 2011.
4. A. B. Morgan and C. A. Wilkie, *Flame Retardant Polymer Nanocomposites,* Wiley, Hoboken, NJ, 2007.
5. E. Eljarrat and D. Barcelo (eds), *Brominated Flame Retardants. Handbook of Environmental Chemistry, Vol. 16*, Springer, Berlin, 2011.
6. P. Mikula and Z. Svobodá, Brominated flame retardants in the environment: their sources and effects (a review), *Acta Vet. Brno*, 2006, **75**, 1587–599.
7. S. L. Schantz, J. J. Widholm and D. C. Rice, Effects of PCB exposure on neuropsychological function in children, *Environ. Health Perspect.*, 2003, **111**, 357–576.
8. D. S. Wikoff and L. Birnbaum, Human health effects of brominated flame retardants, in *Brominated Flame Retardants. Handbook of Environmental Chemistry,* Vol. 16, ed. E. Eljarrat and D. Barcelo, Springer, Berlin, 2011, pp. 19–53.
9. M. Trachsel, *Consensus Platform 'Brominated Flame Retardants,' Final Document*, Swiss National Science Foundation, Berne, 2007; http://www.nrp50.ch/uploads/media/finaldocumentenglish_06.pdf (last accessed 27 December 2012).
10. T. Stoker, R. L. Cooper, C. S. Lambrecht, V. S. Wilson, R. J. Furr and L. E. Gray, *In vivo* and *in vitro* antiandrogenic effects of DE-71,

a commercial polybrominated diphenyl ether (PBDE) mixture, *Toxicol. Appl. Pharmacol.*, 2005, **207**, 78–88.

11. P. Eriksson, H. Viberg, E. Jakobsson, U. Örn and A. Fredriksson, A brominated flame retardant 2,2′,4,4′,5-pentabromodiphenyl ether: uptake, retention and induction of neurobehavioral alterations in mice during a critical phase of neonatal brain development, *Toxicol. Sci.*, 2002, **67**, 98–103.

12. R. J. Law and D. Herzke, Current levels and trends of brominated flame retardants in the environment, in *Brominated Flame Retardants. Handbook of Environmental Chemistry*, Vol. 16, ed. E. Eljarrat and D. Barcelo, Springer, Berlin, 2011, pp. 123–140.

13. P. Lindberg, U. Sellström, L. Häggberg and C. A. De Witt, Higher brominated diphenyl ethers and hexabromocyclododecane found in eggs of peregrine falcons (*Falco peregrinus*) breeding in Sweden, *Environ Sci Technol.*, 2004, **38**, 93–96.

14. A. Schecter, O. Päpke, K. C. Tung, J. Joseph, T. R. Harris and J. Dahlgren, Polybrominated diphenyl ether flame retardants in the U.S. population: current levels, temporal trends and comparison with dioxins, dibenzofurans and polychlorinated biphenyls, *J. Occup. Environ. Med.*, 2005, **47**, 199–211.

15. J. She, A. Holden, M. Sharp, M. Tanner, C. Williams-Derry and K. Hooper, Polybrominated diphenyl ethers (PBDEs) and polychlorinated biphenyls (PCBs) in breast milk from the Pacific Northwest, *Chemosphere*, 2007, **67**, S307–317.

16. R. A. Hites, Polybrominated diphenyl ethers in the environment and in people: a meta-analysis of concentrations, *Environ. Sci. Technol.*, 2004, **38**, 945–956.

17. A. Sjödin, J. R. Patterson, D. G. Erson and A. Bergman, A review on human exposure to brominated flame retardants – particularly polybrominated diphenyl ethers, *Environ. Int.*, 2003, **29**, 829–839.

18. T. Marzi and A. Beard, The ecological footprint of flame retardants over their life-cycle. A case study on the environmental profile of new phosphorus based flame retardants, in *Flame Retardants 2006: Proceedings of the Flame Retardants 2006 Conference*, Interscience Communications, London, 2006, pp. 21–30.

19. A. Beard and T. Marzi, Sustainable phosphorus based flame retardants: a case study on the environmental profile in view of European legislation on chemicals and end-of-life (REACH, WEEE, RoHS), presented at the CARE Innovation Conference, Vienna, November 2006.

20. S. O'Connell and High Density Packaging User Group, *Project Report: Halogen Free Guideline*, High Density Packaging User Group, Austin, TX, 2008; www.hdpug.org (last accessed 27 December 2012).

21. international Electronic Manufacturing Initiative (iNEMI), *iNEMI Low-Halogen Position Statement*, iNEMI, Herndon, VA, 2009; www.inemi.org/node/1970 (last accessed 27 December 2012).

22. European Commission, *Commission Regulation (EC) No. 552/2009 of 22 June 2009 amending Regulation (EC) No. 1907/2006*, European

Commission, Brussels; see also http://echa.europa.eu/chem_data/author-ization_process/candidate_list_table_en.asp (last accessed 27 December 2012).

23. A. Beard and P. E. G. Leonards, Life-cycle and risk assessment of environment-compatible flame retardants: ENFIRO, a prototypical case study, presented at the CARE Innovation Conference, Vienna, 2010.

24. I. van der Veen and J. de Boer, Phosphorus flame retardants: properties, production, environmental occurrence, toxicity and analysis, *Chemosphere*, 2012, **88**, 1119–1153.

25. S. L. Waaijers, D. Kong, H. S. Hendriks, C. A. de Wit, I. T. Cousins, R. H. S. Westerink, P. E. G. Leonards, M. H. S. Kraak, W. Admiraal, P. de Voogt and J. R. Parsons, Persistence, bioaccumulation and toxicity of halogen-free flame retardants, *Rev. Environ. Contam. Toxicol.*, 2012, **222**, 1–71.

26. Regulation (EC) No 1272/2008 of the European Parliament and of the Council of 16 December 2008 on Classification, Labelling and Packaging of Substances and Mixtures, Amending and Repealing Directives 67/548/EEC and 1999/45/EC and Amending Regulation (EC) No 1907/2006, *Off. J. Eur. Union*, 2008, L 353/1.

27. Institute of Electrical and Electronics Engineers, *IEEE 1680.2-2012: IEEE Standard for Environmental Assessment of Imaging Equipment*, and *IEEE 1680.3–2012: Standard for Environmental Assessment of Televisions*, IEEE, New York; www.epeat.net (last accessed 27 December 2012).

MBDC Cradle to Cradle® Product Evaluation and Certification Program

JAY J. BOLUS,* RACHEL PLATIN AND CHRISTOPH SEMISCH

ABSTRACT

Cradle to Cradle® is a design philosophy created by Michael Braungart and William McDonough, two pioneers merging intentional design, chemistry and products for industry. In essence, Cradle to Cradle is about intentional design using Nature as a guide with all materials seen as nutrients for either the technical or biological metabolism, thus eliminating the concept of waste as everything becomes food for another system. Products designed according to this paradigm not only utilize materials that are safe for human and ecological systems, but also utilize manufacturing processes that maximize the use of clean, renewable energy while maintaining, or improving, overall water quality and ensuring social fairness for all involved.

3.1 Introduction

Cradle to Cradle® was developed by Michael Braungart and William McDonough, two pioneers merging intentional design, chemistry and products for industry. Originally used loosely as a term with different meanings as contraindication to 'cradle to grave,'[1] Cradle to Cradle is a beneficial design approach

---

*Corresponding author

Issues in Environmental Science and Technology, 36
Chemical Alternatives Assessments
Edited by R.E. Hester and R.M. Harrison
© The Royal Society of Chemistry 2013
Published by the Royal Society of Chemistry, www.rsc.org

integrating multiple attributes including safe materials, continuous reclamation and reuse of materials, clean water, renewable energy and social fairness.

Michael Braungart formed the Environmental Protection and Encouragement Agency (EPEA) Internationale Umweltforschung GmbH[2] in 1987 and soon afterwards launched the Intelligent Products System (IPS), which defined materials as nutrients with the unique characterization that such materials could be continually reused in biological and technical cycles. The IPS was based on the European precautionary principle and brought a new perspective: that materials can be seen as key parts of technical and biological metabolisms.

At the same time, William McDonough was working as an architect in New York pioneering approaches to building design and concepts—such as '*a building like a tree, a city like a forest*' – which became foundational to the green building movement. His projects included building the first green office in New York for the Environmental Defense Fund in 1984, a solar-powered daycare center operated by children (1989) and a strategy for carbon balance and offset that garnered front-page coverage in the *Wall Street Journal* – 3 years before the Rio Earth Summit. He was a founding member of the American Institute of Architects Committee on the Environment (COTE) and a charter member of the USGBC.

Braungart and McDonough met in 1991 and began to share ideas. Together they merged the concept of materials as nutrients within biological and technical cycles and the concept of intentional design. This later would become the Cradle to Cradle design framework, which presents all of human creation within biological/technical cycles, continuous reuse, renewable power and clean water – all celebrated as a human right.

At the suggestion of Braungart, McDonough was commissioned in 1992 by the city of Hannover, Germany, to prepare the design principles for EXPO 2000, the World's Fair. *The Hannover Principles – Design for Sustainability*[3] were received and honored by Jaime Lerner, Mayor of Curitiba, at the World Urban Forum of the Rio Earth Summit (UNCED) in 1992. They were delivered as a gift from the state of Lower Saxony by McDonough, who attended as a representative of the American Institute of Architects and the International Institute of Architects. In 1995, Braungart and McDonough co-founded McDonough Braungart Design Chemistry, LLC (MBDC).[4]

The *Atlantic* magazine published an article by McDonough and Braungart entitled *The Next Industrial Revolution*[5] in October 1998. This article chronicled the rise of '*eco-efficiency*' (doing more with less) as the main environmental strategy of many leading businesses and introduced the idea of '*eco-effectiveness*' to determine the right thing to do before doing it efficiently. In this article, the terms '*downcycling*' and '*upcycling*' were introduced to show how, by design, we can return product materials with improved, rather than degraded, quality over time.

By 2001, several case studies on the integration of the Cradle to Cradle design principles in product design by leading businesses were made available in video and DVD form by Earthome Productions.[6] Included in this compilation were stories from Designtex (Steelcase), Herman Miller, Ford and Nike. In

2002, the book *Cradle to Cradle – Remaking The Way We Make Things* was published.[7]

MBDC launched the *Cradle to Cradle Certified*[CM] Program[8] in October 2005. As the program has grown worldwide, the desire for an independent review and issuance of certificates was identified to bring the program into the global mainstream. In August 2010, an exclusive, worldwide license was granted to the Cradle to Cradle Product Innovation Institute[9] as a third-party not-for-profit organization to manage the certification program.

Note: Cradle to Cradle®, C2C® and 〉 are registered marks of McDonough Braungart Design Chemistry, LLC. Certified Cradle to Cradle[CM], Cradle to Cradle Certified[CM], and 〉 are registered marks of McDonough Braungart Design Chemistry, LLC used under license by the Cradle to Cradle Products Innovation Institute.

3.1.1 What is Cradle to Cradle® Design?

The Cradle to Cradle® design principles provide a positive agenda for continuous innovation around the economic, environmental and social issues of human design and use of products and services. Specifically, the purpose of the product certification program is to improve the way we make, use and reuse things recognizing two metabolisms, the *biological metabolism* and the *technical metabolism*, with a goal to leave a beneficial footprint for human society and the environment.

The aim is to set a positive course for product and process design and development in a way that will allow natural and technical systems, products and processes to support the diverse living population on Earth. Cradle to Cradle design mirrors the healthy, regenerative productivity of Nature and considers materials as assets not liabilities.

Management theorist Peter Drucker has said that it is a manager's job to do something the right way – to be efficient – but it is an executive's job to do the right thing – to be effective. To date, global efforts by businesses have been focused on becoming more efficient and reducing the (bad) environmental 'footprint' by optimizing existing systems, which may be wrong designs. Cradle to Cradle design is about choosing the right thing to do and then doing that thing the right way to achieve positive outcomes. In other words, to become *'more good'* not just *'less bad.'*

For example, while it makes sense to slow down the use of fossil fuels, this is not the goal. Cradle to Cradle is a continuous improvement process design tool that starts with the positive or beneficial end in mind and executes efficiently towards achieving this goal. In this example, the Cradle to Cradle goal is a move to renewable energy sources.

3.1.2 Long-term Goals – Short-term Actions and Transitions

We start by defining long-term Cradle to Cradle goals and then develop transitional strategies to achieve them. In the short term, we can make successive design-based decisions that will move us to a more sustaining condition. The short-term actions for product development start with complete identification of the materials and chemicals which make up the product and process in order to assess them for human and ecological impacts.

In the medium term, the goal is for designs which are positive or beneficial in terms of cost, performance, esthetics, material health and material (re)utilization potential with continuous use and reuse periods, and additionally, moving renewable energy forward in a cost-effective way, celebrating clean water as a human right and honoring social systems are part of the holistic Cradle to Cradle approach.

The long-term goals can be wholly positive and intended to support 10 billion people and other species. For example, McDonough and Braungart's long term goal is:

> *Our goal is a delightfully diverse, safe, healthy and just world, with clean air, water, soil and power – economically, equitably, ecologically and elegantly enjoyed.*

Cradle to Cradle provides a unique frame of thinking that is based on the precautionary principle and trust in the product supply chain. This is not a framework based on guilt or intended as an opportunity for taking legal actions. Rather, it is the basis of building up a support system.

We work with humility and recognize that checking single chemicals in materials and products does not give the complete picture and there may be unintended consequences, but it is a good start. In focusing attention on chemicals, it is not our intention to promote more animal testing. If a chemical bio-accumulates, we would rather see alternatives substituted.

The question becomes one of design intention and we can ask, what type of products do we want to see? Chemists become designers and designers become chemists. As humans, we accept the limitations of our knowledge and we will make mistakes, but these mistakes need to be reversible by future generations.

The product certification program is a *quality* statement using *quantity* indicators. Each level represents a higher quality indicator using multiple attributes. Today, the program is primarily oriented from a Western cultural perspective. In the longer term, the program is expected to evolve and quality indicators respecting and celebrating cultural diversity are anticipated.

3.1.3   The Cradle to Cradle Principles

In Nature, there is no concept of waste. Everything is effectively food for another organism or system. Materials are reutilized in safe cycles. There are no persistent, bio-accumulative materials that can lead to irreversible changes. The Earth accrues biota grown from the energy of the Sun. We celebrate the diversity of people and of species. We become native to place, celebrating abundance and honoring every child that is born. In short, the design of goods and provision of services can be achieved with three principles in mind:

1. **Eliminate the concept of waste**
   - Nutrients become nutrients again. All materials are seen as potential nutrients in one of two metabolisms – technical and biological metabolisms.
   - Design materials and products that are effectively 'food' for other systems. This means designing materials and products to be used over and over in either technical or biological systems.
   - Design materials and products that are safe. Design materials and products whose nutrient management system leaves a beneficial legacy economically, environmentally and equitably.
   - Create and participate in systems to collect and recover the value of these materials and products. This is especially important for the effective management of scarce materials.
   - Clean water is vital for humans and all other organisms. Manage influent and effluent water streams responsibly, consider local impacts of water use to promote healthy watersheds and ecosystems.
   - Carbon dioxide should be sequestered in soil. Our current practice where carbon dioxide ends up in the oceans and in the atmosphere is a mismanagement of a material.
2. **Use renewable energy**
   - The quality of energy matters. Energy from renewable sources is paramount to effective design.
   - Aligning with Green-e's list of eligible sources, renewable energy sources are solar, wind, hydropower, biomass (when not in competition with food supplies), geothermal and hydrogen fuel cells.
3. **Celebrate diversity**
   - Use social fairness to guide a company's operations and stakeholder relationships.
   - Encourage staff participation in creative design and research projects to enhance your Cradle to Cradle story.
   - Technological diversity is key for innovation; explore different options in looking for creative solutions.
   - Support local biodiversity to help your local ecosystem flourish; strive to have a beneficial social, cultural and ecological footprint.

In the Cradle to Cradle design approach, products that result in materials flowing into the biosphere (from either the product contents or the packaging) are considered to be 'products of consumption.' Materials that are recovered after use can be considered to be 'products of service.' Note that some materials such as paper and bio-plastics are products of consumption as they ultimately return to the biosphere after a number of post-use cycles.

3.1.4   *Complementary Metabolisms*

Cradle to Cradle certification focuses on the characteristics of sustainable materials, products and systems. As a result, this method places a major emphasis on the human and ecological health impacts of a product's ingredients at the chemical level, as well as on the ability of that product to be truly recycled or safely composted. The quality of energy used to create a product, water quantity and quality and social fairness also are essential Cradle to Cradle characteristics and focus areas in this certification process.

Cradle to Cradle design draws on knowledge from the fields of environmental chemistry and material flows management (broadly termed 'industrial ecology') and the fields of industrial and architectural design. It includes the *Intelligent Product System* (IPS) pioneered by chemist Michael Braungart in 1986.

Cradle to Cradle is an innovative approach that models human industry on the processes of Nature's *biological nutrient metabolism* (Figure 3.1) integrated with an equally effective *technical nutrient metabolism* (Figure 3.2) , in which the materials of human industry safely and productively flow within the two metabolisms in a fully characterized and fully assessed way. Products which are designed as services are made from materials which cycle in the technical metabolism at the end of their use cycle. Consumption products, those which naturally end up in the environment (biological cycle) during or post-use, are made from materials which are inherently safe for the biosphere.

Nature's metabolism runs on renewable energy and returns all materials safely in cycles for reuse. Everything can be considered a nutrient with future value. All of our man-made designs exist in this metabolism and many products will result in the nutrients connecting with, and flowing directly into, this

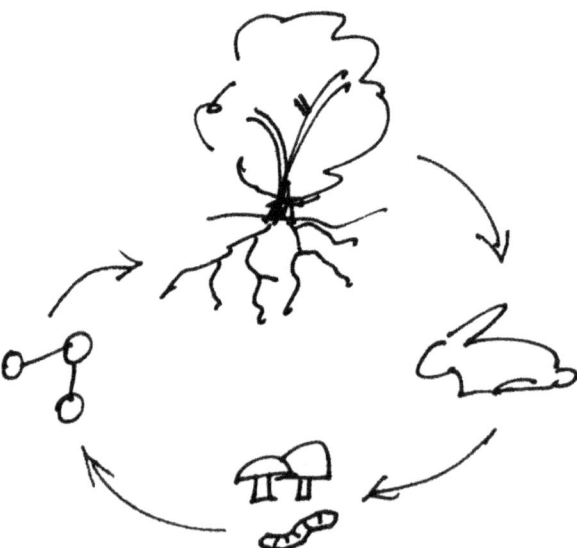

Figure 3.1   Biological metabolism (BM).

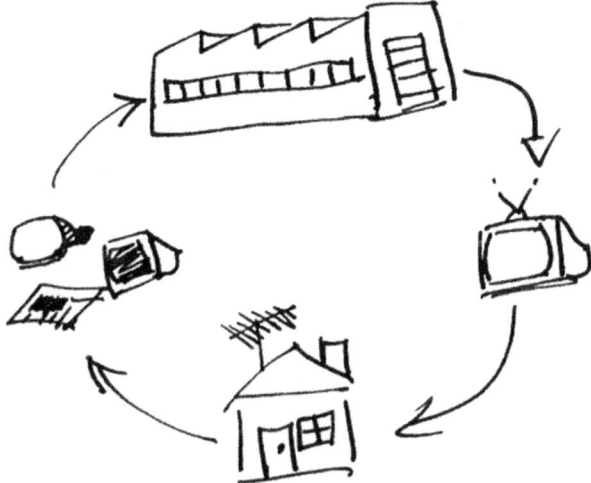

Figure 3.2   Technical metabolism.

system during and after use. These materials need to meet a standard for 'biological nutrients' with the highest level of safety designed in.

Products that have achieved positive design milestones along the continuum of improvement are shown to be suitable for cycling perpetually on Earth, using ingredients that are safe and beneficial – either to biodegrade naturally and restore the soil or to be fully recycled into high-quality materials for subsequent product generations, again and again. This allows a company to eliminate the concept of waste and recover value, rather than creating a future of solid waste liability. Cradle to Cradle design turns contingent liabilities into assets.

3.1.4.1   Effective Material Cycles

3.1.4.1.1   Products of Consumption
A product of consumption is a material or product that is typically changed biologically, chemically or physically during use and therefore enters the biosphere either by nature or by human intention. As a result, products of consumption should consist of biological nutrient materials.

Biological cycle materials and products need to be designed for safe combustion without the need for filters. Biological cycle products such as paper and bio-plastics may go through a series of technical cycles (*e.g.* recycling) before finally going safely into biological systems (*e.g.* composting or incineration for energy recovery).

A biological nutrient product is usable by defined living organisms to carry on life processes such as growth, cell division, synthesis of carbohydrates, energy management and other complex functions. Any material emanating from product consumption that comes into intentional or likely unintentional and uncontrolled contact with biological systems is assessed for its capacity to support their metabolism. Metabolic pathways consist of oxidation, catabolism (degradation, decrease in complexity) and anabolism (construction, increase in complexity), the last two occurring generally in a coupled manner. The classification of products as biological nutrients (or source of nutrients) depends on the biological systems with which they interact. These systems can be more or less complex along the following organizational hierarchy:

- organisms (nutrients for predators);
- organic macromolecules (and combinations thereof) (nutrients for fungi, microorganisms, vegetarian animals; oral, dermal or olfactory nutrients);
- minerals (nutrients for autotrophic plants).

For example: a detergent that is comprised of readily biodegradable materials could be designed such that the material or its breakdown products provide nutrition for living systems. Products such as tires and brake shoes which abrade in use are also products of consumption, but have yet to be designed with biological nutrient materials.

3.1.4.1.2 Products of Service
A product of service is a material or product designed to provide a service to the user without conveying ownership of the materials. Products of service are ideally comprised of technical nutrients which are recovered at the end of use phase.

Technical nutrients (TNs) are products or materials which 'feed' technical systems. While they may or may not be suitable to return to air, soil or water, technical nutrients are never consumed but instead are catabolized (deconstructed) and anabolized (constructed) according to the following hierarchy:

- (dismantle and) reuse;
- (dismantle and) physical transformation (*e.g.* plastic remolding);
- (dismantle and) chemical transformation (*e.g.* plastic depolymerization, pyrolysis, gasification).

Technical nutrients can therefore be managed with service contracts or leasing models so that users benefit from the product service, without owning the materials. In the case of scarce materials, it is especially important to use them in products of service so that they remain available over the long term as useful materials.

3.1.4.1.3 Externally Managed Components (EMCs)
EMCs are a subset of TNs. If the product or any of its assemblies, components or materials fulfills all of the following criteria, the general requirement for full compositional identification will not apply:

1. The supplier of the assembly/component provides the customer with a guarantee for take back and appropriate nutrient management. The supplier may designate a third party or parties for implementation.

2. The supplier provides a guarantee that no subcomponents or chemicals will negatively impact humans and/or the natural environment during the intended and unintended but highly likely use of the end product. This guarantee may be provided via a Cradle to Cradle certification (Gold level or higher) of the assembly/component or other appropriate evidence.
3. Off-gas, migration, leaching, *etc.*, testing of each EMC has proven that no harmful substances are being emitted above detection limits.

Examples of potential EMCs include solar panels, computer motherboards, electric motors and gas lifts for office chairs. Any sealed component with an appropriate take back and management strategy has the potential to be an EMC.

3.2 Product Certification Program Overview

Companies with Cradle to Cradle Certified^{CM} products enjoy increased brand value by achieving product differentiation, building customer retention, facilitating transparency, reducing liability and fostering innovation.

Companies receiving certification will have the opportunity to license the Cradle to Cradle brand certification mark. This mark signifies to the global marketplace that the company has chosen a positive path toward using chemicals, materials and processes for production that are healthy and fit in perpetual use cycles.

This certification program applies to materials, sub-assemblies and finished products, although it is worth noting that materials and sub-assemblies can also be considered 'products' for certification purposes.

This program does not address performance measures associated with any products that qualify for Cradle to Cradle certification. Product compliance with all applicable laws and regulations is assumed. Some rules in the program address activities that are also subject to regulation by local, state or federal authorities. However, nothing contained herein changes legal regulatory requirements or prescribes how compliance is to be achieved. Documentation of compliance with certain key regulations may be included in some sections of the standard, but this in no way changes the underlying regulatory requirements.

3.2.1 Certification Levels

When products qualify for certification at any one of the five levels of Basic, Bronze, Silver, Gold or Platinum, the manufacturer will receive a certificate and a 'scorecard' that can be used to educate consumers on the level of achievement attained in all five categories (as listed in the following section).

In addition, the product and its related certification level and scorecard, will be listed on the Cradle to Cradle Products Innovation Institute's website (www.c2ccertified.org).

3.2.2   Program Categories

Products seeking Cradle to Cradle Certification are evaluated against criteria in the following five categories:

1. **Material Health** – The ultimate goal is for all products to be manufactured using only those materials that have been optimized and do not contain any X or Grey assessed materials/chemicals. As such, products are able to achieve increasingly higher levels of certification as the percentage of optimized materials in the finished product increases.
2. **Material Reutilization** – A key component of Cradle to Cradle design is the concept of technical nutrients and biological nutrients flowing perpetually in their respective metabolisms. Products are evaluated for their nutrient potential and nutrient actualization, in addition to the role the manufacturer plays in product recovery.
3. **Water Stewardship** – Water is a scarce and valuable resource. Product manufacturers are evaluated against their understanding of, and responsibility for, water withdrawals, consumption and releases within the local ecology and are rewarded for innovation in the areas of conservation and quality of discharge.
4. **Renewable Energy and Carbon Management** – Cradle to Cradle products are manufactured in a way that positively impacts our energy supply, ecosystem balance and community and ultimately strives to keep carbon in soil and earth vegetation where it belongs.
5. **Social Fairness** – Cradle to Cradle product manufacturers strive to ensure that progress is made towards sustained business operations that protect the value chain and contribute to all stakeholder interests including employees, customers, community members and the environment.

3.2.3   The Cradle to Cradle Certified^{CM} Marks

These marks may only be used under license and in direct association with the actual certified product or that product's marketing materials. The certification mark may not be used as a general-purpose mark associated with the company and its products. A style guide is available to demonstrate correct usage. An example of the Silver level certification mark is shown in Figure 3.3 as part of a sample scorecard.

3.3   Product Certification Overview

Manufactured products will be reviewed according to the Cradle to Cradle Certification criteria, with the goal of optimization and continuous improvement, in the following five program categories: Material Health, Material Reutilization, Renewable Energy and Carbon Management, Water

	CERTIFICATION LEVEL				
Program Category	BASIC	BRONZE	SILVER	GOLD	PLATINUM
Material Health			√		
Material Reutilization			√		
Renewable Energy and Carbon Management			√		
Water Stewardship				√	
Social Fairness					√
OVERALL CERT LEVEL			√		

Figure 3.3 Certification scorecard.

Stewardship and Social Fairness. The overall certification level for the product will be the lowest level achieved for a given category. An example certification scorecard is shown in Figure 3.3.

3.3.1 Scope

- **Material Health** – The boundaries of review are drawn at the product leaving the direct production facility. The process chemicals associated with the production of certain inputs is included, where applicable (*e.g.* textiles, plated parts, paper, foam).
- **Material Reutilization** – A quantitative measure of a product's design for recyclability and/or compostability. The larger the percentage of a product and/or its components that remains in a technical and/or biological metabolism, the better the score is for this category.
- **Renewable Energy and Carbon Management** – A quantitative measure of the percentage of renewably generated energy that is utilized in the manufacture of the product. Scope 1, Scope 2 (as determined by the Greenhouse Gas Protocol) and Supply Chain are considered depending on the level of certification.
- **Water Stewardship** – A quantitative and qualitative measure of water usage and water effluent related directly to the manufacture of the certified product.
- **Social Fairness** – A qualitative measure of the impact a product's manufacture has on people and communities and includes some measures of general environmental impacts. Requirements apply to the facility or facilities where final product is manufactured unless otherwise noted.

Since this is an evaluation scheme for products only, processes, buildings, cities, *etc.*, are outside the scope and cannot be certified. In addition, there are a number of product attributes that may exclude a manufacturer from certifying. In general, products counter to the intention of Cradle to Cradle design are prohibited from applying. The following are some examples:

- products from rare or endangered species (*e.g.* ivory);
- products with ethical issues (*e.g.* weapons, tobacco, alcohol);
- products leading to or including animal abuse;
- products with apparent safety concerns related to physical and chemical characteristics;
- nuclear power.

3.3.2   Continuous Improvement and Optimization

It is expected that certification holders will make a good faith effort towards optimization in all five categories. Program conformance requires that all applicants outline their intention for the eventual phase-out/replacement of problematic substances (*i.e.* those materials or chemicals with poor ratings) as part of certification. The plan constructed is meant to lay the foundation for prioritizing the phase-out of problematic product inputs in order to move along the Cradle to Cradle continuum.

3.3.3   Material Health

3.3.3.1   Intent.   The review for Material Health generates material assessment ratings based on the hazards of chemicals in products and their relative routes of exposure during the intended (and highly likely unintended) use and end of use product phases. Chemical composition data for materials are needed down to the 100 ppm level (0.01%) to generate full assessment ratings. Exceptions to this include colorants and finishes that might be less than 0.01% of a material but will need to be assessed. Materials deemed 'Externally Managed Components' are exempt from the 100 ppm characterization. The following rating system has been developed to identify those substances that pose the greatest risk to those that pose little to no risk, for a defined product use scenario. The purpose is to give product designers the opportunity to see those substances that pose the greatest risk in order to pick safer alternatives.

Certain substances are inherently problematic in the biosphere and are therefore banned from use in Cradle to Cradle Certified products. In general, these substances are those that have the potential to accumulate in the biosphere or find their way into breast milk and lead to irreversible negative health impacts. Examples include certain phthalates, certain brominated flame retardants and other halogenated organic compounds.

Figure 3.4 illustrates the assessment ratings, which combine the impacts a substance has on human and environmental health within a specific product use scenario, with the potential for the substance to be a technical or biological nutrient.

A	The ingredient is ideal from a Cradle to Cradle perspective for the product in question.
B	The ingredient supports Cradle to Cradle objectives for the product.
C	Moderately problematic properties of the product in terms of quality from a Cradle to Cradle perspective are traced back to the ingredient. The ingredient is still acceptable for use.
X	Highly problematic properties of the product in terms of quality from a Cradle to Cradle perspective are traced back to the ingredient. The optimization of the product requires phasing out this ingredient.
GREY	This ingredient cannot be fully assessed due to either lack of complete ingredient formulation, or lack of toxicological information for one or more components.
Banned	BANNED FOR USE IN CERTIFIED PRODUCTS This material contains one or more Banned substances and cannot be used in a certified product.

Figure 3.4 Material assessment ratings.

3.3.3.2 Summary of Requirements – Material Health
Basic
A product will achieve the level of Basic if the following conditions are met:

- The product does not contain any banned substances based on supplier declarations.
- The product is 100% characterized by its generic materials (*e.g.* aluminium, polyethylene, steel) and/or product categories and names (*e.g.* coatings).
- The appropriate metabolism (*i.e.* technical or biological) has been identified for the product and its components.

Bronze
A product will achieve the level of Bronze if the following conditions are met:

- Meets Basic level requirements.
- The product is at least 75% assessed (by weight) using A, B, C, X ratings. If a product contains EMCs, they will be considered assessed and will contribute to the overall percentage of the product that is assessed.
- For products that are entirely BN in nature (*e.g.* cosmetics, personal care, soaps, detergents) they must be at least 100% assessed.
- A phase-out/optimization strategy has been developed for those materials with an X rating.

Silver

A product will achieve the level of Silver if the following conditions are met:

- Meets Bronze level requirements.
- The product is at least 95% assessed (by weight) using A, B, C, X ratings. If a product contains EMCs, they will be considered assessed and will contribute to the overall percentage of the product that is assessed.
- For products that are entirely BN in nature (*e.g.* cosmetics, personal care, soaps, detergents) they are 100% assessed.
- The product contains no substances known or suspected to cause cancer, birth defects, genetic damage or reproductive harm after the A, B, C, X assessment has been carried out.
- A phase-out/optimization strategy has been developed for those materials with an X rating.

Gold

A product will achieve the level of Gold if the following conditions are met:

- Meets Silver level requirements.
- The product is 100% assessed and contains only non-X materials. All EMCs are considered to be assessed as non-X.
- Products designed for indoor use meet the Cradle to Cradle emissions standard.

Platinum

A product will achieve the level of Platinum if the following conditions are met:

- Meets Gold level requirements.
- All process chemicals are assessed and none are assessed X.

3.3.4   Material Reutilization

3.3.4.1   Intent.   A significant focus of Cradle to Cradle as a product design framework is to promote the creation of an optimized materials economy that eliminates the concept of 'waste.' This category of the program is intended to create incentives for industry to eliminate the concept of 'waste' by designing products with materials that may be perpetually recycled to retain their value. The Program challenges companies to take more responsibility for creating the infrastructure and systems necessary for recovering and recycling materials as the nutrients necessary to fuel our global economies. There are many opportunities for companies to use products as part of the services they offer their customers.

The goal of this program category is to increase the material reutilization potential of a product determined by using the Nutrient Reutilization Scoring method. This method uses a weighted average of the recycled/renewable content and recyclable/compostable content. From a Cradle to Cradle perspective it is more desirable to have a product/material that is truly recyclable/compostable rather than have one that contains a high recycled content but cannot then be recycled or safely composted after use. Therefore, recyclability/compostability is weighted twice compared with recycled/renewable content, according to the following formula:

$$\frac{[(\% \text{ recyclable or compostable}) \times 2] + (\% \text{ recycled or renewable})}{3}$$

3.3.4.2 Summary of Requirements – Material Reutilization

Basic

A product will achieve the level of Basic if the following conditions are met:

- Each generic material in the product is clearly defined as an intended part of a biological or technical cycle.

Bronze

A product will achieve the level of Bronze if the following conditions are met:

- Meets Basic level requirements.
- Material Reutilization Score is between 35 and 49.

Silver

A product will achieve the level of Silver if the following conditions are met:

- Meets Basic level requirements.
- Material Reutilization Score is between 50 and 64.

Gold

A product will achieve the level of Gold if the following conditions are met:

- Meets Basic level requirements.
- Material Reutilization Score is between 65 and 99.
- Company has documented 'nutrient management' strategy for the product including scope, timeline and budget

Platinum

A product will achieve the level of Platinum if the following conditions are met:

- Meets Basic level requirements.
- Material Reutilization Score is = 100.
- Product is actively being recovered and cycled in either a technical and/or biological metabolism.

3.3.5   Renewable Energy and Carbon Management

3.3.5.1   Intent: Eco-effective Energy Production.   Cradle to Cradle envisions a future in which industry and commerce positively impact energy supply, ecosystem balance and community. This is a future powered by current solar income and built on circular material flows. The Renewable Energy and Carbon Management category is a combination of these core principles of Cradle to Cradle design: *produce and use renewable energy* and *eliminate the concept of waste.*

Renewable energy displaces energy produced from fossil fuels, which emit carbon. Changing the quantity and quality of energy used affects the balance of carbon in the atmosphere and ultimately the climate. Ideally, emissions are simply eliminated and renewable energy is produced in excess to be supplied to local communities. When emissions do occur, they are managed as biological nutrients and balanced with an equivalent uptake by natural systems. If we are to reach the ultimate goal of net positive impact, it is critical to measure energy use and emissions accurately. By obtaining these measurements, we can identify and carry out effective plans for transitioning to renewable energy use and achieving a balance of carbon in the atmosphere and as food for building healthy soil.

3.3.5.2   Summary of Requirements – Renewable Energy and Carbon Management

Basic
- Energy use and carbon emissions relevant to the product are characterized (product allocated scope 1 and 2 emissions).

Bronze
- Basic level requirements are met.
- A renewable energy use and carbon management strategy is developed.

Silver
- Basic level requirements are met.
- 5% of emissions as estimated at the Basic level are covered by on-site renewable energy production, RECs and/or offsets.

Gold
- Silver level requirements are met.
- 50% of emissions as estimated at the Basic level are covered by on-site renewable energy production, RECs (Renewable Energy Certificates), and/or offsets.

Platinum
- Gold level requirements are met.
- > 100 % of emissions as estimated at the Basic level are covered by on-site renewable energy production, RECs, and/or offsets. The embodied

energy associated with the product from Cradle to Gate is characterized and quantified, and a strategy to optimize is developed.

- 5% + of embodied energy have been covered by offsets or otherwise addressed.

3.3.6   Water Stewardship

3.3.6.1   Intent: Treating Clean Water as a Valuable Resource.   Water Stewardship creates awareness and drive towards the treatment of water as a valuable resource by encouraging effective management and use strategies. Every product manufacturer has an important responsibility to care for this vital resource and would be wise to manage water resources effectively. These goals are addressed within the program by encouraging an understanding of, and responsibility for, water withdrawals, consumption and releases within local ecosystem(s) and rewarding innovation in the areas of conservation, quality and social equity.

3.3.6.2   Summary of Requirements – Water Stewardship
Basic
- All applicable effluent quality regulatory compliance thresholds are met.
- Local and business specific water-related issues are characterized (for example, the manufacturer will determine if water scarcity is an issue and/ or if sensitive ecosystems are at risk due to direct operations).
- A statement of water stewardship intentions describing what action is being taken for mitigating identified problems and concerns is provided.
- Note: for re-application, progress is demonstrated against any action plans generated under this category.

Bronze
- Basic level requirements are met.
- A facility-wide water audit is competed.

Silver
- Bronze level requirements are met.
- Product relevant effluent chemistry is characterized.
  OR
- Supply chain-relevant water use is quantified and a positive impact strategy is developed (required for facilities with no product relevant effluent).

Gold
- Silver level requirements are met.
- Product-relevant effluent chemistry is optimized (effluents identified as problematic are kept flowing in systems of nutrient recovery; effluents leaving the facility do not contain chemicals assessed as problematic).
  OR

- Demonstrated progress against the strategy developed for the Silver level requirements (required for facilities with no product-relevant effluent).

Platinum
- Gold level requirements are met.
- All water leaving the manufacturing facility meets drinking water quality standards.

3.3.7   Social Fairness

3.3.7.1   Intent: Positive Support for Social Systems.   Social Fairness ensures that progress is made towards sustained business operations that protect the value chain and contribute to all stakeholder interests including employees, customers, community members and the environment. It is important for business ethics to go beyond the confines of the corporate office and permeate the supply chain, engaging responsible manufacturing, enforcing fair treatment of workers and reinvesting in natural capital.

3.3.7.2   Summary of Requirements – Social Fairness
Basic
- A streamlined self-audit is conducted (to be developed) to assess the protection of fundamental human rights.
- Management procedures aiming to address any identified issues are provided.
- Note: demonstration of *progress against the management plan is required for re-application.*

Bronze
- Basic level requirements are met.
- A full social fairness self-audit is complete and a positive impact strategy is developed (based on UN Global Compact Tool or B-Corp).

Silver
- Bronze level requirements are met, plus *one* of the following.
- Material specific and/or issue-related audit or certification relevant to the product is complete (FSC Certified, Fair-trade, *etc.*; relevant to 25% + of product).
  OR
- Supply chain-relevant social issues are fully investigated and a positive impact strategy is developed.
  OR
- The company is actively conducting an innovative social project that positively impacts employee's lives, the local community, global community, social aspects of the product's supply chain or recycling/reuse.

Gold
- Silver level requirements are met, plus *two* of the following.
- Material specific and/or issue-related audit or certification relevant to the product is complete (FSC certified, Fair-trade, *etc.*; relevant to 25% + of product).
  OR
- Supply chain-relevant social issues are fully investigated and a positive impact strategy is developed.
  OR
- The company is actively conducting an innovative social project that positively impacts employee's lives, the local community, global community, social aspects of the product's supply chain or recycling/reuse.

Platinum
- Gold level requirements are met.
- A facility level audit is completed by a third party against an internationally recognized social fairness program (SA8000, B-Corp, *etc.*).
- Material-specific and/or issue-related audit or certification relevant to the product is complete (FSC certified, Fair-trade, *etc.*; relevant to 25% + of product).
- Supply chain-relevant social issues are fully investigated and a positive impact strategy is developed.
- The company is actively conducting an innovative social project that positively impacts employee's lives, the local community, global community, social aspects of the product's supply chain or recycling/reuse.

3.3.8   Certification Program Summary

In general terms, this is what the various achievement levels mean:

3.3.8.1   Basic

- Provisional certification – maximum of 2 years.
- Beginning to understand the product from a Cradle to Cradle perspective (understand emissions/energy footprint, water flows, *etc.*).
- Product's materials are fully characterized, but not assessed (*i.e.* generic material types are known but not all ingredients are known).
- Identifying appropriate metabolism for product and it's components.

3.3.8.2   Bronze

- Officially 'in the game' from a Cradle to Cradle perspective.
- The majority of the product has been assessed.
- Optimization plans are in development for any X-assessed materials.
- Creating a plan to introduce renewable energy into the manufacturing process.

- Gaining a better understanding of water and energy flows.
- Performing a self-audit for social fairness.

3.3.8.3 *Silver*

- Product at least 95% assessed by weight with no highly problematic substances identified (X substances may still be present but none considered CMR (Carcinogenic, Mutagenic, or Reproductively Toxic)).
- Optimization of problematic substances under way.
- More significant amount of renewable energy used to manufacture product.
- Expanding water stewardship and social fairness review to supply chain.

3.3.8.4 *Gold*

- Product is fully assessed and does not contain any X-assessed substances.
- At least 50% renewable energy or offsets for scope 1 and 2 emissions.
- Continuing to expand scope of water and social issues to supply chain and local community where product is manufactured.

3.3.8.5 *Platinum*

- Cradle to Cradle 'holy grail.'
- Waste = food.
- Energy positive.
- Supply chain fully engaged.

The overall intention is to inspire and promote innovation in quality products in a way that supports 10 billion people on Earth without increasing the natural background level of materials.

Notes and References

1. The term 'cradle to cradle' was used in the 1980's by Walter Stahel and Michael Braungart. The term was first used in a limited way to counter the prevailing 'cradle to grave' paradigm in Germany related to manufacturing processes. Braungart and McDonough expanded this to a more holistic, design led framework.
2. Braungart's organization Environmental Protection and Encouragement Agency (EPEA) Internationale Umweltforschung GmbH was formed in 1987 and subsequently created the 'Intelligent Products System' which described materials as nutrients flowing in either biological or technical cycles. M. Braungart and J. Engelfried, An 'intelligent product system' to replace 'waste management,' *Fresenius' Environ. Bull.*, 1992, **1**, 613–619; http://epea.com (last accessed 22 December 2012).
3. *The Hannover Principles – Design for Sustainability* was first published in 1992 and republished as a 10th Anniversary edition in 2002 and a 20th Anniversary edition in 2012; http://www.mcdonough.com/principles.pdf.

4. Michael Braungart and William McDonough co-founded McDonough Braungart Design Chemistry, LLC (MBDC) in the USA in 1995 to help companies learn and implement the Cradle to Cradle design framework; http://MBDC.com (last accessed 22 December 2012).

5. In October 1998, the *Atlantic* magazine published an article entitled 'The NEXT Industrial Revolution.' This article posited the idea that humans could incorporate positive intentions for equity, economy and ecology though product design; http://www.theatlantic.com/magazine/archive/1998/10/the-next-industrial-revolution/4695/ (last accessed 22 December 2012).

6. In 2001, the documentary film *The Next Industrial Revolution* was released by Earthome Productions. This chronicled several active Cradle to Cradle projects; http://www.earthome.org/designfuture.html (last accessed 22 December 2012).

7. In 2002 the book W. McDonough and M. Braungart, *Cradle to Cradle: Remaking the Way We Make Things*, Farrar, Straus and Giroux, New York, was published.

8. In October 2005, MBDC launched the Cradle to Cradle Certified[CM] Program. By 2010, over 400 products from over 100 companies had achieved certification.

9. In August 2010, The Green Products Innovation Institute (the original name of the Cradle to Cradle Products Innovation Institute) was granted a free, exclusive worldwide license by MBDC to independently manage the product certification program Version 2.1.1.

China's Implementation of Alternatives Assessment in the Building Industry: GIGA

BRANDON ZANG, RAEFER K. WALLIS* AND RYAN D. DICK

ABSTRACT

China has been known as the world's factory for decades. It is synonymous with cheap goods, heavy pollution, copyright infringement and innumerable breaches of trust. Despite its infamous recent history, this chapter explores how China is poised to become a global leader in the demand for the assessment of chemical alternatives. In order to understand how the world's most populous country can undergo this radical transformation, one must understand its cultural history and drivers of change. This chapter begins with China's long relationship with personal health and explains the impact of newly found wealth and access to information. It continues by exploring how the combination of social media and China's tenuous relationship with trust can move the demand for chemicals assessment into the hands of everyday consumers. Finally, it ends by giving an example of how these lessons are being applied locally through an online platform.

It is important to note that the authors of this chapter are neither toxicologists nor pioneers in the development of green chemistry. Rather, they are architects and chemists who have spent the past four and a half years researching and implementing ways to create demand for green chemistry and alternatives assessment. This chapter is

---

*Corresponding author

Issues in Environmental Science and Technology, 36
Chemical Alternatives Assessments
Edited by R.E. Hester and R.M. Harrison
Published by the Royal Society of Chemistry, www.rsc.org

not a scientific piece, but rather a report on how the authors are attempting to harness developments and trends in both the IT and building industries to support research in the assessment of chemical alternatives.

4.1   Introduction

Over the course of the next few years, China will emerge as a leader in the field of assessment for chemical alternatives. It will also emerge as a leader in the development and production of healthy chemicals and products. Whereas most of the country's recent growth has been driven by foreign demand, the demand for healthy chemistry will come from China itself.

Although these statements may seem lofty for a country whose history of tragic health scandals seems to have no end, this chapter explains how the stage is now set for what promises to be a perfect storm of change. Of course, change is nothing new to China. The transformations that the country has undergone over the past three decades alone have been called miraculous by many – and they are far from being complete. The depth, speed and impact of these future transformations will be defined by how effectively they are guided. In view of this, the first part of this chapter serves to identify the drivers of change, so that the impact of these drivers can be catalyzed and magnified.

Architects by profession, we have spent the last decade working with designers, consumers, developers and manufacturers in China to define healthy building standards that are locally relevant. Starting at a time and place when the topic was still mostly unknown, public awareness of health hazard sources was devoid and the culture for disclosure was lacking, the challenge has been enormous. However, behind this challenge lies an equally enormous opportunity: that of leap-frogging the West in terms of impact by deploying innovative systems to measure and communicate product chemistry while creating market pull for healthy products. Rather than limit ourselves to established Western standards (which have little meaning in China), our strategy has been to define an ideal set of standards and principles and then look for points of overlap with existing cultural values in order to drive implementation. The resulting points of overlap have been significant enough to suggest that our targets are both feasible and highly desirable.

Whereas most global standards focus on minimizing the negative impact of manufacturing and construction, our focus for China has been on leveraging these activities to maximize environmental regeneration. As we shall see, this approach relies heavily on the rapid development of 'green' chemistry. The two biggest obstacles to achieving this goal are creating consumer demand for chemical assessment and substitution and finding scalable ways to meet that demand. Overcoming these obstacles is the subject of the second part of this chapter.

4.2 China Context

4.2.1 Health

Without a doubt, China's first and most important driver of change is personal health, particularly with respect to children. Health is a deeply rooted value that can be traced back thousands of years.

At first glance, the above statement may seem counter-intuitive to the Western reader. For decades, China has produced inexpensive goods for the rest of the world at tremendous costs to ecology and human health. The country's recent history is marked by a tragic string of high-profile health scandals that include baby formula tainted with melamine,[1,2] children's toys laced with lead,[3] drinking water contaminated with lead and cadmium,[4,5] ... and the list goes on. On the surface, these events indicate a culture for which health has little to no value, when in fact exactly the reverse is true. Yet, if health is such a fundamental concern for China, how could these incidents be tolerated for so many years? Although the answer is complex and largely outside the scope of this chapter, we have sketched out the key barriers to health below.

4.2.2 Barriers to Health

Following the creation of the People's Republic of China in 1949, the country entered a nationwide period of poverty. Combined with the priorities of the new government – equal distribution of resources to citizens and the maintenance of social stability – this period was also marked by the standardization of manufacturing and the control of information. This virtually eliminated the ability of individuals to make health-related choices for almost 40 years.

This early period of information control also greatly crippled public opinion and China's rich traditional network of social relationships. Especially acute during the Cultural Revolution, the chaos of accusations and persecutions decimated fundamental social relationships amongst communities, families, friends and peers.[6]

Few cultures are as famous as China's for the importance placed on relationships and their prioritization within a network or community. In a community where relationships are strong, questions and concerns are answered by a trusted network of family, friends and peers. The more close knit the community, the more the members look out for the health and well-being of one another and the less the chance of having breaches of trust across various sectors. Conversely, when social relationships and communities collapse, the primary priority becomes personal survival and consideration for the health and well-being of others is eliminated.

At the risk of grossly oversimplifying a complex period in history, the importance of networks and social relationships is spelled out here in view of the critical role it must play in the growth of chemical assessments in China – a topic which is further developed below.

It is important to note that during this time of great internal transformation, where critical cultural values were heavily challenged and the topic of health was virtually starved, China's traditional relationship with the value of personal health has remained relatively unchanged. With the removal of the barriers mentioned above, the recent resurgence of health as a critical priority is a testament to this idea.

4.3 Resurgence of Health

The publication of Rachel Carson's *Silent Spring* in 1967 marked the beginning of the ecological movement in the West. The book documented the death of an ecosystem. In other words, the trigger to Western awareness was ecological health. To this day, the topic is still central to any Western conversation about sustainability, regeneration, responsibility and stewardship.

In China, the trigger was entirely different. Despite an influx of Western influence, the topic of ecological health did not foster a significant movement. Rather, the trigger to public awareness for ecological health was human health.

In September 2008, news of melamine-contaminated milk swept the nation, provoking panic and outrage. By November, China reported an estimated 300 000 victims, with 860 babies hospitalized and six infants dying from kidney stones and other kidney damage. The crime was so significant that two of the main culprits were executed, another was given a suspended death sentence and three others received life imprisonment.[7]

Although the milk scandal was health related, what it actually triggered was public interest in greater environmental awareness. In many ways, it marked the tipping point in a series of environmentally related events in 2008, both dramatic and high profile: the Sichuan Earthquake in May,[8] the floods of Southern China in June[9] and the 'pollution-free' Beijing Olympics in August.[10] By September, the seeds of environmental awareness had already been planted nationwide, yet the event that was to transform public awareness into public engagement would prove to be health related. In a country famous for its one-child policy and emperor children, the milk scandal could not have hit on a more sensitive topic.

Made nationally infamous by melamine, product chemistry had suddenly entered the language of an entire population. It helped galvanize awareness of other chemicals such as formaldehyde, lead and *para*-xylene, which had previously appeared through localized incidents and would later be joined by the likes of aluminium and cadmium.

The list of health-related breaches of trust by manufacturers is long and ongoing: the 2011 lead poisoning incident in Shanghai, attributed to Johnson Controls, and the 2012 cadmium discharge into the Liu river are just two of many examples. Both were primarily reported as threats to human health, with the first focused on levels of lead in children's blood and the second on contamination of drinking water for 3.2 million people.[5,10,11]

As a driver of change, human health is far more powerful than ecological health. Simply put: human health is personal whereas ecological health is not. Although the end result is the same, China's approach to environmental issues requires a faster, more targeted response and it could be argued that it will most likely be more enduring and further reaching than anything the West has seen. Newly found access to wealth and information, in addition to China's tenuous relationship with trust, are serving as the main catalysts for change towards ecological health *via* personal health.

4.3.1   Trust as a Catalyst

Since 2008, the scandals associated with Chinese manufacturing have contributed to a near collapse of consumer trust, both towards manufacturers and in the ability of the government to serve as a watchdog. With nowhere else to turn to, Chinese consumers have been steering their newly found spending power towards foreign products, a trend catalyzed by the milk scandal. In this particular case, consumers boycotted domestic brands and the market for expensive imported milk products swelled overnight.[12]

Amidst all these events, it is also important to note that the future is not necessarily better for foreign companies. The Chinese are acutely aware that most of the world's foreign companies are still manufacturing their products in China, either in part or in whole. Recent breaches of trust by Apple and its suppliers,[13] and also the Johnson Controls case study mentioned above, serve as potent examples.

Ultimately, loss of consumer trust translates directly into loss of business. With new wealth, Chinese consumers now have the ability to choose between products and they have shown that if the pain is acute enough, they will pay a premium. This is a critical driver of change which did not exist a decade ago.

For Chinese manufacturers, finding ways to earn trust has become a necessary priority. With chemistry now part of popular language, chemical assessment will have a growing role to play in this respect.

4.3.2   Wealth as a Catalyst

Over the past decade, China's GDP has grown at nearly 10% p.a. and as a result it is now home to more then 1 110 000 millionaires.[14] Although the economic gap between the wealthy and poor continues to widen, the amount of disposable income per capita has been trending upwards for over a decade, giving the general population access to a new standard of living.

As mentioned above, disposable income is accompanied by the power of choice, yet even wealth is worthless without access to the information that is needed to guide choices. Fortunately, access to information has exploded during recent decades and now underpins the topics of health, wealth and trust.

4.3.3   Access to Information as a Catalyst

There are few countries in which the Internet revolution has been as deep as it has been in China. The number of individuals using the Internet to access information is staggering: 564 000 at the time of writing.[15] However, the extent of the revolution is far more evident when looking at the number of users who are actively communicating and creating information, primarily through social media. China boasts 450 000 000 users of instant messaging and 300 000 000 users of Weibo, a micro-blogging platform similar to Twitter. The Chinese numbers make the USA look small not only in terms of volume, but also in terms of activity. Whereas only 24% of web users create and post information in the USA, 44% are doing so in China.[16] Power users of Chinese social media are followed by millions and have often used their influence to guide market trends.

The popularity of social media in China has grown for obvious reasons: it has given a voice to a population that has been denied one for decades. Most importantly, it has allowed news and opinions to spread faster than they can be edited. This has been beneficial not only for the average individual, but also for China's central government, where it serves as a barometer for social stability and public opinion. In many ways, social media also serve as a whistle-blowing system for local breaches of trust, an area that the central government has long struggled to control. This creates an 'eyes on the street' effect which could eventually play a major role in keeping manufacturers under scrutiny and creating a culture of transparency.

However, the real influence of social media is in the way they allow individuals to radically expand their circles of trust and focus their spending power.

4.4   Social Media and the Rebuilding of Trust

In a traditional (offline) social network, individuals build relationships with a relatively small number of like-minded and trusted allies: family, friends and close peers. Whenever a choice needs to be made and information is lacking, individuals rely heavily on the experience and opinions of the network. Social media (online) simply take this process and expand it exponentially. The result is that more often than not decisions are made on perceived trust and the opinions of highly vocal thought leaders. Yet when transparent and validated information is made easily available, it tends to trump all else.

Without a doubt, the single best example of transparency and trust validation is Taobao, China's own version of Amazon. Here, over 370 million users peruse and rank the vendors of more than 800 million product listings according to their customer service using a simple system of diamond icons.[17] Seeking the best, buyers only purchase from the vendors who are ranked highest and sport the highest number of diamonds. Consequently, providing unparalleled service with a view to earning diamonds has become an issue of economic life or death for vendors. In a country infamous for its complete lack of customer service, this transformation was nothing short of radical.

This example underscores the role of social media in terms of building transformative networks of trust. Now, what if this was applied to green design, green chemistry and alternatives assessment?

4.4.1   Perfect Storm of Change

China has reached an unprecedented time in history and the stage is now set for what could be a perfect storm of change. On the one hand, access to information has moved human health back on to center stage, particularly with respect to children. Meanwhile, consumers have lost trust in domestic manufacturers, impacting both national and international brands.

On the other hand, newly found wealth and access to information have created consumer choice. Similarly, the web and the behavior of online users have given rise to powerful social tools that have proven to be transformative, particularly in terms of building trust in vendors. The key to determining the direction of the storm will be in figuring out if these elements can be aligned to drive change. However, before then, it is important to define the direction of change.

4.5   Minimization *Versus* Regeneration in China

Over the course of the past four decades, the focus of green design and green chemistry has been to minimize our negative impact on the environment. As a society, we have developed myriad ways of minimizing our carbon emissions, waste water emissions, consumption of resources, toxic chemical emissions and so on. Although this has proven to be an effective first step, it is now turning into an extremely dangerous one.

To illustrate this, we only need to imagine discharging lethal doses of cadmium directly into a river as a byproduct of manufacturing, thereby rapidly killing thousands of people downstream. Recognizing the problem, the fastest solution, with minimal impact to economy and society, is to minimize the discharge of cadmium to an acceptable level – in other words, one that is not immediately lethal. Un-noticed, the cadmium is allowed to travel further and contaminate more people. Instead of killing thousands of people quickly, we end up killing millions of people slowly.

Another way to visualize this is to study emergency dynamics. Imagine being in an accident and bleeding profusely. Without any intervention, death follows within 5 min. The first and most effective response is to minimize the negative impact of the accident by applying pressure to the wound and thereby minimize the bleeding. Although this is a necessary first step, the end effect is still the same: we still bleed to death. The only difference is that it takes 30 min as opposed to 5 min. Minimization's only effect is to buy us time, allowing us to fix the damage and support regeneration.

Within a culture of minimization, the goal is to build less, consume less and live less. In short, the goal is to limit growth heavily or eliminate it. Applying such a strategy to China is setting oneself up for assured failure. China is growing and will continue to do so for the next few decades. The only way to ensure impact is by enabling regenerative growth.

4.5.1   Regenerative Chemistry

In terms of architecture and urban planning, regeneration refers to buildings and cities that produce clean water, clean air and healthy topsoil, sequester carbon, increase biodiversity and are both carbon and energy neutral. Although these goals may sound far-fetched, the reality is that most of them are achievable today. The Riverhouse China – an eco-retreat on the Yangtze River – is an example of a completed project that achieves many of these targets already.

In terms of interior design, regeneration refers to spaces that help repair or maintain the good health of occupants. Again, this is a target that is easily achievable today.

Without a doubt, the greatest challenge for regeneration lies in the chemistry of materials. In many ways, toxicology is founded on principles of minimization, with one of the core activities of green chemistry being that of setting thresholds for chemical compounds. At first glance, the logic of regeneration seems to fall apart at the chemical level. What is a regenerative chemical? How is it defined?

In its simplest definition, green chemicals are those that are conducive to creating life and enabling it to thrive. Within the biosphere, resources perpetually travel in closed cycles and within defined time periods. Therefore, green, regenerative chemicals are those that do not compromise this ability and/or foster it. Of course, this eliminates both bio-accumulative and persistent chemicals and their compounds.

In terms of minimization, green chemistry sets a cap on the dosage of chemicals. There, green chemicals are most commonly defined as composed of those that are of low threat to human and greater ecological health.

However, the greatest interest in green chemistry is identifying those chemicals that help us optimize the existing activities of our living systems. In its simplest form, this is as obvious as cooking with an iron skillet as opposed to a Teflon-coated one, allowing the beneficial iron to leach into the food and supplement our diets. In its more complex form this means creating pro-microbial materials that help keep indoor environments safe or materials that help fight cancer cells. In short, green chemicals need to be redefined as those which are beneficial to our health.

Referring to the above as a challenge is somewhat of an under-statement. However, the beauty of working in China is that there is no precedent. Unlike the West, it is not yet entrenched in a culture of minimizing negative impact. For this reason, we have spent the past four and half years working on an open-source database designed to catalyze a culture of regeneration.

4.6   GIGA (Green Ideas, Green Actions)

GIGA is a research entity whose purpose is to catalyze regeneration in China across the ecological, social and economic environments. Driven by architects and building materials experts, the scope of research covers the impact and life-cycle of chemicals, building materials, interior design and architecture.

To quantify and measure the progress of buildings and their materials appropriately, GIGA has developed a set of targets called RESET™: an acronym for Regenerative, Ecological, Social and Economic Targets. RESET serves as an open-source genome project for the manufacturing and construction industry, created to identify the DNA of how Nature designs, *i.e.*, in closed cycles of regeneration.

Reflecting the current state of the industry, GIGA has also developed a system to document and recognize achievements in the building industry that are focused on minimizing negative impact. For clarity, these standards have been grouped under the title MNI: an acronym for Minimizing Negative Impact.

GIGA ranks building materials and projects according to both MNI and RESET standards within a public online database. This allows the database to highlight what are currently the world's greenest chemicals, materials and buildings and lead them from minimization to regeneration. As described by the example of emergency dynamics, this process enables us to support minimization while buying us time to develop and implement principles of regeneration. GIGA is the only system in the world to use this approach.

4.6.1   GIGA: the User's Perspective

From the user's perspective, GIGA is simply an online tool designed to help save time and money over the course of a working day. Designers use it to search for materials to be specified for projects, vendors and case studies. They also use it to communicate with colleagues and publicize their work. Manufacturers use it to search for project leads and promote their products. Contractors use it to search for vendors, and developers use it to search for service providers. The key is that the results of each one of these searches are prioritized according to green attributes, with chemical assessment carrying the most weight. For designers, manufacturers, contractors and developers, the critical aspect is that these activities are already part of their current routines. In other words, the idea is that the demand for alternatives assessment will be most successful if it becomes part of people's existing habits and activities.

4.6.2   GIGA: Green Algorithm

Similarly to Google, GIGA uses an algorithm to rank search results. However, whereas Google's algorithm is secret, GIGA's is transparent and based on green attributes. The greener the product, the higher it ranks and the easier it is found by designers. By making it an integral part of their daily tools, designers, manufacturers, vendors, developers and contractors create market demand for environmental regeneration simply by using the database, regardless of whether or not they are environmentally focused. Within the database, the most effective way for a product to raise its ranking is by understanding and assessing its chemistry. As the database considers the entire life-cycle of materials and chemistry cuts across all parts of the cycle (sourcing, production, installation, use and renewal), it is *de facto* the factor that carries the most weight.

4.6.3 GIGA: Unpacking Alternatives Assessment

Over the past few years, GIGA's greatest challenge has been in creating public demand for a topic as heady as chemistry and alternatives assessment. After struggling with the topic for a couple of years, we launched our first alternatives assessment pilot in 2010.

We began by inviting manufacturers to participate without charge. Not surprisingly, only a handful of early adopters, including Haworth, Biohouse and Desso, responded. For most manufacturer's the offer of free chemical analysis was viewed as a risk rather than an opportunity. We knew that properly aggregating demand would take time, so we worked with our early adopters to build an infrastructure that could be replicated at scale. Participating manufacturers selected products to analyze and we collectively broke each product down into components, raw materials and finally chemical compounds (CAS), manually screening them against the chemicals listed in the REACH SIN List. The aim was simply to identify hazards and unknowns.

Obviously, assessments are only as good at the data points they measure against. Although the SIN List includes incredibly useful information (we still use it today), its coverage was not wide enough to differentiate products adequately or identify the majority of chemical unknowns. The process also took our chemist up to a few weeks to analyze each product, depending on complexity. At best we were on pace to analyze around 40–50 products per year whereas a simple interiors project might include 200 or more products. Even at maximum capacity, our efforts had no hope of making a dent and were anything but scalable. In order to solve this, we began working with Dr Lauren Heine, Science Director of Clean Production Action, in order to include her Green Screen Lite benchmarking system as a quick screen for chemicals. By using the Green Screen, our chemical hazard list instantly grew to include the SIN List in addition to nearly 240 other global regulations. Designed as a software, this tool also allowed us to test products at scale to ensure meaningful impact in China. We launched a pilot to test several real products and create baselines for several commonly used materials such as powder coatings and polyurethanes. These baselines were communicated to our early adopters to identify effective ways to communicate the scientific results of the screening process.

Most of our users have not studied chemistry since high school, if at all. For them, chemical hazards must first be comprehensible, transparent and consistent before there is value in having a manufacturer disclose product chemistries.

The following is the result of over 4 years of standards development and experimentation with the market, four generations of online databases, dozens of communication strategies tested on both designer and manufacturer pilot groups and chemical collaboration with numerous manufacturers – simply to figure out ways to communicate and finance sustainably such a complex, sensitive but critical subject. In order to achieve this, we broke the process of developing regenerative products through alternatives assessments into simple and marketable steps, leveraging every major driver of change that was available to us in China:

- *Step 1:* Create a product. Keep it simple.
- *Step 2:* Make the product part of an existing choice/activity.
- *Step 3:* Reward/prioritize manufacturers who use the product.
- *Step 4:* Communicate the results in a way that is honest and non-threatening.
- *Step 5:* Develop a road map.
- *Step 6:* Build trust by communicating the road map's implementation.

4.6.3.1   Step 1: Create a Product. Keep It Simple.   The first part of this chapter served to identify an acute consumer need for the transparency of product chemistry and for alternatives assessment. Stating the obvious, the first step in supplying demand is to create an actual product that fulfills a need. Given the complexity of the topic, our key challenge was to keep the product extremely simple.

In this case, the product was Green Screen Lite, an automated chemical quick screen developed by Dr Lauren Heine. Our strategy was to represent it using a simple icon, placed on the summary of each material within GIGA's online database, thereby separating the 'haves' from the 'have-nots'. This simple strategy creates a choice for consumers, between a material that has been screened and one that has not. More importantly,and as we shall see below, the product is influential in defining how easy it is to find the project.

4.6.3.2   Step 2: Make the Product Part of an Existing Choice/Activity.   Choosing materials to be used in projects is the bane of every designer's existence. If materials could magically take the form, color and texture of a designer's imagination, the process of design would be perfect. Unfortunately, each material comes with strengths and limitations. Designers spend days sifting through materials to find the most suitable one while considering innumerable variables including, shape, size, finish and desired performance. Finally, adding in green requirements makes the process even more time consuming, particularly when information is scarce.

GIGA was created to save designers' time sourcing and specifying materials. As mentioned above, the database prioritizes search results according to green attributes.

The end effect is that regardless of whether or not a designer is looking for a material that prioritizes green attributes, it is the first thing he or she inevitably finds. By making the chemical screen part of an existing design process, no added effort is required on the part of the user. Yet, as these materials appear first and have a greater chance of being specified into projects, manufacturers suddenly have an economic incentive to screen their materials.

4.6.3.3   Step 3: Reward/Prioritize Manufacturers Who Use the Product.   Within the material ranking system, the best way to earn points is to undertake a chemical quick screen. This is because the results of the screen allows us to satisfy several criteria across the entire life-cycle of a material.

The critical challenge is that a product which is subjected to a chemical quick screen will always look worse that one that is not. GIGA solves this problem in two ways.

The first is by placing equal value on unknowns and highly toxic chemicals (those to be immediately eliminated). This means that it is impossible for a material that has been screened to look worse than one that has not. It also creates necessary pressure to eliminate all unknowns from the onset.

The second is by granting points simply for doing the chemical quick screen, regardless of the results. Therefore, a screened material will always appear above one that has not been screened. The better the chemistry of a product, the more points are earned, the higher is the ranking and the easier it is for the designer to find and specify the material.

This approach recognizes that most manufacturers have no idea what chemicals go into their materials, let alone what the health impact of those chemicals are. In many cases, the complexity of supply chains make the process of identifying chemicals extremely difficult. Since it is impossible to fix a problem without knowing what the problem is, rewarding manufacturers who take the first step of simply identifying their product chemistry is extremely important.

Of course, one critical element of concern is knowing whether or not the chemicals entered into the quick screen are accurate. To get manufacturers engaged in the process and lower the barrier to entry, GIGA allows for self-entry of chemical lists for screening. However, this level of participation assumes unknowns which, as we have seen above, are considered to be on a par with chemicals of very high concern. The impetus for manufacturers is that they receive a base level of points and are rewarded over those whose chemistry is completely ignored. For manufacturers to score above the base level, GIGA requires their material chemical lists to be verified by approved third parties.

The reward and ranking process ensures that materials which have undergone the chemical quick screen float to the top of search results, thereby bringing market value to the screen itself and providing economic incentive to document the chemical composition of products.

4.6.3.4 Step 4: Communicate the Results in a Way That Is Honest and Non-threatening. Given the current state of the industry, the results of most chemical quick screens look threatening to both manufacturers and consumers. Yet another tremendous challenge is that of communicating the results publicly and transparently, without creating a negative impact for the manufacturer. In fact, is it possible to communicate the results of the screen in such a way that creates positive impact, encouraging improvement and further transparency?

First and foremost, it is important to note that users and manufacturers have different needs. Therefore, the results of the screen must be communicated to them in different ways.

For the manufacturer, the results need to be quantitative. They need to known exactly which chemical is of highest concern and across how many indicators.

For the user, the results need to be qualitative. All they want to know is how safe the product is. At this level, too much information is confusing and counter-productive.

In order to illustrate this, here is how chemicals are scored within the database and communicated to the user, following both RESET and MNI criteria:

- *RESET: Ranks materials based on chemical disclosure and chemical safety:*
  - ○ RESET 1: Material chemistry is fully known and only includes Benchmark 4 chemicals. (5 points).
- *MNI: Ranks materials based on chemical disclosure and chemical safety:*
  - ○ A: Material chemistry is fully known and includes Benchmark 3 chemicals and above. (3 points).
  - ○ B: Material chemistry is fully known and includes Benchmark 2 chemicals and above. (2 points).
  - ○ C: Material chemistry is fully known includes Benchmark 1 chemicals and above.
    Or Material chemistry is fully known but self-documented.
    Or Material chemistry contains unknowns.
    (1 point)

Note that Benchmarks are defined as follows:

- *Benchmark 1:* Eliminate. Chemicals of very high concern to be eliminated immediately.
- *Benchmark 2:* Substitute. Chemicals of high concern to be replaced as soon as possible.
- *Benchmark 3:* Improve. Chemicals of moderate concern that can be improved.
- *Benchmark 4:* Prefer. Chemicals of low concern or of benefit. Note that with research, this category will need to be split into two, distinguishing between chemicals of low concern and those that are beneficial.

The final step is to communicate these results as compared with an industry average. In this way, if a material only scores as MNI B but the industry average is MNI C, this particular material would still be understood as an industry leader. Alternately, a material scoring at MNI C would be cast in neutral light, eliminating a direct threat and giving it a chance to clean up its chemistry. Users would not know whether the score was granted because the material contained chemicals of very high concern, because the information was self-documented or because it contained unknowns. In the meantime, the database would steer consumer priority towards MNI B.

The above system creates enough transparency to provide users with the information they need, rewards manufacturers in the process and protects their proprietary information.

4.6.3.5 Step 5: Develop a Road Map. Upon receipt of the results, the first question posed by manufacturers is what to do next. In most cases, the results expose issues they never even knew they had. In virtually all cases, expectations need to be managed and a road map created. The chemical quick screen is both enlightening and misleading. It takes a process that used to require weeks and deep pockets and reduces it to a few minutes and petty cash. It is an absolutely critical tool if alternative chemical assessment is to become scalable. However, in order to be quick, it misses out on many details. As the name implies, it is a screen, nothing more and nothing less.

Therefore, the first step following the screen is to identify the areas that need more research. At this stage, the task is to introduce manufacturers to a set of resources, depending on their needs. Typically, the first point of contact would be a consultant or toxicologist who would help define a road map according to a series of questions. Which chemicals need more research? Which ones can be replaced easily and affordably? Which ones require new chemicals to be developed? Most importantly, the consultant would help define goals in addition to the time and cost required to achieve those goals.

Once complete, GIGA's role would be to help communicate, bring market value to the road map and help accelerate its deployment. In other words, GIGA's role is to create demand and market value for assessment.

4.6.3.6 Step 6: Build Trust by Communicating the Roadmap's Execution. Some of the world's most progressive brands have long demonstrated the value of communicating progress with consumers. Manufacturers such as Interface, Haworth and Desso have built a loyal following by communicating what their products do and do not contain, in addition to their goals and their progress. Simply said: transparency and communication build trust. This is equally true whether at the brand or the retailer level, with Walmart serving as an example of the latter.

GIGA currently serves as a platform for designers to 'discover' green brands and connect with them. Combining social media tools with the deployment of a green road map allows designers to follow, influence and accelerate the development of these leaders. Interaction builds loyalty, trust and ultimately added market value for the greenest brands.

In an age of instant access to information, full transparency is becoming a basic expectation. Although the average consumer does not understand the chemistry behind the products that they buy, there is an expectation to know what hazards and risks are inherent to the products they use. As before, the challenge lies in simplifying communication. To this effect, GIGA has developed a simple graph that tracks a manufacturer's progress towards achieving green chemistry. Although the graph displays the proportion of chemicals with low, medium, high and very high health impacts contained in a material, it does so over time (Figure 4.1). Therefore, the emphasis is placed on progress and trends over time.

4.7 Impact: Process

The goal of GIGA's process is to turn alternatives assessment into a product that is market driven and easily understood by the consumer. In summary, the

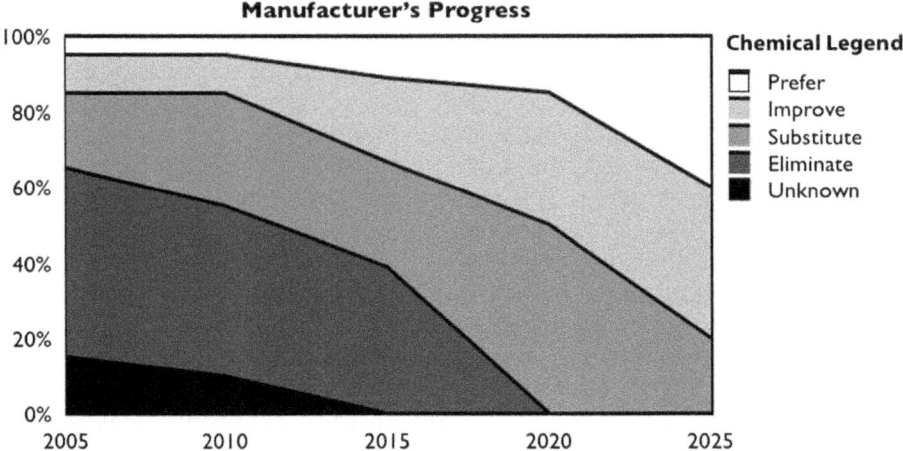

Figure 4.1   Tracking current and projected manufacturer progress towards green
chemistry.

process guides manufacturers through the following steps, primarily driven by
consumer demand:

- *Step 1:* Identify all product chemistries.
- *Step 2:* Screen chemicals against Green Screen Lite Benchmarks to
  identify known hazards, unknown chemicals and known benign
  chemicals.
- *Step 3:* Create a research roadmap with toxicologists to improve product
  chemistry.
- *Step 4:* Incrementally substitute hazardous or unknown chemicals with
  known safer ones.
- *Step 5 (if needed):* Develop new formulations.
- *Step 6 (ongoing):* Communicate progress with consumers to build trust
  and a loyal consumer following.

   This process is nothing new, and variations of it have existed for years. What
is of interest is the automation of the process, making it simple and accessible
for manufacturers to participate. Most importantly, the process enables alternatives assessments to be conducted and communicated at scale. In this way,
GIGA serves as a platform to aggregate and create scale on both the demand
and the supply sides.
   It should also be noted that in the process of using the database to search for
building materials, the user is not expected to dig down to the chemical level for
each material choice. That is unrealistic and undesirable. In most cases, the user
focuses simply on the key indicators. First and foremost would be the overall
rank. Second would be price, lead time, performance or whatever other indicator is important to the user for their particular application. However, as the

first contact is driven by the ranking, it is important that the mathematics, chemistry and process behind the ranking remains transparent at all times. Consequently, a new user typically starts with the ranking and drills all the way down to the certifications that underpin the chemical assessments. This process builds trust in the system and then allows them to pull out and focus on the indicators.

4.8   Alternatives Assessment at Scale

Traditionally, creating demand for alternative assessments has been the business of government regulations, particularly in the Western world. This is exemplified by the exponential growth in health regulations over the last century (Figure 4.2). As regulations increase, chemical disclosure becomes motivated by the risk of direct fines or loss of business. As an example of the former, Walmart's US$5.5 million fine in 2001 for storm water contamination is just one of many.[18] In terms of the latter, Sony also serves as one example amongst many. In 2001, 1.3 million Playstation game consoles were found to contain levels of cadmium that exceeded local regulations. The consoles were blocked by Dutch officials, costing Sony an estimated $100 million in direct sales alone.[19]

However, the reality of government regulations is that they are difficult to implement. When rules are broken and problems arise, the responsibility needs to be carried by the entity which is at fault. However, the extreme fragmentation of our current manufacturing and distribution chains is so complex that it can be impossible to assign fault. Most manufacturers are actually better described as assemblers who create end products by combining dozens to hundreds of component materials. As assemblers, these components are sourced from any number of suppliers depending on availability, price, location, lead time, *etc*. Often, the end product finds its way to the consumer through a distribution chain that is equally complex. Unwanted information is easily lost along the way. As a result, the exponential increase in government regulations has not created enough leverage to generate a widespread adoption of alternatives assessments. Government regulations are the equivalent of a beating stick. However, as Walmart would show, when combined with a carrot, the donkey of change starts to move a lot faster.

In 2009, Walmart announced the alternatives assessment work that it was pioneering within its supply chain. Leveraging its global purchasing power, Walmart made it compulsory for suppliers and vendors to disclose their product chemistry. In other words, Walmart notified suppliers around the world that if they wanted to sell to the world's largest retailer, they needed to disclose. In order to empower this, Walmart also enabled a third-party disclosure system. For a supplier, selling at volume through Walmart's international distribution network represents an economic carrot that is large enough to start digging through the chemistry of their own supply chain.

Of course, Walmart implemented this system as a result of a growing number of fines and lawsuits. However, the lesson learned is how regulatory pressure was transformed into economic pull – a far more effective means of implementation.

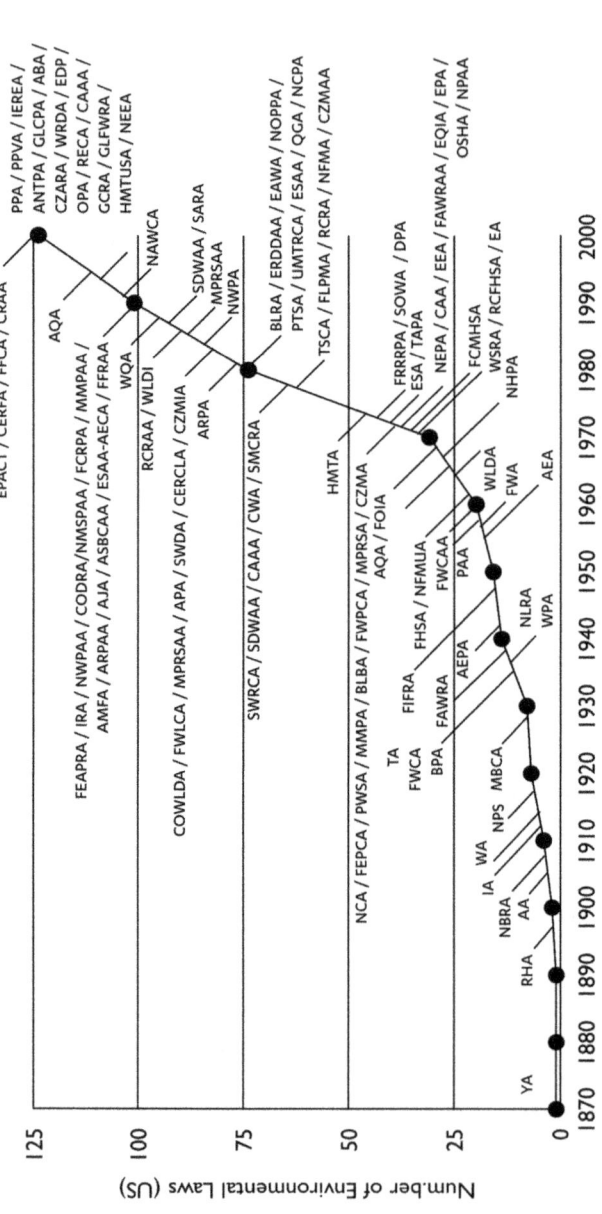

Figure 4.2   Cumulative growth in federal environmental laws and amendments.
Source: D. T. Allen and D. R. Shonnard, *Green Engineering: Environmentally Conscious Design of Chemical Processes*, Prentice Hall, Upper Saddle River, NJ, 2002.

Replicating this model across other sectors will require aggregating scale. For instance, although a few large retailers of building materials exist, most materials are sourced from distributors or from the brands themselves. This decentralized process currently removes any economic potential to create demand for alternatives assessment.

However, once aggregated, the market for building materials in China reached 1.9 trillion RMB ($310.9 billion) in 2011, about 3/4 the size of Walmart's reported gross revenue of $418.9 billion for the same year.[20,21] Although these numbers are not directly comparable seeing as they cover different industries, they do give a good indication of magnitude. In terms of direct comparison, the 15.4 trillion RMB ($2.45 trillion) market for consumer goods in China is 5.8 times that of Walmart's.[22]

In a country where the consumer watchdog is fed by social media and growing infinitely faster than the government watchdog, the adoption of alternatives assessments by market forces is far more plausible than anywhere else.

4.9 Conclusion

China is poised to become a global leader in the demand for the assessment of chemical alternatives – all the elements are currently in place for this revolution to occur. The depth, speed and impact of this transformation will be defined by how successfully these elements are assembled in a way that makes the process simple and scalable for both consumers and manufacturers.

Over the past two decades, the combination of newly found wealth and access to information has opened the doors for consumer choice, decimated trust in locally manufactured goods and highlighted the threat of manufacturing on China's culture of personal health. This has led to an acute consumer need for information in order to empower health-related choices.

In the absence of such information, the market has trended towards imported goods and/or the advice of trusted peers within a social network. As a result, the need which has now emerged for manufacturers is that of rebuilding trust with consumers. Currently, efforts to address these needs are isolated and non-systemic.

The key challenges to solving these needs are scale and simplicity. On the one hand, demand must be aggregated in order to reach an influential market scale. Similarly, tools, resources and a process must be in place and equally scalable to meet demand. On the other hand, the results must be simple enough for a consumer to understand within the context of ubiquitous daily activity. In other words, the results must be simple enough to guide consumer choice. Similarly, the process for manufacturers must be simple enough to enable them to respond to demand.

Part of the solution lies in social media. Not only does it enable networks of trust to be rebuilt, it also can aggregate demand at a national and global scale.

The other part of the solution lies in developing a simple language and process to leverage social media, making alternatives assessment comprehensible, interactive and accessible. GIGA is demonstrating one such approach with respect to the construction industry, whereby green attributes and chemical screening are woven into the activities of everyday users. Techniques of web-gamification focus further attention on the greenest materials. Social tools allow users to communicate the features of these materials to others and follow brands. Finally, users can dig through the validation of manufacturers' claims in order to help rebuild credibility and trust, down to the source. This approach gives Chinese consumers both information to understand product safety and the means to push manufacturers to produce healthier materials. More importantly, it helps drive market demand.

GIGA's system has evolved tremendously over the past 4 years, as a result of user validation and simplification. It will continue to evolve and, in so doing, it aims to remove all barriers to participation. More importantly, although the user facing part of GIGA's platform caters to the building industry, the back-end system of information validation and chemical screening processes can be applied to any industry.

If properly combined and catalyzed, the drivers of change currently active in China could easily forge the country into becoming the world's biggest market for green chemistry.

References

1. CCTV News, Guo jia zhi jian zong ju gong bu jian chu san ju qing an ying you er pei fang nai fen qi ye ming, 16 September 2008.
2. Xin Hua News, Wei sheng bu: Xian you 1.2 wan yu ming ying you er yin shi yong nai fen zhu yuan zhi, 21 September 2008.
3. E. S. Lipton, and D. Barboza, *New York Times*, 19 July 2007.
4. Occupational Knowledge International, Global Village Beijing and Institute of Public and Environmental Affairs, *Health and Environmental Impacts from Lead Battery Manufacturing and Recycling in China*, 2011, revised August 2012; http://www.ipe.org.cn/Upload/Report-Battery-II-EN.pdf. (last accessed 26 February 2013).
5. Caixin Online, *Cadmium Contamination Raises Concerns*, 1 February 2012, http://english.caixin.com/2012-02-01/100352441.html (last accessed 26 February 2013).
6. P. Clark, *The Chinese Cultural Revolution: a History*, Cambridge University Press, Cambridge, 2008.
7. China dairy products found tainted with melamine, *BBC News*, July 2010.
8. Sichuan earthquake, *Times Topics*, 6 May 2009; http://topics.nytimes.com/topics/news/science/topics/earthquakes/sichuan_province_china/index.html (last accessed 28 December 2012).
9. Death toll from floods rises to 171 in south China, *China.org.cn*, 18 June 2008; http://www.china.org.cn/environment/news/2008–06/18/content_15846517.htm (last accessed 28 December 2012).
10. B. Wen, After Beijing's 'green' Olympics, *Chinadialogue*, 5 September 2008; http://www.chinadialogue.net/article/show/single/en/2367-After-Beijing-s-green-Olympics- (last accessed 28 December 2012).

11. J. Watts, Cadmium spill threatens water supplies of major Chinese city, *Theguardian*, 30 January 2012; http://www.guardian.co.uk/environment/2012/jan/30/cadmium-spill-china-river (last accessed 28 December 2012).

12. Jiefang ri bao: 'Yang nai fen' jin kou 3 nian zeng 3 bei duo, guo chan nai fen hui fu shang xu shi ri, 16 September 2011.

13. Friends of Nature, Institute of Public and Environmental Affairs, Green Beagle, Envirofriends and Green Stone Environmental Action Network, *The Other Side of Apple – Pollution Spreads Through Apple's Supply Chain*, 31 August 2011; http://www.ipe.org.cn/Upload/Report-IT-V-Apple-II.pdf (last accessed 28 December 2012).

14. F. Balfour, China's Millionaires Leap Past 1 Million On Growth, Savings, *Bloomberg*, 1 June 2011.

15. China Internet Network Information Center (CNNIC), *Statistical Report on Internet Development in China*, January 2013, http://www1.cnnic.cn/IDR/ReportDownloads/201302/P020130221391269963814.pdf (last accessed 26 February 2013).

16. B. Cao, Sina's Weibo Outlook Buoys Internet Stock Gains: China Overnight, *Bloomberg*, 28 February 2012.

17. Alexa, *Taobao.com*, http://www.alexa.com/siteinfo/taobao.com (last accessed 2 August 2012).

18. Underground Construction, *Walmart Violates Storm-Water Rules*, August 1 2001, http://www.highbeam.com/doc/1G1-77930585.html (last accessed 26 February, 2013).

19. The Economist, *Electronics, Unleaded*, 10 March 2005, http://www.economist.com/node/3714013 (last accessed 26 February 2013).

20. *China - Construction Materials*, MarketLine, 11 September 2012.

21. Walmart 2011 Annual Report, http://www.walmartstores.com/sites/annualreport/2011/financials.aspx (last accessed February 26, 2012).

22. LFRC, IFTE, *Blue Book of China's Commercial Sector (2011)*, Cengage Learning Asia, 1 May, 2011.

A Collaborative Industry and University Alternative Assessment of Plasticizers for Wire and Cable

GREGORY MOROSE* AND MONICA BECKER

ABSTRACT

There is increasing regulatory and market pressure on companies to eliminate known chemicals of concern from their products. These companies can face significant challenges in identifying alternatives that are safer and effective. Switching to a new chemical can cost up to millions of dollars and no company wants to switch to a new chemical that later becomes a regulatory target. To assist these companies, the Green Chemistry and Commerce Council (GC3) launched a project in 2010 to develop and pilot a new model for companies and universities to evaluate collaboratively safer alternatives, with the aim of yielding more robust and cost-effective results through pooling of knowledge and funding for evaluations. The pilot focused on alternatives to toxic plasticizers in PVC and non-PVC wire and cable applications. The project workplan included assessing the relative hazard of the alternatives using the GreenScreen™, and also technical performance and cost. Chemical manufacturers, plastics compounders, original equipment manufacturers (OEMs), a retailer and the University of Massachusetts Lowell successfully collaborated to assess nine alternative plasticizers and created a group process that can be repeated for other chemicals and applications. This chapter describes the process developed and lessons learned, including strategies used to assess plasticizers with proprietary ingredients.

---

*Corresponding author

Issues in Environmental Science and Technology, 36
Chemical Alternatives Assessments
Edited by R.E. Hester and R.M. Harrison
© The Royal Society of Chemistry 2013
Published by the Royal Society of Chemistry, www.rsc.org

5.1 Background

There is increasing regulatory, market and consumer pressure for the development and adoption of safer alternatives to chemicals and materials with known toxic effects. Examples of toxic chemicals in products abound: product recalls of toys contaminated with lead, the presence of endocrine disruptors in food containers, the detection of flame retardants in household dust and humans and recent toxicology studies showing asbestos-like health effects associated with carbon nanotubes. These concerns have led to increasing demands for 'greener' and 'safer' products.

Companies today face the challenge of identifying which chemicals are problematic, determining whether they are present in their products and then developing strategies to eliminate these chemicals and materials. There are several approaches and many tools available to help companies address these challenges, including risk assessment, life-cycle assessment and alternatives assessment.

Risk assessment is a systematic approach to evaluate the potential health risks associated with exposure to toxic chemicals or materials. Life-cycle assessment is a process to evaluate the environmental burdens associated with a product and may include such categories as human toxicity, energy consumption, global warming potential, ozone depletion, water usage and ecotoxicity. Life-cycle assessment examines all stages of the product life-cycle, including raw material extraction, materials processing, manufacturing, assembly, transportation, product use and disposal.

Alternatives assessment involves a comprehensive evaluation of available alternatives to a chemical of concern in a particular application to determine which alternatives are safer, technically feasible and affordable. These alternatives may be chemical or material substitutes or modifications to processes or product designs that facilitate a shift to safer chemicals.

Companies seeking safer alternatives to chemicals of concern want to be sure that the alternatives they choose are indeed safer to avoid 'regrettable substitutions.' However, the use of available approaches, such as alternatives assessment, can be challenging because these methods are resource intensive and their use may be constrained by a lack of toxicological data for the chemicals and materials being assessed.

To respond to these and other challenges that slow the advancement of safer chemistries and products, the Lowell Center for Sustainable Production at the University of Massachusetts Lowell created the Green Chemistry and Commerce Council (GC3), a business-to-business forum that promotes the application of green chemistry and design for environment across supply chains. Launched in 2005, the mission of the GC3 is to promote and support green chemistry practices in research and application nationally and internationally among companies and other governmental and non-governmental entities by:[1]

1. Implementing green chemistry, green engineering and design for environment throughout supply chains and sharing successful case studies and strategies to overcome barriers.

2. Promoting education and information on safer chemicals and products that can increase demand by a broad range of consumers.
3. Identifying existing and needed information on toxics hazards, risks, exposures and safer alternatives to promote green chemistry as defined in the '12 Principles of Green Chemistry.'

GC3 members include leading companies from a range of industry sectors, including Nike and Timberland in the footwear and apparel sector, Hewlett-Packard, EMC and Dell in the electronics sector, Dow, DuPont and BASF in the chemical sector and Staples and Target in the retail sector.

The GC3 holds annual roundtables to bring GC3 members together from around the world to share best practices and new ideas. The concept for developing partnerships between GC3 companies and between companies and academic institutions was first raised during the 2009 annual GC3 Innovators Roundtable. GC3 companies were interested in collaborating to conduct evaluations of safer alternative chemicals/materials to:

1. Pool their knowledge of chemicals of concern and their potential alternatives.
2. Share the costs of conducting the chemical and material evaluations.
3. Enable conversations between companies up and down a supply chain to work toward solutions that meet the needs of all participating companies.
4. Allow companies to initiate chemical and material changes to their products and within their supply chain in a timely and cost-efficient manner ahead of anticipated regulations.
5. Leverage the technical expertise of universities and their ability to bring competitors together to work on common sustainability challenges.

The GC3 has several ongoing projects, including: *Green Chemistry Education*; *Business and Academic Partnerships for Safer Chemicals*; *Engaging Retailers in the Adoption of Safer Products*; and *Facilitating Chemical Data Flow Along Supply Chains*.

In 2010, the GC3's *Business and Academic Partnerships for Safer Chemicals* project group began in earnest to develop and pilot an alternatives assessment partnership to create a model that future partnerships might follow and build upon. The goals for the project were to:

1. Provide valuable information on the relative hazards of chemicals of concern to GC3 companies and their supply chain partners.
2. Develop a repeatable process for identifying potentially safer chemical alternatives to chemicals of concern, evaluating their toxicity, assessing their technical performance and evaluating the economic implications of selecting one alternative over another.

3. Demonstrate project success and form a basis for the development of a model for continued university/business partnerships.
4. Share pilot project results with the public in an effort to lead to the more rapid adoption of safer chemicals and materials in supply chains.

Barbara Hanley from Hewlett-Packard, Roger McFadden from Staples and David Levine from the Environmental Health Fund comprised a Steering Group that helped to establish the initial vision and workplan for the project. Initially, Greg Morose from the Massachusetts Toxics Use Reduction Institute (TURI) and Melissa Coffin of the GC3 coordinated the project. Their time was provided as an in-kind contribution by the GC3 and the Toxics Use Reduction Institute. Both organizations are part of the University of Massachusetts Lowell. Midway through the project, the role of project coordination and facilitation was transitioned to Monica Becker, a Lowell Center Fellow and GC3 contractor.

Funding for the project was provided by GC3 companies interested in the results of the project and the establishment of a model for future business/ academic partnerships to address other chemicals of concern. The scope of the project was drawn to match the level of funding available.

This chapter describes the approach taken in this collaborative alternative assessment project, accomplishments to date, a critique of the model created and recommendations for future collaborative efforts of this kind.

5.2 Project Workplan

The overall approach used to conduct the pilot project is summarized in the steps below, with more detail provided for some steps in the next few sections.

1. Select a chemical/chemical class of concern.
2. Select an application to be the primary focus.
3. Assemble a pilot group, consisting of companies in the supply chain for the selected product application.
4. Develop a preliminary list of potential alternatives.
5. Obtain input from the pilot group on preliminary list of potential alternatives.
6. Screen potential chemical alternatives against a list of chemicals of concern (*i.e.* a 'red list' of chemicals).
7. Select priority alternatives for further screening with the Quick Chemical Assessment Tool (QCAT).
8. Conduct chemical hazard screening using the QCAT.
9. Select priority alternatives based upon QCAT results.
10. Conduct detailed chemical hazard evaluation using the GreenScreen™ assessment tool.
11. Conduct collaborative technical performance and economic evaluations on top-performing alternatives.

12. If none of the alternatives are acceptable, commence efforts to develop new chemicals/materials.
13. Publish the results of the evaluation.

5.3 Selection of a Chemical/Chemical Class of Concern and Application

First, a chemical or class of chemicals of concern and application was needed to focus the pilot. The aim was to select a chemical and application that had the broadest applicability to GC3 members. Based on input from members, phthalates used as plasticizers were selected for several reasons, including: GC3 members consider certain phthalates to be a concern, some members are under pressure to eliminate certain phthalates from their products and replacing phthalates poses a challenge in many different industry sectors.

Phthalates are used as plasticizers in a variety of products such as flooring, wire and cables, footwear, adhesives, toys, medical devices, roofing, wall coverings, upholstery, films and car under-coatings. Over 90% of the phthalates produced are used specifically for their plasticizing function, giving plastics, primarily poly(vinyl chloride) (PVC), enhanced flexibility and durability. Phthalate ester plasticizers are usually colorless and are mostly odorless. Although various plasticizers in use today have structural similarity, each one is different in the way it performs. Di(2-ethylhexyl) phthalate (DEHP) is the most commonly used phthalate plasticizer. DEHP is the international standard for PVC and properties of other plasticizers are usually reported relative to its properties. As a plasticizer for PVC, DEHP generally offers excellent compatibility, desirable fusion properties and a set of performance properties that often require little modification with other types of plasticizers.[2]

DEHP is classified by the US Environmental Protection Agency (EPA) as a Class B2 carcinogen. This determination is based on liver cancer in rats and mice. In 2000, the International Agency for Research on Cancer (IARC) changed its classification for DEHP from 'possibly carcinogenic to humans' to a Class 3 carcinogen, 'cannot be classified as to its carcinogenicity to humans,' because of the differences in how the livers of humans and primates respond to DEHP compared with those of rats and mice.[3] No studies are currently available that directly indicate reproductive effects in humans after oral exposure of humans to DEHP; however, many animal studies have been conducted. Studies of rodents exposed to doses in excess of $100 \, \text{mg kg}^{-1} \, \text{d}^{-1}$ of DEHP indicate that the testes are a primary target organ, resulting in decreased testicular weights and tubular atrophy. Also, weights of the seminal vesicles, epididymis and prostate gland in rats and mice are reduced by oral exposure to DEHP.[4,5]

Since phthalate-based plasticizers are used in a variety of products, the project group sent a survey to GC3 members to determine which product category was of most interest to them. Based on this member input, and also input from EPA's Design for Environment program regarding their chemical action plans for some phthalates, the Steering Group decided to focus on wire and cable used in the electronics sector.

5.4 Pilot Project Team Formation

Given the focus on plasticizers for wire and cable applications, a project group was formed consisting of additional personnel from Hewlett-Packard and also representatives from EMC, Dow Chemical, Dell, HallStar, BASF, Teknor Apex, PolyOne, Clean Production Action and the Pacific Northwest Pollution Prevention Center. The following parts of the wire and cable supply chain were represented within the project group:

- chemical manufacturers: Dow Chemical, HallStar, BASF
- plastics compounders: Teknor Apex, PolyOne
- original equipment manufacturers (OEMs) for electronics products: Hewlett-Packard, EMC, Dell
- retailers for electronics products: Staples.

The next step taken was to develop an inventory of potential plasticizer alternatives. Based on a review of the available literature, an inventory was developed listing more than 100 commercially available plasticizers that could potentially be used for wire and cable applications.

A survey was created and distributed to GC3 members and to other interested companies asking respondents to identify which of the 100 commercially available plasticizers were of most interest. The survey also requested the level of priority for each plasticizer and whether the plasticizers were used for PVC and/or non-PVC materials. Respondents were also prompted to tell the GC3 about any additional plasticizers they were interested in having evaluated. In total, surveys were received from ten companies and one trade association. The responses were consolidated and presented without attribution to members of the project group.

5.5 Screening Against the Red List

A 'red list' of chemicals is a group of chemicals that are known or suspected to be carcinogens, mutagens, reproductive toxins (CMRs), persistent, bio-accumulative or aquatically toxic (PBT), or otherwise toxic or hazardous to humans or the environment. The red list used in this project was developed by Clean Production Action and the Healthy Building Network[6] and contains approximately 2500 substances that appear on authoritative hazardous chemicals lists developed by a body established by one or more government entities, addressing one or more chemical hazards.

Project staff compared the plasticizer inventory developed by the project group against the red list. Any chemicals on the chemical inventory appearing on the red list were removed from consideration for further investigation. A short list of plasticizers that did not appear on the red list and that also scored well on the survey were presented to the project group for review. After several conference calls, the project group selected the 10 chemicals listed in Table 5.1

Table 5.1 Plasticizers selected for assessment.

Plasticizer Name	Brand Name	Reasons Selected for QCAT
Di(2-ethylhexyl) isophthalate	Flexol 380	• Isophthalates are potential replacement for traditional phthalates • Listed as a high priority by survey response
COMGHA – acetylated monoglycerides of fully hydrogenated castor oil	Grindstead Soft n' Safe	• Listed as a high priority by survey response • Represents a chemical family of plasticizer alternatives
DEHP – di(2-ethylhexyl) phthalate		• Chosen as benchmark for evaluation of other plasticizers • Listed as a high priority by survey response
DGD – dipropylene glycol dibenzoate	Benzoflex 9-88	• Used in both PVC and non-PVC applications • Listed as a high priority by survey response • Represents a chemical family of plasticizer alternatives
DIDP – diisodecyl phtha-late and di(C_{10}-rich branched C_9–C_{11}-alkyl phthalate)	Jayflex	• Used in wire and cable • Likely candidate to replace DEHP in PVC applications • Listed as a high priority by survey response
DINCH – diisononylcy-clohexane-1,2-dicarboxylate	Hexamoll® DINCH®	• Likely candidate to replace DEHP in PVC applications • Listed as a high priority by survey response • Represents a chemical family of plasticizer alternatives
DOZ – di(2-ethylhexyl) azelate	Edenol 9058, San-socizer DOZ	• From partially renewable content • Likely candidate to replace DEHP in PVC applications • Represents a chemical family of plasticizer alternatives • Listed as a high priority by survey response
LPLAS-1100 Series	Ecolibrium	• New to the market • Bio-based • Listed as a high priority by survey response
TEHTM – tri(2-ethyl-hexyl) trimellitate	Palatinol TOTM, TOTM-I	• Listed as a high priority by survey response • Used in wire and cable, likely replacement for DEHP • Used in PVC/non-PVC applications • Represents a chemical family of plasticizer alternatives
TXIB – 2,2,4-trimethyl-1,3-pentanediol diisobutyrate	Eastman TXIB, Texanol	• Represents a chemical family of plasticizer alternatives • Found in wire/cable products already on the market • Listed as a high priority by survey response

for assessment using the Quick Chemical Assessment Tool (QCAT). The table also provides the major reasons that the plasticizers were selected.

5.6 Using the Quick Chemical Assessment Tool (QCAT) to Screen Chemicals

The GC3 contracted with ToxServices to conduct reviews, using the QCAT, of the 10 plasticizers. ToxServices is a scientific consulting firm with expertise in toxicology, risk assessment, comparative hazard assessment and the application of science to promote human health and environmental sustainability. ToxServices provides services as an approved third-party profiler for the US EPA's Design for the Environment (DfE) Formulator Program, Green Blue's CleanGredients database and Clean Production Action's GreenScreen™. The QCAT is a rapid screening tool developed by the Washington State Department of Ecology. The QCAT rates chemicals and their degradation byproducts with a letter grade based on the following endpoints:

- environmental fate: persistence and bio-accumulation
- ecotoxicity: acute aquatic toxicity
- human health effects: carcinogenicity, mutagenicity/genotoxicity, reproductive/developmental toxicity, endocrine disruption, acute mammalian toxicity.

The QCAT draws data for chemical hazard assessments primarily from a set of hazardous chemical lists and databases developed by authoritative government bodies. Where data gaps exist, the QCAT utilizes estimated data from models such as the US EPA's PBT Profiler, and also measured data from peer-reviewed scientific journals and risk assessments. The QCAT does not use advanced scientific modeling or apply advanced, 'expert' review of other available data sources.

The definitions of QCAT grades are as follows:

Grade A: Few concerns, *i.e.* safer chemical
Grade B: Slight concern
Grade C: Moderate concern
Grade D: High concern
Grade F: Toxic chemical.

When a grade is determined from modeled data, the letter grade carries an asterisk. The QCAT grades for the plasticizers ranged from A ('Few concerns, *i.e.* safer chemical') for DOZ, to F ('Toxic chemical') for several plasticizers. A summary of the QCAT results are provided in Table 5.2.

In addition to the toxicity assessment, the QCAT results enabled the team to assess the availability of data for a more detailed assessment. Based on these results and an assessment of technical feasibility and commercial availability,

Table 5.2 Summary of QCAT grades.

Plasticizer Name	Grade Given Available Data	Available Data as the Basis for this Grade	Data Gaps	Data Gap Grade
Flexol 380 Di(2-ethylhexyl) isophthalate	B*	Modeled data for persistence and bio-accumulation	• Carcinogenicity • Reproductive/ developmental toxicity • Endocrine disruption	F
COMGHA Acetylated monoglycerides of fully hydrogenated castor oil	C*	Modeled data for persistence and bio-accumulation	• Acute mammalian toxicity • Carcinogenicity • Reproductive/ developmental toxicity • Genotoxicity/ mutagenicity • Endocrine disruption	F
DEHP Di(2-ethylhexyl) phthalate	F	Present on lists for • Carcinogenicity • Endocrine disruption • Reproductive/developmental toxicity	None	F
DGD Dipropylene glycol dibenzoate	D*	• Present on the Canadian DSL list for environmental toxicity • Modeled data for persistence and bio-accumulation • Acute aquatic toxicity	• Carcinogenicity • Reproductive/ developmental toxicity • Genotoxicity/ mutagenicity • Endocrine disruption	F
DIDP Diisodecyl phthalate and di(C_{10}-rich branched C_9–C_{11}-alkyl) phthalate	F	Present on lists for • Endocrine disruption • Reproductive/developmental toxicity Very high/High • Acute aquatic toxicity • Bio-accumulation	• Carcinogenicity	F
DINCH Diisononylcyclohexane-1,2-dicarboxylate	B*	Modeled data for persistence	• Acute mammalian toxicity • Carcinogenicity • Reproductive/ developmental toxicity • Genotoxicity/ mutagenicity • Endocrine disruption • Acute aquatic toxicity	F*

Table 5.2   Continued.

Plasticizer Name	Grade Given Available Data	Available Data as the Basis for this Grade	Data Gaps	Data Gap Grade
DOZ Di(2-ethylhexyl) azelate	A	Rated of low concern for • Acute mammalian toxicity • Reproductive/developmental toxicity • Genotoxicity/mutagenicity • Aquatic toxicity • Persistence and bioaccumulation	• Carcinogenicity • Endocrine disruption	C
TEHTM Tri(2-ethylhexyl) trimellitate	D	• Acute aquatic toxicity • Persistence	• Carcinogenicity • Reproductive/ developmental toxicity • Endocrine disruption	F
TXIB 2,2,4-Trimethyl-1,3-pentanediol diisobutyrate	D	Present on Canadian DSL list for environmental toxicity	• Carcinogenicity • Endocrine disruption	D

Ecolibrium LPLAs 1100 series (proprietary chemistry):
Non-disclosure agreements necessary for the evaluation of this plasticizer are not yet in place. A QCAT assessment will be performed as soon as possible to enable the chemical to be considered for a later GreenScreen™ should funding become available.

the following three plasticizers were selected to receive full GreenScreen™ assessments:

- COMGHA (acetylated monoglycerides of fully hydrogenated castor oil)
- TEHTM [tri(2-ethylhexyl) trimellitate]
- DINCH (diisononylcyclohexane-1,2-dicarboxylate).

DOZ [di(2-ethylhexyl) azelate] and TXIB (2,2,4-trimethyl-1,3-pentanediol diisobutyrate) were selected as alternatives in the event that the three plasticizers listed above had insufficient data for a robust GreenScreen™ assessment.

5.7   Applying the GreenScreen™

The GreenScreen™ for Safer Chemicals is a screening tool developed by Clean Production Action (CPA) to assess and compare the environmental, health and safety profiles of chemicals. The GreenScreen™ was selected as the primary method for evaluating chemicals for this project because it provides a consistent, repeatable, transparent and scientifically robust method for comparison.

The GreenScreen™ defines four Benchmarks, 1–4, with 4 being the safest:

- *Benchmark 1:* Toxic chemical, avoid – chemical of high concern
- *Benchmark 2:* Use but search for safer substitutes
- *Benchmark 3:* Use but still opportunity for improvement
- *Benchmark 4:* Prefer – safer chemical.

The Green Screen assesses 16 human and environmental endpoints, grouped into the following four categories:

- Environmental Fate: persistence and bio-accumulation
- Ecotoxicity: acute aquatic and chronic aquatic toxicity
- Human Health Effects: carcinogenicity, mutagenicity/genotoxicity, reproductive/developmental toxicity, endocrine disruption, neurotoxicity, acute mammalian toxicity, skin sensitization, respiratory sensitization, immune system effects and systemic toxicity/organ effects
- Physical/Chemical Properties: explosivity and flammability.

The GreenScreen™ utilizes a wide range of data sources such as toxicity lists and databases from authoritative bodies, estimated data (*e.g.* from the US EPA's PBT Profiler), measured data from peer-reviewed scientific journals and risk assessments, advanced scientific modeling and expert review of data sources. Each endpoint is rated as high, medium or low and then entered into a summary table, called the 'hazard matrix.' Once a chemical is fully characterized with all the available data, the Benchmarks are applied, data gaps are noted and a GreenScreen™ report is completed including references to all data sources utilized.

The plasticizers that were ultimately assessed using the GreenScreen™ differed from the original plan. There were several reasons for this departure. One reason was the lack of data availability. COMGHA was chosen as a target plasticizer. However, upon initial efforts to conduct a GreenScreen™ for this chemical, it was determined that there were not enough publicly available to data to characterize the endpoints. Therefore, DOZ was determined to have greater data availability and was therefore selected to replace COMGHA.

Additionally, after the target list of plasticizers was developed, companies in the GC3 project group were asked to contribute financially to help cover the costs of the GreenScreen™ assessments and were given the option to choose which chemical they wanted to have assessed. BASF, Dell, Dow, HallStar and Hewlett-Packard provided funding to cover the cost of the GreenScreens™. Some companies chose chemicals that were not part of the targeted list. One company wanted to fund the assessment of Flexol 380 [di(2-ethylhexyl) isophthalate] but because publicly available data for this chemical were found to be significantly lacking, the company requested DEHT in its place. This company was interested in having either an iso- or terephthalate assessed and, in particular, they wanted to learn more about DEHT because DEHT was being detected in components that it sources from its supply chain. DEHT had a full SIDS Dossier and was considered relatively data rich; therefore, it was believed to be a good candidate for a GreenScreen™ assessment. The plasticizers assessed with the GreenScreen™ are listed in Table 5.3.

5.8 Conducting the Chemical Hazard Assessment Portion of the GC3 Project

The GreenScreens™ were conducted by the firm ToxServices, under contract to the GC3. ToxServices is licensed by CPA to conduct GreenScreens™.

5.8.1 Obtaining Data for GreenScreens™

GreenScreens™ are data intensive. The sources of data for the GreenScreens™ in this project varied according to a number of factors that are best described by categorizing the GreenScreens™ that were completed into two categories:

- *Category 1* – The data used for GreenScreens™ were in the public domain or provided by the supplier without an non-disclosure agreement (NDA), plus some data derived from modeling by ToxServices.
  - **Plasticizers:** DEHT, DINP, DOZ, Hexamoll® DINCH®, TEHTM.
  - **Formulation:** Plasticizers consisting of a single chemical/single CAS number.
  - **Transparency:** GreenScreen™ assessments were 'transparent,' *i.e.* the GreenScreen™ assessment reveals (1) the chemical identity of the plasticizer (chemical name and CAS number), (2) the hazard matrix which indicates the scoring of each hazard endpoint, (3) the Benchmark scores, (4) the toxicological information used to determine the hazard endpoints and benchmark scores and (5) references indicating the source of the toxicological data and other data used.
- *Category 2* – The data used for GreenScreens™ were provided by the suppliers that were part of the project group under an NDA between the supplier (*i.e.* the manufacturer of the plasticizer) and ToxServices.
  - **Plasticizers:** Ecolibrium®, Dioplex®, Paraplex®.
  - **Formulation:** Plasticizers are formulations, consisting of more than one chemical/CAS number.
  - **Transparency:** GreenScreens™ were performed by ToxServices under an NDA with the manufacturer, the manufacturer provided the toxicological data to ToxServices and the manufacturer had the ability to review and comment on drafts in confidence with ToxServices prior to the final draft of the GreenScreen results being released to the project group. Draft GreenScreen™ reports are redacted, *i.e.* the chemical identify of the formulation is not disclosed and some or all of the toxicological data are not disclosed in the assessment report.

Table 5.3 Plasticizers assessed with the GreenScreen™.

Chemical Name	Abbreviation, Trade Name	Chemical Type	CAS No.	Manufacturer(s)	GreenScreen™ Transparent or Redacted?
Di(2-ethylhexyl) terephthalate	DEHT Eastman 168®	Phthalate	6422-86-2	Eastman Chemical	Transparent
Diisononyl phthalate	DINP	Phthalate	68515-48-0	Exxon Mobil, BASF, *etc.*	Transparent
Di(2-ethylhexyl) azelate	DOZ	Azelate	103-24-2	Various	Transparent
Di(2-propylheptyl) phthalate	DPHP, Palatinol® DPHP (or Palatinol® 10P in Europe)	Phthalate	53306-54-0	BASF	Transparent
Modified vegetable oil derivatives	Ecolibrium® – four formulations	Confidential	Confidential	Dow	Redacted
Polymeric adipate	Dioplex®	Adipate	Polymer exempt	HallStar	Redacted
Polymeric adipate	Paraplex®	Adipate	Polymer exempt	HallStar	Redacted
Diisononyl cyclohexanedicarboxylate	Hexamoll® DINCH® (BASF)	Aliphatic ester	166412-78-8 (outside the USA), 474919-59-0 (inside the USA)	BASF	One full toxicological study shared confidentially with GreenScreen™ Profiler, otherwise transparent
Tri(2-ethylhexyl) trimellitate	TEHTM, Palatinol® TOTM	Trimellitate	3319-31-1	Various	Transparent

For GreenScreens™ in both categories, in cases where data were lacking and a suitable chemical analog was available, ToxServices filled relevant data gaps with data for structurally similar chemicals. ToxServices followed the GreenScreen™ guidelines for choosing surrogates.[†] This was the case for the plasticizer DOZ. ToxServices chose bis(2-ethylhexyl) adipate (CAS No. 103-23-1) as a surrogate. When data for the surrogate were used in the GS for DOZ, ToxServices indicated it as such in the assessment report.

5.8.2   Reviewing Draft GreenScreens™

In this project, we aimed to maximize learning about chemical hazard assessment and the GreenScreen™ method and to be as open and transparent as possible in the way we conducted the work. When ToxServices prepared a draft of a GreenScreens™ assessment, it was circulated to the project group by email and was posted on a password-protected website set up specifically for this project. Group members were asked to review and comment on drafts. The comments received were also posted on the webpage so that other project group members could see all comments submitted. When changes were made to the GreenScreens™ as a result of comments received, the Group discussed the reasons for those changes during periodic conference calls with GC3 team members.

ToxServices considered all comments submitted by project group members and made changes that they believed were relevant and appropriate. In cases where there was disagreement between the expert judgment of the project team reviewer and the expert judgment of ToxServices, ToxServices' expert judgment prevailed in the draft GreenScreens™ reports. The project facilitator (Monica Becker) requested that ToxServices compile a report summarizing the comments received on the draft GreenScreens™ and describe whether these comments led ToxServices to modify the GreenScreen™ assessments and to describe the changes made. This document was also posted on the website as were all interim GreenScreen™ drafts.

5.9   Results

5.9.1   Results of the GreenScreen™ Assessments of
Alternative Plasticizers

Although the GreenScreens™ method is robust and assessments are done by a Licensed GS Profiler, Clean Production Action (CPA), and other organizations that have been involved in the development and use of the GreenScreens™ have decided that GreenScreens™ need to be viewed as draft assessments until after the GreenScreens™ have been peer reviewed by a CPA-designated, independent, third-party expert reviewer. This peer review or 'validation' process, as it

---

[†]GreenScreen Guidance, refer to section on choosing surrogates.

is referred to in the GreenScreen™ Program, is designed to bring to light additional, or emerging, relevant toxicological studies that the profiler was not aware of or did not have access to, that could legitimately influence the GreenScreen™ assessment; and to have other experts weigh in on expert judgments, particularly to harmonize judgments where there are conflicting results.[‡]

CPA is in the process of developing the validation process, including the identification of experts who would be qualified to do validation. While the exact timing for finalizing the validation program is still being worked out, the goal is to complete a pilot of the process by the end of 2012.

Although the draft GreenScreen™ assessments for all of the targeted plasticizers in this project have been completed, the GC3 project group decided that they did not want to release the individual chemical results until the GreenScreens™ have undergone validation. Overall, it can be noted that for the six transparent plasticizers evaluated, the following GreenScreen™ Benchmark values were determined: one plasticizer received a Benchmark 1 score, one plasticizer received a 2 and three plasticizers received a U for insufficient data. The data gaps identified for these three plasticizers were primarily for cancer and endocrine activity. The ingredients for the proprietary polymers received Benchmark scores of 2 and 3.

During the GC3's 2012–2013 project year, the project group will support the work of CPA to develop, pilot and finalize the validation process for the GreenScreen™ program and to offer the draft GreenScreens™ from this project for the piloting effort.

Once the GreenScreen™ results are validated through the CPA validation process, the GC3 will publish and disseminate the results publicly. This will enable GC3 members and companies throughout the wire and cable supply chain to make more informed decisions about alternative chemicals that could be used as plasticizers in their products.

5.9.2   Technical and Cost Evaluation of Plasticizers

After finalizing the draft GreenScreens™, the project group began to discuss the technical and cost evaluation portion of the project. The original aim of including technical and cost evaluations was to allow potential adopters of safer alternatives to determine more easily whether the plasticizers will be appropriate for their applications or, as one plastic compounder stated, 'information to let us know whether it was worth going to the next step of laboratory bench evaluation,' and whether they would be cost-effective.

The group quickly came to the decision that it would not be fruitful to dedicate the time and resources of this group to the cost evaluation. It was concluded that the cost of a plasticizer is highly dependent on the application and that the cost of these products is a 'moving target.' For example, newer

---

[‡]Today, CPA does not permit companies to make marketing claims based on GreenScreen™ Benchmark scores. CPA is considering a process for allowing marketing claims. While the process is not yet defined, CPA has stated that claims will only be allowed if a licensed GreenScreen™ profiler conducted the GreenScreen™ assessment and the assessment is validated.

products may be expensive at first but their cost typically drops when demand and production scale increase. Therefore, the cost evaluation was removed from the project scope.

The group is currently discussing whether and how to conduct a technical performance evaluation of the plasticizers assessed for hazard. Similarly to the cost evaluation of plasticizers, performance is highly dependent on the specific application, specific formulation of the plastic compound (i.e. polymer plus additives) and how the plastic is compounded (i.e. the equipment used, process steps and techniques applied to create a homogeneous application-specific plastic compound from a number of different raw materials). Some members of the group believe that it is not useful to provide generalized data on plasticizer performance. A proposal being considered now is simply to provide web links to technical information on the plasticizers on the manufacturers' websites with a disclaimer stating that each application is different and the GC3 should not compare performances of plasticizers.

5.9.3   Lessons Learned in the Project

5.9.3.1   Assessing Single Chemical Products with Known Ingredients Versus Proprietary Formulations.
Companies that are choosing plasticizers for formulated compounds or specifying plasticizers for their finished products need information to support their decision-making. The initial aim of this project was to provide a consistent, robust set of information upon which to evaluate and compare plasticizers, to include: the chemical identity(ies) of single ingredient or formulated plasticizers; a full, unredacted report containing a Benchmark score; a hazard table; and detailed information on the toxicology test results that formed the basis of the GreenScreens™ assessment.

However, it was not stated as a requirement at the outset of the project that the suppliers participating and contributing financially to the project would be expected to reveal the formulations of their products. Dow and HallStar, the suppliers of the multi-ingredient plasticizers evaluated in this project, informed the project group that they would not be willing to reveal their formulations, citing their need to protect their intellectual property. They were willing to provide the formulations and toxicological data to ToxServices under an NDA and then to have ToxServices generate redacted GreenScreens™ for the project under the condition that the redacted versions did not reveal the identity of their formulations.

In response to this development, the group developed an approach for structuring redacted GreenScreen™ reports for this project, with the aim of meeting the needs of the companies that did not want to disclose their formulations while encouraging them to provide as much information as possible in the GreenScreens™. The protocol sought to harmonize the approach used in the GC3 project with the approach taken in similar projects conducted by the US EPA's DfE partnership programs. In these projects, EPA and their partner companies reach an agreement on a generic description for the chemical ingredients that need to be concealed. For confidential toxicological studies,

EPA summarizes the data but references the study as a submitted confidential study.

The protocol that was developed for the GC3 project had three elements, as follows:

1. The company seeking to conceal the identity of the chemical(s) in their product enters into an NDA with ToxServices.
2. The company and ToxServices jointly decide on a descriptive, generic name(s) for the chemical ingredients to be masked. The naming protocol followed will be that developed by the US EPA DfE Program. An example is provided below from the Furniture Flame Retardancy Partnership report (page 4-2).[§] The descriptive, generic name of the chemical will be used in the ToxServices GreenScreen™ report and any summary statements developed in the project.
3. Any studies provided to ToxServices by a company seeking to conceal a chemical name will be referenced as a submitted confidential study. If it is a private, unpublished study, the company must provide information to ToxServices on the organization that funded the study since that information will need to be included in the GreenScreen™ report (per Green-Screen™ guidelines).

Despite the project group agreeing to this protocol, when the redacted GreenScreens™ were released by ToxServices, the descriptors were not in line with the EPA DfE model. Rather, the chemical ingredients were indicated only by their company Trade Names (in the case of the Dow product) and as 'monomer A' and 'monomer B' by HallStar. Despite efforts to encourage these companies to conform to the EPA model, they were not willing to do so, again citing their need to protect intellectual property.

In contrast to the process used to generate and review the unredacted GreenScreens™, the process for the redacted GreenScreens™ was much less transparent and produced less information. ToxServices obtained the chemical information and toxicological test data from the manufacturer under an NDA and generated a GreenScreen™ report. The report was then sent to the manufacturer for their review and comment. ToxServices then determined if and how to address the comments from the manufacturer, *i.e.* to change the GreenScreen™ results or not on the basis of these comments. This process was hidden from the other project group members.

The redacted reports for the proprietary products had varying levels of disclosed information and redaction, but in all cases it is possible to know the Benchmark score and to see the hazard table for the ingredients. For some companies, this information may be sufficient for decision-making purposes; others may insist on knowing the full identity of chemical ingredients prior to adopting a substitute product such as a plasticizer. In the latter case, the prospective customer could ask the supplier to provide chemical identity

---

[§]Available at http://www.epa.gov/dfe/pubs/flameret/altrep-v1/altrepv1-f1c.pdf.

information under an NDA, although there is no guarantee that the supplier would be forthcoming.

At least one project group member objected to the inclusion of these proprietary GreenScreens™ in this project. This member expressed a desire for the GC3 to promote transparency by rewarding companies that are transparent with participation in the project and barring companies from participating if they are not forthcoming with chemical ingredient information.

The GC3 is considering the possibility of repeating this type of project to expand the number of plasticizers or to assess another category of products (*e.g.* flame retardants). If this comes to pass, the GC3 will consider the question of whether it should require full transparency as a condition of participation.

5.9.3.2 Fewer Toxicology Data Exist for Newer Plasticizers and Plasticizers from Smaller Suppliers.
GreenScreens™ are data intensive and toxicology testing is expensive and time consuming. As discussed earlier, some data gaps can be filled by modeling or by obtaining data for suitable analogous chemicals. In the GC3 project, we found that data were more available for single chemical products, particularly when the chemicals have been in commercial use for some time and are produced by larger companies with greater in-house capabilities or financial resources to do testing.

In contrast, products with multiple chemical ingredients and those that are being marketed by smaller companies with fewer testing capabilities and resources have fewer data. This may create an uneven playing field for newer products made by smaller companies. In particular, products that are less data rich may receive a Benchmark U, indicating that there was insufficient data for a GreenScreen™ Benchmark determination.

5.9.3.3 Need to Provide Clear Guidance at the Beginning of the Project.
Companies that participated in the project by paying for their plasticizers to be evaluated were not given information at the beginning of the project, prior to their commitment, on the data requirements for GreenScreen™ assessments. As a result, in some cases when the GreenScreen™ profiler indicated that there were insufficient data to reach a GreenScreen™ benchmark score and acceptable analogs or modeling could not be used to fill data gaps, the companies were unpleasantly surprised. A GreenScreen™ result of U is clearly less desirable than a GreenScreen™ Benchmark 3 or 4 and may send a signal to the marketplace that the product has not been adequately tested. Therefore, if this project were to be repeated, the data requirements for a GreenScreen™ should be clearly communicated at the outset.

5.9.3.4 Participating in a Collaborative Assessment Process Carries Some Risk for Suppliers.
When a supplier privately contracts a GreenScreen™ profiler to conduct a GreenScreen™ assessment of a product and the product receives a GreenScreen™ Benchmark score of 1, a U or a score that is lower

than the supplier expected or hoped for, the supplier does not need to publicly release the results.

In contrast to this private transaction, when the profiler completed the draft GreenScreens™ results in the GC3 project, the results were made public within the GC3 project group community. Although this is a closed group, nevertheless, once suppliers agreed to participate in the project, the Green-Screens™ results for their products were slated to be released to competitors and potential customers within the group; whether or not the supplier was happy with the results. This caused consternation among some suppliers when their products either received a score that was lower than they expected or received a U rather than a numerical score. It is not clear that these companies had thought this through completely prior to engaging in the project. If this project is to be repeated, this should be clearly articulated at the outset.

5.9.3.5 Funding for Project Management. In addition to the funding required to carry out the environmental, health and safety evaluations of the alternative chemicals using tools such as QCAT and GreenScreen™, there is a need to obtain funding for project management and coordination efforts and to streamline the funding process to make it less time consuming for administrative staff.

An alternative model to that used in this project, *i.e.* asking each participating company to pay for a GreenScreen™ of one or more chemicals, has been discussed. This would entail deciding on the total number of alternative chemicals to be evaluated, creating a budget for the entire project, including project management costs, and dividing the total cost among the companies participating.

5.10   Conclusion

The GC3 Pilot project succeeded in demonstrating that collaborative chemical assessments – involving companies at different points in a product supply chain, academic organizations and non-governmental organizations – can be done and that participants will share in the costs.

The dialog that took place during monthly conference calls and the collective review of draft documents yielded robust chemical hazard assessments of nine alternative plasticizers for wire and cable applications. In addition, there was a significant amount of learning on topics such as the application of QCAT and GreenScreen™ assessment processes, chemical hazard assessment methods in general, the need for and availability of data for these types of assessments, key challenges to managing chemicals of concerns within products and supply chains, and insight into how different companies assess and decide on chemical substitutes for their products. In addition, the project was a cost-effective way for the companies in the group to gain access to

GreenScreen™ results for multiple chemicals without having to pay for each assessment.

The GC3 would consider engaging in another project of this kind, to develop further the model for doing collaborative chemical assessments, if a few conditions are met. First, the chemical category needs to be one that is under significant regulatory pressure, such as chemical bans or restrictions. Otherwise, chemical users (such as formulated material suppliers and OEMs) would not have sufficient incentive to participate and contribute financially. Second, the project needs to be structured to build on the collaborative model that we have begun to create in the pilot project described in this chapter. Although the GC3 may be a credible forum for convening this type of collaborative process, it is unlikely that the organization would repeatedly undertake this type of effort. The ultimate goal of the GC3 is to create a model for how these collaborations can take place and then to have the GC3 and other organizations carry on this type of work. This is necessary so that the initiation of these collaborations can scale to meet the demand necessary for industry to respond effectively to regulatory requirements such as REACH and the California Green Chemistry Initiative.

Finally, the GC3 would require at least one influential OEM to commit to participate in the project to ensure that it can recruit suppliers and other companies in the supply chain to participate and to contribute financially.

Acknowledgements

The authors would like to acknowledge the contributions to the pilot project described in this chapter by the following individuals and organizations: Joel Tickner, GC3; Melissa Coffin, (formerly with the GC3 but now with the Healthy Building Network; Barbara Hanley, Cory Robertson and Helen Holder, Hewlett-Packard; Roger McFadden, Staples; David Levine, Environmental Health Fund; Meg Whittaker and Chris Schlosser, ToxServices; Shari Franjevic, Clean Production Action; and individuals from the companies whose plasticizer products were assessed, in addition to the plastics compounders and OEMs that participated in the project.

References

1. Green Chemistry and Commerce Council, *Moving Business Toward Safer Alternatives*; http://www.greenchemistryandcommerce.org/index.php (last accessed 5 September 2012).
2. Toxics Use Reduction Institute, DEHP, in *Five Chemicals Alternatives Assessment Study*, Toxics Use Reduction Institute, University of Massachusetts Lowell, Lowell, MA, 2006, Chapter 7, pp. 7-1–7-84.
3. Agency for Toxic Substances and Disease Registry, *Toxicological Profile for Di(2-ethylhexyl) Phthalate (Update)*, US Department of Health and Human Services, Public Health Service, Atlanta, GA, 2002.

4. T. J. Gray and K. R. Butterworth, Testicular atrophy produced by phthalate ester, *Arch. Toxicol.*, 1980, Suppl. 4, 452–455.
5. J. C. Lamb IV, R. E. Chapin, J. Teague, A. D. Lawton and J. R. Reel, Reproductive effects of four phthalic acid esters in the mouse, *Toxicol. Appl. Pharmacol.*, 1987, **88**, 255–269.
6. Clean Production Action and Healthy Building Network, *Chemicals of High Concern – List of Lists ('Red List of Lists')*, 2009; www.cleanproduction.org/library/CPA-HBN_Red_List_26jan09.doc (last accessed 28 December 2012).

Chemical Hazard Assessment and the GreenScreen™ for Safer Chemicals

LAUREN G. HEINE* AND SHARI A. FRANJEVIC

ABSTRACT

Whether driven by market pressures or increasing global regulations, businesses are seeking to understand better the hazards associated with the chemicals in their supply chain and to move toward the use of inherently safer alternatives to substances of high concern. In order to do so, they need reference tools that will allow them to compare options. Chemical hazard assessment (CHA) is an approach that allows one to compare chemicals based on inherent hazard and to make more informed decisions. Clean Production Action, a non-profit organization, has developed the first freely available and fully transparent CHA methodology called the GreenScreen™ for Safer Chemicals (Green-Screen™). The method builds on national and international precedents for hazard classification including the Globally Harmonized System of Classification and Labelling of Chemicals (GHS) and the US EPA Design for the Environment (DfE) Program Alternatives Assessment Criteria for Hazard Evaluation, and includes a structured decision logic in the form of Benchmarks. Clean Production Action has published all of the supporting documentation and is in the process of developing infrastructure to enable broader adoption of the GreenScreen™. This chapter traces the origins of the GreenScreen™, describes the process of applying the method, highlights how it can be and is being used to make more informed decisions by organizations including Hewlett-Packard, Staples,

*Corresponding author

Issues in Environmental Science and Technology, 36
Chemical Alternatives Assessments
Edited by R.E. Hester and R.M. Harrison
© The Royal Society of Chemistry 2013
Published by the Royal Society of Chemistry, www.rsc.org

Royal DSM and The Wercs and outlines the future direction of the GreenScreen™ program.

6.1   Introduction

Whether driven by the increasing number of global regulations that specify substance restrictions or by sustainability initiatives, businesses are moving toward improving their knowledge of chemicals in their supply chains and toward better understanding the hazards associated with those chemicals. These drivers create both threats and opportunities for innovation. One of the most compelling areas of potential innovation is Green (and Sustainable) Chemistry. Green chemistry is 'the utilization of a set of principles that reduces the use and/or generation of hazardous substances in the design, manufacture and application of chemical products.'[1] The 12 Principles of Green Chemistry call for the design of chemicals that are fully effective and inherently safer – such chemicals will have little or no toxicity, use innocuous or, better, avoid the need for solvents and auxiliaries in manufacturing, break down to innocuous substances that do not accumulate in the environment and minimize the potential for chemical accidents including explosions, fires and releases to the environment.

Chemists can use the 12 Principles to guide their practices. However, for most companies, activities such as procurement or product design and development involve the *selection* of chemicals and materials rather than the creation of new molecules. These companies are chemical and material 'choosers' who depend on suppliers for providing options. A significant challenge, with respect to procurement or designing safer and more sustainable products, is to identify the safest and most sustainable chemical and material options that provide the needed functionality. Chemical hazard is one attribute that can be used to compare options and to advance green chemistry through chemical hazard assessment (CHA). CHA is based in the risk paradigm whereby risk is a function of hazard and exposure. An important premise behind green chemistry is that in many cases, chemical risk can be effectively managed by reducing hazard, rather than by controlling exposure. Exposure controls can and do fail and products are used in ways that were never intended. The primacy of hazard reduction as the preferred option for reducing risk is established in the Pollution Prevention Act of 1990, which defines pollution prevention (also known as source reduction) as any practice that 'reduces the hazards to public health and the environment.'[2] However, even if one were to focus solely on the exposure side of the equation, the practices of elimination and substitution of toxic chemicals are considered top priority in global occupational and health and safety exposure control prioritization schemes.[3]

CHA allows one to compare chemicals based on their inherent hazard and to make more informed decisions. Informed substitution, a term coined at the United States Environmental Protection Agency (US EPA), promotes the considered transition from a chemical of concern to a safer chemical or non-chemical alternative. Informed substitution builds on the best available information and reduces business risk by minimizing the opportunity for unintended consequences that may occur by switching from a chemical of concern to a chemical with unknown characteristics. Using CHA early in the material selection process allows one to compare directly alternatives for a particular functional use based on their inherent hazards that include human toxicity, physical hazards and environmental toxicity and fate.

CHA is particularly effective when comparing alternative chemicals or materials with similar functional uses, because differences in life-cycle use patterns and exposure potential are minimized. CHA is necessary but not sufficient for sustainability. It addresses one impact category (hazard) that can be used along with other indicators of sustainability and/or product performance. However, it is a critical attribute that has too often been overlooked, in part due to the lack of a robust and systematic way to compare chemicals based on hazard. It is difficult for organizations to take action to improve their chemical inventories without a common reference. Based on this need, the GreenScreen™ for Safer Chemicals (GreenScreen™) was developed as the first fully transparent, freely and publicly accessible method to assess and evaluate chemicals based on inherent hazard.

6.1.1 Origins of the GreenScreen™ for Safer Chemicals

The GreenScreen™ was developed by Clean Production Action (CPA), a non-profit research and education organization with the mission to advance green chemistry and sustainable materials management. GreenScreen™ version 1.0 emerged from the confluence of two initiatives – the US EPA Design for Environment (DfE) Program and state regulatory initiatives seeking to ban certain flame retardants in the USA.

DfE hosts Alternatives Assessment Partnerships to identify and evaluate alternatives to chemicals of concern. The first two DfE Partnerships focused on alternatives to flame retardants – pentabromodiphenyl ether (pentaBDE) used in polyurethane foam in furniture and tetrabromobisphenol-A (TBBPA) used in printed circuit boards.[4,5] The DfE Partnerships are voluntary initiatives designed to leverage knowledge and experience from the US EPA's New Chemicals Program, which has over 30 years of experience implementing the Toxic Substances Control Act (1976) and assessing chemical hazard and risk for new chemical substances based on test data, structure–activity relationship modeling and expert judgment.[6] DfE Partnerships target alternatives to chemicals of concern identified by stakeholders or by the US EPA, including chemicals for which there are Action Plans.[7]

Each DfE Chemical Alternatives Assessment addresses a comprehensive set of hazard endpoints including carcinogenicity, mutagenicity, reproductive and developmental toxicity, neurotoxicity, acute and chronic mammalian toxicity, sensitization, irritation/corrosion and environmental toxicity and fate. Each chemical, whether as a pure substance or in a mixture, is assigned a hazard classification (*e.g.* high, moderate or low) for each hazard endpoint based on test data, estimated values and/or expert judgment. The results are presented in an easy to read hazard table (Table 4-1)[8] that has become a hallmark of CHA because it provides a wealth of hazard information about a chemical in a comprehensive, clear and meaningful format.

While the DfE hazard table approach laid out chemical hazard information in a clear and comprehensive format, DfE could not recommend, beyond general guidelines, which chemical alternatives were more or less preferable. Participants were left wondering how best to use the hazard table to make decisions. Some users considered simple algorithms such as counting the number of High or Low hazard classification values and/or applying weighting factors to certain endpoints.

Around the same time as the initial DfE Partnerships, certain US states (*e.g.* Washington) were looking to ban the use of decaBDE as a flame retardant. Before doing so, they wanted to ensure that safer alternatives were available. Independently, CPA evaluated and compared the inherent hazard profiles of three flame retardants, including decaBDE. To do this, CPA adopted and made modifications to the hazard assessment and classification approach used in the DfE Alternatives Assessment Partnerships. The results were published in a report that became GreenScreen™ version 1.0.[9] The modifications included the following:

- Addition of endocrine activity, flammability and reactivity as hazard endpoints considered in the assessment.
- Addition of criteria from the Globally Harmonized System of Classification and Labelling of Chemicals (GHS).[10]
- Addition of a set of authoritative hazard lists to assist with hazard classification.
- Modification of criteria for persistence and bio-accumulation potential to identify better chemicals with low persistence and low bio-accumulation potential.
- Inclusion of transformation products from the parent chemical as part of the assessment.
- Addition of Benchmarks to help guide decision-making toward the use of substances with inherently lower hazard.

The development of GreenScreen™ Benchmarks enhanced the utility of the DfE approach by providing structured decision logic to the results of the hazard classifications. The benchmarking system was developed based on principles underlying regulations in the USA, Canada and Europe. For example, chemicals of concern (Benchmark 1) are those chemicals with hazard attributes that include carcinogenicity, mutagenicity, reproductive and developmental toxicity, persistence and/or bio-accumulation potential in combination with toxicity and substances of 'equivalent concern.' These attributes are consistent with the definition of a substance of very high concern (SVHC) under REACH.[11] Preferred chemicals (Benchmark 4) are those with inherently low hazard across all hazard endpoints. Benchmarks 2 and 3 are based on the same logic but with lesser degrees of hazard. Much has changed since the first DfE Partnership and since the publication of GreenScreen™ version 1.0, but movement continues toward even greater alignment of the hazard criteria between GreenScreen™ and the DfE Program's Alternatives Assessment Criteria for Hazard Evaluation.[12]

6.2 How it Works

Applying the GreenScreen™ method involves three broad steps:

1. Assess and classify hazards.
2. Apply the Benchmarks.
3. Make informed decisions and drive innovation.

6.2.1 Assess and Classify Hazards

Assessment and classification of hazards are accomplished by:

1. Identifying the chemical constituents to be assessed along with their relevant and feasible transformation products.
2. Comprehensively reviewing all available data (including hazard lists, summary data sets, scientific and toxicology literature, models and structural analogs).
3. Assigning a hazard level for each endpoint based on GreenScreen™ criteria.
4. Determining the level of confidence.
5. Populating the hazard table.

6.2.1.1 Identifying Chemical Constituents and Transformation Products. It is sometimes challenging to identify accurately (by CAS number) the chemicals to be assessed. Some chemicals are identified by multiple CAS numbers, while other chemicals share the same CAS number even though they vary in structure (*i.e.* more or less highly branched) in ways that may affect hazard classification for endpoints such as persistence. Using the GreenScreen, all intentionally added chemicals[13] should be identified and evaluated. In addition, any chemicals not intentionally added but present at or above *de minimus* (100 ppm) should be included in the assessment. These chemicals are typically identified based on life-cycle knowledge, particularly of upstream manufacturing processes.

Best practice for applying the GreenScreen™ entails reporting of all chemicals known to be present in a substance, material or mixture. However, the key is to ensure that the *de minimus* value used is reported to ensure that chemicals or materials can be compared on an equivalent basis.[14] Polymeric materials present some special considerations. Polymeric materials can comprise many constituents, including catalysts, residual (unreacted) monomers, oligomers and processing aids. While these constituents are not considered intentionally added, their use or presence must be reported along with their concentration. A standardized reporting template for GreenScreen™ assessments is provided to ensure consistent reporting.[15]

In some cases, it is necessary to identify chemicals beyond the primary chemical being assessed. For example, when data are not available on the primary chemical for a particular endpoint, data from a suitable analog may be used to estimate the hazard of the parent chemical for that endpoint. When assessing inorganic chemicals, identification and assessment of key moieties may provide the most relevant information regarding the hazard of the parent chemical. A moiety is defined in Canada's Domestic Substance List Guidance[16] as 'a discrete chemical entity that is a constituent part or component of a substance.' Identification and assessment of moieties are relevant because they reflect the actual species present based on dissolution, dissociation, speciation or other transformations that occur in environmentally and/or biologically relevant pathways.

To apply the full GreenScreen™ method, it is important to identify and assess feasible and relevant transformation products because in some cases, the transformation products are more toxic and persistent than the parent compound. Feasible means that based on the use pattern and life-cycle of the substance or product, the chemical is likely to undergo a specific transformation pathway such as hydrolysis, biodegradation or combustion. Relevant means that the transformation pathway is likely to generate a particular transformation product with hazardous characteristics in sufficient quantities or with persistent and/or bio-accumulative characteristics that could potentially result in increased risk from the use of the parent chemical across its life-cycle. The intention is *not* to identify transformation products that are transient or that occur naturally and are found at ambient concentrations. Feasible transformation products include degradation products and metabolites that may be formed during use or upon release. For example, if a flame retardant is being evaluated for use in an electronic product that may be burned after use to recover materials, then combustion products are feasible transformation products. Likewise, if a personal care product is designed to go down the drain, then aquatic degradation is considered.

Estimating feasible and relevant transformation products requires expertise and expert judgment. The following resources may support the identification of feasible and relevant transformation products based on chemical and biological processes.

1. Literature from peer-reviewed scientific journals on environmental science, monitoring, toxicology and chemistry.
2. Published risk assessments from governmental and other authoritative sources.
3. Online search, modeling and prediction tools such as the SRC Fate-Pointer[17] or the University of Minnesota Pathway Biocatalysis Biodegradation Prediction Program.[18]

4. Comprehensive, publicly accessible toxicology databases such as the Hazardous Substances Data Bank (HSDB) of the US National Library of Medicine or eChemPortal maintained by the Organisation for Economic Co-operation and Development (OECD) that may contain information on metabolism and metabolites.[19,20]

6.2.1.2 Hazard Classification. The GreenScreen™ hazard endpoints and associated criteria reflect those used by government agencies and authoritative bodies in their chemical initiatives including the US EPA, the OECD,[21] the Canadian Government's categorization scheme,[22] the International Joint Commission (a commission established by the USA and Canada to protect transboundary waters), the European Union's REACH[23] and CLP Regulation,[24] the Stockholm Convention on Persistent Organic Pollutants and the Globally Harmonized System of Classification and Labeling of Chemicals (GHS). Every effort was made to harmonize with national and international precedents in order to leverage the usefulness of data and other hazard information. In the GreenScreen™, a chemical's hazard characteristics are defined by its potential to cause acute or chronic adverse effects in humans or wildlife, its fate in the environment and certain physical/chemical properties that relate to worker safety.

The GreenScreen™ hazard endpoints are divided into the following groups:

1. Group I (Human Health)
    (a) Carcinogenicity (C)
    (b) Mutagenicity/Genotoxicity (M)
    (c) Reproductive Toxicity (R)
    (d) Developmental Toxicity including Neurodevelopmental Toxicity (D)
    (e) Endocrine Activity (E)
2. Group II (Human Health)
    (a) Acute Mammalian Toxicity (AT)
    (b) Systemic Toxicity/Organ Effects – Single Exposure sub-endpoint (ST-single)
    (c) Neurotoxicity – Single Exposure sub-endpoint (N-single)
    (d) Irritation/Corrosivity – Eyes (IrE)
    (e) Irritation/Corrosivity – Skin (IrS)
3. Group II* (Human Health)
    (a) Systemic Toxicity/Organ Effects – Repeated Exposure sub-endpoint (ST-repeated)
    (b) Neurotoxicity – Repeated Exposure sub-endpoint (N-repeated)
    (c) Respiratory Sensitization (SnR)
    (d) Skin Sensitization (SnS)
4. Ecotoxicity
    (a) Acute Aquatic Toxicity (AA)
    (b) Chronic Aquatic Toxicity (CA)
5. Fate
    (a) Persistence (P)
    (b) Bio-accumulation (B)

6. Physical/Chemical Properties
   (a) Flammability (F)
   (b) Reactivity (Rx)

Group I endpoints include Carcinogenicity, Mutagenicity/Genotoxicity, Reproductive Toxicity, Developmental Toxicity (including Neurodevelopmental Toxicity) and Endocrine Activity. Group I endpoints are consistent with priorities reflected in national and international governmental regulations. These endpoints address hazards that can lead to chronic or life-threatening effects or adverse impacts that are potentially induced at low doses and transferred between generations.

Group II (and II*) endpoints are additional endpoints that are necessary for understanding and classifying hazards. Group II endpoints tend to be threshold based or dose dependent. Group II and Group II* are differentiated in the benchmarking system because Group II endpoints are based on testing with single doses and have four hazard levels (vH, H, M and L) whereas Group II* endpoints are based on testing with repeated doses and have three hazard levels (H, M and L). Systemic Toxicity/Organ Effects and Neurotoxicity endpoints can belong in either Group II or Group II* depending on whether the data are generated from single exposure (acute) or repeated exposure (sub-chronic or chronic) studies. Results from single and repeated exposures are not considered as separate endpoints but rather sub-endpoints. If data exist for both single and repeated exposure studies, and if both are of good quality, then data from repeated exposure studies should be prioritized.

Endocrine disruption is not considered an adverse effect *per se*, 'but rather a potential mechanism of action,'[25] particularly for developing organisms. However, changes in hormone levels and/or disruption of hormonally regulated processes such as those caused by endocrine-disrupting chemicals can lead to severe health effects. The GreenScreen™ hazard criteria for Endocrine Activity distinguish between substances that are endocrine active (Moderate) and substances that are endocrine active and demonstrate a related adverse effect (High). This distinction is reinforced in a recent publication by the Danish Centre on Endocrine Disruptors.[26] It is currently beyond the scope of the GreenScreen™ method to specify a minimum set of tests required to define clearly low hazard for endocrine activity.

6.2.1.3 Defining Levels of Concern for Each Hazard Endpoint. Each GreenScreen™ hazard endpoint is defined by a set of criteria. The set of criteria defines the three primary classification levels (High, Moderate or Low) for that hazard endpoint. Some endpoints such as Persistence (P) and Bio-accumulation Potential (B) also have very High (vH) or very Low (vL) classification options. An excerpt of the GreenScreen™ version 1.2 hazard criteria for Developmental Toxicity is shown in Table 6.1 and the full set of criteria are available online.[27]

The GreenScreen™ criteria may include thresholds based on standardized OECD or US EPA test methods, categorization criteria described in GHS,

Table 6.1 GreenScreen™ criteria for developmental toxicity.

Information Type	Information Source		High (H)	Moderate (M)	Low (L)
Developmental Toxicity (D)	Data	**GHS Criteria & Guidance** *Note: GHS Reproductive Toxicity includes both reproductive and developmental effects, while the Green Screen separates them into two distinct hazard endpoints. This classification must be based on developmental effects alone.*	GHS Category 1A (Known) or 1B (Presumed) for any route of exposure	GHS Category 2 (Suspected) for any route of exposure or limited or marginal evidence of developmental toxicity in animals (See Guidance)	Adequate data available, and negative, no structural alerts, and GHS not classified.
	A Lists	EU H-statements	Authoritative	H360FD, H360D, H360Df, or H362	H360Fd, H361d, H361fd	
		EU R-phrases	Authoritative	R61 or R64	R63	
		NTP-OHAaT	Authoritative	Clear Evidence of Adverse Effects – Developmental		Clear Evidence of No Adverse Effects - Developmental
		Prop 65	Authoritative	Known to the state to cause developmental effects		
	B Lists	G&L	Screening	Developmental Neurotoxicant		
		Boyes-N	Screening		Developmental Neurotoxicity Effects	
		MAK	Authoritative	Pregnancy Risk Group A or B	Pregnancy Risk Group C	
				Pregnancy Risk Group D		
		NTP-OHAaT	Authoritative	Limited Evidence of Adverse Effects - Developmental or Some Evidence of Adverse Effects - Developmental	Limited Evidence of No Adverse Effects - Reproductive or Some Evidence of No Adverse Effects - Developmental	
				Insufficient Evidence for a Conclusion - Developmental Toxicity		

evidence from the scientific and toxicological literature or references to regulatory or hazard lists developed by authoritative bodies or organizations with relevant expertise. Whereas GHS criteria are used wherever they apply, GHS criteria are not available for all of the endpoints specified in the Green-Screen™ including Persistence, Bio-accumulation Potential and Endocrine Activity. In addition, for some endpoints, GHS does not provide toxicity thresholds, rather it classifies hazard based on evidence that a chemical either does, or does not, demonstrate a specific hazard, regardless of the potency of that hazard.

Establishing criteria for some hazard endpoints was more challenging than for others. Establishing criteria for P and B was particularly challenging for two reasons. First, there is significant national and international variation in threshold values used to define persistence and bio-accumulation potential. For example, the high level of concern for bio-accumulation used by the US EPA [bioconcentration factor (BCF) > 5000] was the very high level of concern used by the European Union (EU) and the Stockholm Convention on Persistent Organic Pollutants (POPs). Second, governments typically establish threshold criteria to regulate problematic chemicals. There is less precedent for identifying thresholds for chemical hazards of low concern. In the GreenScreen™, criteria are included to classify chemicals with low hazard such as rapid degradability and low potential for bio-accumulation.

Regulatory and authoritative hazard lists can be used to support hazard classification and have been incorporated into the GreenScreen™ criteria. The lists leverage the expertise of the scientific and regulatory communities for chemicals that are well characterized for particular hazards. Examples of chemical listings that support classification of carcinogenicity include the International Agency for Research on Cancer (IARC),[28] National Toxicology Program,[29] California Proposition 65,[30] European Union CLP Regulation[31] and the US EPA Integrated Risk Information System.[32]

The GreenScreen™ includes specified hazard lists defined as follows:

1. Authoritative Lists – Listing is based on a comprehensive expert review by a recognized authoritative body.
2. Screening Lists – Lists are identified as Screening Lists if they were developed using a less comprehensive review, if they were compiled by an organization that is not considered to be authoritative, if they were developed using exclusively estimated data or if the chemicals are listed because they have been selected for further review and/or testing.
3. A lists – Each category in the list translates directly to a single level of concern for a single GreenScreen™ hazard endpoint or a single Benchmark. Authoritative A lists typically represent a high level of confidence.
4. B Lists – Lists meet one or more of the following: (i) each category in the list incorporates a single GreenScreen™ hazard endpoint and does not translate directly to a single level of concern or benchmark; or (ii) each category in the list refers to more than one hazard endpoint.

A supporting resource called the GreenScreen™ List Translator™ maps authoritative and screening A and B lists to hazard endpoints and where feasible, to hazard classifications and Benchmarks.[33]

GreenScreen™ carries inherent weightings for the use of different types of data/information. In general, measured data from valid studies trump estimated values from sources such as models and analogs. Typically, a literature review will uncover conflicting results for chemical hazards. When lists conflict, the most authoritative list is used and, of those, the most conservative results are used. When scientific literature and test results conflict, the assessor will use all available information to make a professional judgment based on weight of evidence. The GreenScreen™ method adopted the DfE convention of using a bold, colored capital letter (*e.g.* a red **H**) to indicate when hazard classification was high confidence *i.e.* based on measured data, and a black italic letter (*e.g.* *H*) to indicate lower confidence, *i.e.* when hazard classification was based on estimated values.

GreenScreen™ clearly distinguishes between low hazard and unknown hazard when classifying each hazard endpoint. When test data are not available on the chemical or its reasonable analogs, when hazard lists do not specify classifications and when specific hazards cannot be modeled or estimated, then the notation 'DG' should be used to designate a data gap for that specific endpoint.

6.2.2   Apply the Benchmarks

The GreenScreen™ defines four benchmarks on the path to safer chemicals, with each benchmark defining a progressively safer chemical. When data are considered insufficient to establish a Benchmark score, the Benchmark is indicated as 'U' for 'Unspecified.' The Benchmarks were developed to discriminate between chemicals based on individual hazards and combinations of hazards. Figure 6.1 illustrates how the hazard classifications are integrated into Benchmarks.

Each benchmark is defined by a set of criteria based on the hazard classification scheme. A chemical, along with its feasible transformation products, must pass each of the criteria within a Benchmark to pass that Benchmark. Starting with the criteria for Benchmark 1, if **any** of the criteria statements (*i.e.* a–e for Benchmark 1) are true, then the chemical fails to pass that Benchmark. The hazard criteria for Benchmark 1 are as follows:

(a) PBT = High P + High B + very High T (Ecotoxicity or Group II) or High T (Group I or II*)]

(b) vPvB = very High P + very High B

(c) vPT = very High P + [very High T (Ecotoxicity or Group II) or High T (Group I or II*)]

(d) vBT = very High B + [very High T (Ecotoxicity or Group II) or High T (Group I or II*)]

(e) High T (Group I)

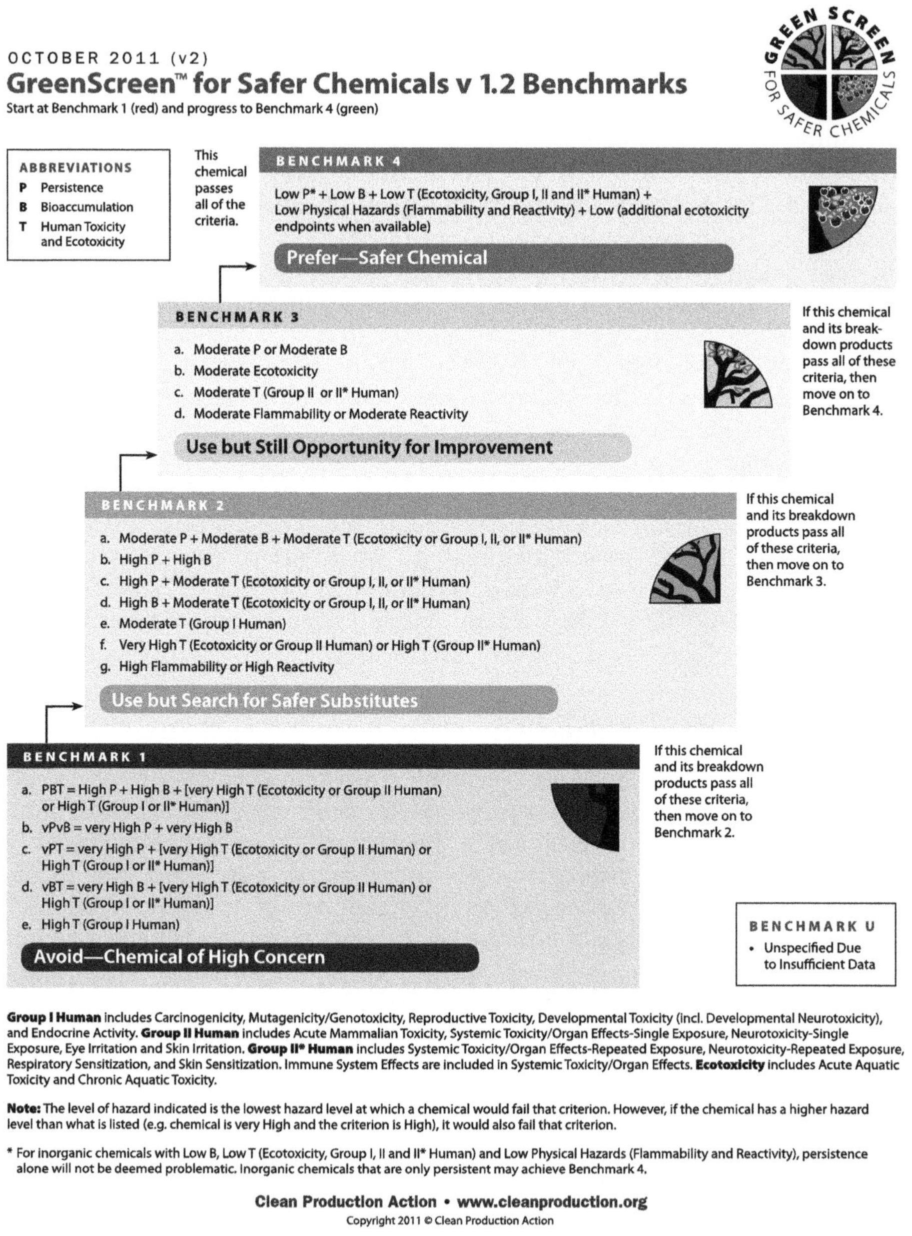

OCTOBER 2011 (v2)
GreenScreen™ for Safer Chemicals v 1.2 Benchmarks
Start at Benchmark 1 (red) and progress to Benchmark 4 (green)

ABBREVIATIONS		This chemical passes all of the criteria.
P	Persistence	
B	Bioaccumulation	
T	Human Toxicity and Ecotoxicity	

BENCHMARK 4

Low P* + Low B + Low T (Ecotoxicity, Group I, II and II* Human) + Low Physical Hazards (Flammability and Reactivity) + Low (additional ecotoxicity endpoints when available)

Prefer—Safer Chemical

BENCHMARK 3

a. Moderate P or Moderate B
b. Moderate Ecotoxicity
c. Moderate T (Group II or II* Human)
d. Moderate Flammability or Moderate Reactivity

Use but Still Opportunity for Improvement

If this chemical and its breakdown products pass all of these criteria, then move on to Benchmark 4.

BENCHMARK 2

a. Moderate P + Moderate B + Moderate T (Ecotoxicity or Group I, II, or II* Human)
b. High P + High B
c. High P + Moderate T (Ecotoxicity or Group I, II, or II* Human)
d. High B + Moderate T (Ecotoxicity or Group I, II, or II* Human)
e. Moderate T (Group I Human)
f. Very High T (Ecotoxicity or Group II Human) or High T (Group II* Human)
g. High Flammability or High Reactivity

Use but Search for Safer Substitutes

If this chemical and its breakdown products pass all of these criteria, then move on to Benchmark 3.

BENCHMARK 1

a. PBT = High P + High B + [very High T (Ecotoxicity or Group II Human) or High T (Group I or II* Human)]
b. vPvB = very High P + very High B
c. vPT = very High P + [very High T (Ecotoxicity or Group II Human) or High T (Group I or II* Human)]
d. vBT = very High B + [very High T (Ecotoxicity or Group II Human) or High T (Group I or II* Human)]
e. High T (Group I Human)

Avoid—Chemical of High Concern

If this chemical and its breakdown products pass all of these criteria, then move on to Benchmark 2.

BENCHMARK U
• Unspecified Due to Insufficient Data

Group I Human includes Carcinogenicity, Mutagenicity/Genotoxicity, Reproductive Toxicity, Developmental Toxicity (incl. Developmental Neurotoxicity), and Endocrine Activity. **Group II Human** includes Acute Mammalian Toxicity, Systemic Toxicity/Organ Effects-Single Exposure, Neurotoxicity-Single Exposure, Eye Irritation and Skin Irritation. **Group II* Human** includes Systemic Toxicity/Organ Effects-Repeated Exposure, Neurotoxicity-Repeated Exposure, Respiratory Sensitization, and Skin Sensitization. Immune System Effects are included in Systemic Toxicity/Organ Effects. **Ecotoxicity** includes Acute Aquatic Toxicity and Chronic Aquatic Toxicity.

Note: The level of hazard indicated is the lowest hazard level at which a chemical would fail that criterion. However, if the chemical has a higher hazard level than what is listed (e.g. chemical is very High and the criterion is High), it would also fail that criterion.

* For inorganic chemicals with Low B, Low T (Ecotoxicity, Group I, II and II* Human) and Low Physical Hazards (Flammability and Reactivity), persistence alone will not be deemed problematic. Inorganic chemicals that are only persistent may achieve Benchmark 4.

Clean Production Action • www.cleanproduction.org
Copyright 2011 © Clean Production Action

Figure 6.1 GreenScreen™ Benchmark criteria.

If the chemical has high persistence, high bio-accumulation potential and very high acute aquatic toxicity (Ecotoxicity) then it would not pass beyond Benchmark 1. Benchmark 1 criteria define the highest inherent hazard. The criteria in Benchmarks 2, 3 and 4 define progressively more benign chemicals.

Benchmark 1 – Avoid – Chemical of High Concern. The criteria associated with Benchmark 1 are consistent with the hazard criteria used by leading governments to identify substances of very high concern (SVHCs). For example, chemicals that are persistent, bio-accumulating **and** toxic to humans or aquatic species (PBTs) are considered especially problematic because their concentrations in the environment may increase over time. Governments and authoritative bodies worldwide have targeted PBTs through vehicles such as the Oslo–Paris Convention for the Protection of the Marine Environment of the Northeast Atlantic (OSPAR) and the Stockholm Convention on POPs. Canada is prioritizing chemicals for further assessment that are not only PBTs, but also P + T or B + T. The European Union's REACH legislation addresses PBTs and very persistent and very bio-accumulating substances (vPvBs).

Benchmark 2 – Use But Search for Safer Substitutes. Benchmark 2 chemicals do not have the characteristics of substances of very high concern. But they have hazardous properties that should be treated with caution before using the chemical in a specific application.

Benchmark 3 – Use But Still Opportunity for Improvement. Benchmark 3 chemicals have inherently lower hazard across all endpoints. These chemicals are relatively benign but specific hazards should always be evaluated to make sure that they are not problematic for the desired application.

Benchmark 4 – Prefer Safer Chemical. Benchmark 4 reflects a chemical that has low inherent hazard across all hazard endpoints. While the chemical is considered inherently benign, proper risk management practices should always be applied. In GreenScreen™ version 1.2, inorganic chemicals that are highly persistent but that have low hazard for all other endpoints may achieve Benchmark 4.

Benchmark U – Unspecified Due to Insufficient Data. Based on guidance described in the next section, Benchmark U chemicals have data gaps that preclude applying a Benchmark score.

6.2.2.1 Benchmarking with Data Gaps. A GreenScreen™ assessment is only as good as the data and information available for hazard classification. In addition, performing a comprehensive hazard assessment can require resources and expertise not commonly found in all organizations. In the ideal scenario, comprehensive hazard data and knowledge of feasible transformation products would be available for all chemicals. Unfortunately, the ideal data scenario is seldom attained because comprehensive hazard data are the exception rather than the norm. To date in the USA, chemical manufacturers have not been required to generate comprehensive test data before providing the US EPA with pre-manufacturing notification about a new chemical intended for commercial use,[34] with the outcome that the vast majority of the more than 80,000 chemicals on the market have limited to no publicly available test data.[35] The situation is changing in Europe with the implementation of REACH. But for now, we live in a world of imperfect and incomplete chemical hazard data. This creates a challenge to using the GreenScreen™ to benchmark chemicals with limited or no experimental test data.

One approach to filling data gaps is to supplement test data with estimated data from analog and structure–activity relationship (SAR) analyses. Combining the results of experimental data with the use of modeling tools and expert judgment-based SAR analyses to address hazard endpoints is common practice at the US EPA, Environment Canada and other government agencies. However, not all endpoints are amenable to estimating hazard, depending in part on the type of molecule, the availability and quality of modeling tools and the availability of reasonable analogs and associated data. In some cases, it is not possible to estimate one or more of the hazards using these methods.

The GreenScreen™ method provides a clear and transparent distinction between a chemical with unknown hazard (data gap) and low hazard. When there are insufficient measured and estimated data to provide any classification for a hazard endpoint, the endpoint is assigned a data gap (DG). The data gap indicates that measured data and authoritative and screening lists have been reviewed and expert judgment and estimation such as modeling and analog data have been applied and there is still insufficient information to assign a hazard level.

The importance of transparency with data gaps is maintained through benchmarking. When rolling up the hazard information into a Benchmark score, the GreenScreen™ method incorporates both the number and type of data gaps into the Benchmark decision logic. The foundational premise is that chemicals lacking important data cannot be assumed to be safe. Incorporating the number and severity of hazard endpoints with unknown hazard into the benchmarking decision logic reduces the chance of selecting a poor substitute, reflects the level of uncertainty with the benchmarking decision and provides incentive for companies to share chemical hazard data to support informed decision-making.

Data requirements are established for each Benchmark and the GreenScreen™ method includes guidance on how to Benchmark chemicals with DGs. Table 6.2 depicts the minimum data requirements to achieve a given Benchmark score. Chemicals missing important data are either downgraded to a lower Benchmark or assigned 'U' for 'Unspecified'. For example, a chemical that receives an initial score of Benchmark 3 based on available data but only meets the Benchmark 2 data requirements is assigned Benchmark 2_{DG} to indicate that data gaps are driving the final score.

Data requirements become more stringent with higher Benchmark scores. Whereas one can confidently assign a Benchmark score of 1 with solid evidence of high hazard for a Group 1 hazard, many more data are needed to assess a chemical and confidently assign it a higher Benchmark score (see Table 6.2).

6.2.2.2 Benchmarking Mixtures. The GreenScreen™ provides guidance for assessing mixtures and materials and reporting the results. Each chemical in a mixture or material is assessed separately and receives an individual Benchmark score. Users are guided to report the constituents in the material or

Table 6.2 Data requirements to achieve each Benchmark score.

Benchmark Score	Data Requirements and Permissible Data Gaps by Hazard Endpoint Category			
	Group I Human	Group II Human	Ecotoxicity and Fate	Physical Properties
Benchmark 1	A chemical may be assigned Benchmark 1 with data on as few as one endpoint. To achieve a higher score, a chemical must meet the minimum data requirements as described for Benchmark 2			
Benchmark 2	Data required for 3 out of 5 endpoints. Permissible data gaps include: 1. Endocrine Activity 2. Reproductive or Developmental Toxicity	Data required for 4 out of 7 endpoints. Permissible data gaps include: 1. Skin OR Respiratory Sensitization 2. Skin OR Eye Irritation 3. One other hazard end-point (unrestricted)	Data required for 3 out of 4 endpoints. Permissible data gaps include: 1. Acute OR Chronic Aquatic Toxicity	Data required for all 2 endpoints
Benchmark 3	Data required for 4 out of 5 endpoints. Permissible data gap is: Endocrine Activity	Data required for 5 out of 7 endpoints. Permissible data gaps include: 1. Skin OR Respiratory Sensitization 2. One other hazard end-point (unrestricted)	Data required for all 4 endpoints	Data required for all 2 endpoints
Benchmark 4	Data required for all 18 endpoints			

Table 6.3   Reporting Benchmark scores for mixtures.

Chemical	CAS	% by weight	Benchmark	BM by %
Calcium carbonate	1317-65-3	30–45%	4	30–45%
Acetone	67-64-1	5–20%	2	5–20%
Petroleum distillates	64742-89-8	5–20%	1	
Toluene	108-88-3	5–20%	1	
Dichloromethane	75-09-2	0–5%	1	10–46%
Methyl ethyl ketone	78-93-3	0–1%	1	

mixture in a consistent and transparent way to allow for informed decision-making. The reporting requirements include the following:

- Report all constituents by name and CAS number.
- Report the Benchmark score for each constituent.
- Report the percentage of material at each Benchmark score.

An example of reported results for a GreenScreen™ mixture assessment are shown in Table 6.3. The GreenScreen™ method focuses on transparency in reporting rather than on providing an overall score. This is an area of future discussion with the Technical Advisory Committee. One obvious approach would be to give a mixture an overall score based on weighted averages of the Benchmark scores of the chemical constituents. However, to provide an overall score using weighted averaging as an algorithm creates the possibility of masking the presence of highly hazardous (Benchmark 1) substances by dilution with low hazard (Benchmark 3 or 4) substances. Transparency allows users to make their own judgments and to assess their own risks.

6.2.3   *Make Informed Decisions and Drive Innovation*

The final step in the GreenScreen™ method is to use both the hazard table and Benchmark results to make more informed decisions and to drive innovation. Organizations use tools such as CHA for diverse purposes specific to their needs. The use of CHA can support an organization in product development, procurement and in chemicals management. We have observed the use of CHA, and specifically the GreenScreen™, in three distinct organizational stages that illustrate a progression from being passive to active in response to chemical restrictions based on hazard:

1. In stage one, organizations are driven by regulations or significant customer pressures to eliminate chemicals of concern. The term 'aversion chemistry' can be applied to describe the practice of identifying chemicals of concern and eliminating them from use.

2. In stage two, organizations look beyond the elimination of chemicals of concern to ask the next obvious question – how to identify safer alternatives. There are business risks in simply moving away from a regulated chemical of concern to another chemical based solely on performance and the fact that it is currently unregulated. If the replacement also turns out to have hazardous properties, then it too could eventually become regulated. Examples of this occur when a particular chemical is targeted for regulation and a less known alternative from the same chemical class is used to replace it. It is not unusual soon to discover that the related chemical has similar hazard characteristics. Organizations that carefully assess alternatives and opt for those with lower hazard are practicing 'safer chemistry.' They are also reducing their business risk.

3. Some organizations are driven by sustainability principles and aspire to practice 'green chemistry.' They seek chemicals that are fully assessed, that have inherently low hazard and that provide life-cycle benefits. They may seek to develop new chemicals themselves or to work with their suppliers to develop new chemicals that perform without unnecessary hazard.

Some organizations practice aversion chemistry, safer chemistry and green chemistry concurrently – depending on where their products are in the product cycle and how well developed their infrastructure is for chemicals management.

6.2.3.1 GreenScreen™ Users and Applications. CHA and specifically GreenScreen™ assessments are being used in diverse ways. Government agencies have used CHA to justify chemical bans and to guide companies toward replacement with safer alternatives. Regulatory action to eliminate the use of certain chemicals is probably the most important driver for CHA in industry. Examples of government agencies that have leveraged and promoted use of the GreenScreen™ include Washington State Department of Ecology and the US EPA DfE Program. Washington State has also incorporated the GreenScreen™ and a simplified version of it into their state pollution prevention and technical assistance programs.[36] They are training individuals on their pollution prevention staff who then support Washington companies in finding alternatives to chemicals of concern to the State. The US EPA DfE points to the GreenScreen™ as a resource that can be paired with information from their Alternatives Assessment Partnership programs. Through the Partnerships, DfE evaluates alternatives to chemicals of concern (*e.g.* flame retardants in furniture foam, insulation and printed circuit boards) and publishes CHAs of the chemical of concern and its available alternatives. Because there is so much alignment between the hazard classification schemes in the GreenScreen™ and in the DfE Alternatives Assessment Criteria, the GreenScreen™ benchmarking process can be easily applied to the results of the EPA hazard assessments to provide a Benchmark score for each alternative and to support decision-making.

Businesses use GreenScreen™ assessments for product development, in support of marketing efforts, for materials selection and as part of their chemicals management in response to regulations and other market pressures. Examples of businesses integrating the GreenScreen™ into their decision-making frameworks include Staples, DSM, Clariant, Hewlett-Packard, the Green Chemistry and Commerce Council and The Wercs. One factor that affects how businesses leverage the GreenScreen™ is where they lie in the supply chain. Suppliers of chemicals and formulated products such as Clariant use the GreenScreen™ in product development and marketing to support claims about improved hazard profiles. Manufacturers of polymeric materials such as DSM have used the GreenScreen™ to evaluate safer additives for use in formulating polymers. Specialized polymer suppliers with lines of halogen-free and phthalate-free products have developed PVC replacements for wire and cable applications that meet stringent performance and regulatory requirements. The Green-Screen™ has been demonstrated to be a valuable tool for evaluating potential additives and supporting the development of inherently safer formulations in product development.

Downstream users of chemicals such as OEMs and retailers may use the GreenScreen™ in material and purchasing specifications. Hewlett-Packard is the world's leading user of the GreenScreen™ (see Chapter 7). It piloted the use of the GreenScreen™ as an alternatives assessment tool for PVC replacements and other substances and found it to be extremely useful in identifying preferable options to chemicals facing restrictions in their supply chain and for communicating materials goals and criteria to suppliers. Hewlett-Packard has expanded its adoption of the GreenScreen™ to be one of the primary tools used to support its commitment to find replacements for restricted materials that will not adversely impact human health or the environment.

The Green Chemistry and Commerce Council (GC3) is a business-to-business forum that advances the application of green chemistry and design for environment across supply chains. As part of their Business and Academic Partnerships for Safer Chemicals, GC3 coordinated a consortium of businesses interested in evaluating the inherent hazards and performance characteristics of plasticizers used in polymers for wire and cable applications (see Chapter 5). The group evaluated the plasticizers using the GreenScreen™. Future efforts include filling key data gaps identified through the CHA and conducting performance testing. GC3 is also planning to support CPA in developing a verification program by submitting fully transparent assessments for verification.

Environmental advocacy organizations and environmental labeling schemes (*i.e.* standards and ecolabels) have used GreenScreen™ assessments to prioritize and support advocacy efforts. The US Green Building Council currently includes credit options in their proposed LEED v4 rating system for disclosing Benchmark 1 chemicals in products and additional credits for the use of products that do not contain Benchmark 1 chemicals. While the fate of the GreenScreen™ Benchmark 1 credits in the LEED rating system is to be determined, its use as a normative reference sets a useful precedent.

6.2.3.2 Applying the GreenScreen™ with Common Sense. The GreenScreen™ methodology provides a framework for comprehensive hazard assessment of a chemical and decision support to rank chemicals along a continuum from least to most preferred based on inherent hazard. It has its strengths and limitations and, as with any tool, it should be used appropriately. The GreenScreen™ is not intended to address all of the attributes of sustainability. It is just one tool in the toolbox. For example, it does not consider social equity or important life-cycle impacts such as energy quantity and quality. However, it can be used in combination with other indicators and metrics, including the results from life-cycle assessment to promote sustainability.

GreenScreen™ assessments focus on the inherent hazards of a chemical and its feasible and relevant transformation products of a chemical in a product as used and at the end-of-life stage of its life-cycle. It may not capture chemicals associated with upstream processing such as feedstock, reagents, auxiliaries and catalysts unless they are present in the final product. However, the Green-Screen™ can be used to assess the inherent hazards associated with any chemical used during any life-cycle stage.

Toxicological data are continually being generated and *in vitro*, *in vivo* and *in silico* test methods are being developed and refined to better characterize existing and newly developed chemicals. In addition, the state of the science and our understanding of how chemicals cause toxicity are continuing to advance. As with any hazard assessment, the GreenScreen™ assessment is only as good as the data used for the assessment. The accuracy of the assessment results is time limited. Users are advised to update assessments periodically, particularly for hazard endpoints where the science is quickly evolving and for chemicals for which new data are emerging.

Performing a comprehensive GreenScreen™ assessment requires expertise. GreenScreen™ assessments include review of data from standard toxicological tests, estimation models, analogs, emerging science and regulatory and authoritative hazard lists that can be challenging to interpret. Performing a high-quality assessment requires a high level of technical expertise including toxicology and chemistry to interpret conflicting results, to weigh the evidence and to classify hazards. Hence the quality of GreenScreen™ assessments depends on the expertise of the assessor. CPA has developed a program to train and certify experts to perform GreenScreen™ chemical profiling services.[37] CPA is also working to create publicly accessible repositories of GreenScreen™ assessments.

Finally, it is important to remember that although the GreenScreen™ can be used to identify the least hazardous chemical options, risk assessment and risk management still need to be applied to ensure that even the safest chemicals are used properly. After all, even water is hazardous if you drink too much, too quickly (hyponatremia). Users should consider the importance of potential exposure routes and data gaps based on knowledge of how the chemical or product will be used across its life-cycle before making material selection decisions. For example:

1. If no data are available for a hazard based on a specific exposure route when it is known that exposure is likely to occur by that route, then

regardless of the Benchmark score, the assessment should not be considered complete. For example, a GreenScreen™ Benchmark may be generated in part based on data for acute mammalian toxicity by oral administration. However, if inhalation is a primary route of exposure for workers and there are no inhalation data, then the GreenScreen™ report should be used to identify the data gap rather than to default to the Benchmark score.

2. For chemicals with the same Benchmark score, one may be more appropriate for a particular application than another based on the specific hazards and the use pattern. For example, a chemical with potential for moderate eye irritation would not be a good candidate for an eyewash. The hazard table and the full GreenScreen™ report can be used to determine what is known and what is not known about a chemical. One should not rely solely on the Benchmark score when making important decisions. Additional data may need to be generated.

6.3   Continual Improvement

The GreenScreen™ method will continue to evolve based on user feedback, advances in science and understanding of hazard as a criterion for sustainable chemicals, materials and products. CPA is developing infrastructure to support the adoption and use of the method.

6.3.1   Advances to the GreenScreen™ Method

Changes to the GreenScreen™ are developed in consultation with a Technical Advisory Committee. The committee includes a diverse group of experts from academia, industry, and non-government and government organizations. A few key modifications under development include:

- Refinement of hazard criteria to align better with updates to the DfE Alternative Assessment Criteria.[38]
- Refinement of the Benchmarking decision logic for inorganic chemicals such as mineral oxides.
- Updating testing protocols to classify endocrine disruption.
- Expanding guidance to support
  - assessing and benchmarking polymers
  - identifying feasible and relevant transformation products
  - applying the GreenScreen™ to formulated materials or products.

6.3.2   Developing Infrastructure for the GreenScreen™ Program

To support broader adoption of the GreenScreen, CPA is developing much-needed infrastructure to support the program. CPA develops educational materials and provides training through workshops and webinars. However, the other elements of the program are outsourced to service and software providers better suited to provide effective business services. The full Green-Screen™ program includes:

1. Certified GreenScreen™ profilers.
2. Education and training.
3. Software/automation.
4. Verification (under development).
5. Shared GreenScreen™ repositories (under development).

Qualified GreenScreen™ profilers are organizations that have demonstrated expertise in toxicology and chemistry and that have the capacity to provide GreenScreen™ assessments on a consulting basis. The qualified profilers are kept up-to-date on all method revisions and CPA audits their application of the method to ensure that it is being applied as intended. Currently certified GreenScreen™ profilers include ToxServices, Inc.[39] and NSF International in the USA.[40] Additional US profilers are currently in the process of qualifying and several European consulting firms have expressed initial interest.

Education and training are important elements of the GreenScreen™ Program. CPA has partnered with the State of Washington, the Business NGO Working Group for Green Chemicals and Sustainable Materials, the US EPA and the National Pollution Prevention Roundtable to provide GreenScreen™ training workshops across the USA and with a focus on the Great Lakes region. The method is also included as part of the training program offered by the Substitution Support Portal (SUBSPORT) project in Europe. The goal of SUBSPORT is to create a state-of-the-art Internet portal to resources on safer alternatives to the use of hazardous chemicals. In addition to information on alternative substances and technologies, it also provides tools and guidance for substance evaluation and substitution management.[41]

Currently, GreenScreen™ assessments are performed manually and elements of the method, such as the searching of authoritative and screening hazard lists, can be monotonous and time intensive. CPA has developed the GreenScreen™ List Translator that maps authoritative and screening hazard lists to hazard classifications for relevant endpoints. By automating the search for chemicals on those lists and linking them directly to hazard classifications, automation can reduce the time needed to perform a full GreenScreen™ assessment by several hours. Automation of the List Translator represents only a portion of the full GreenScreen™ method. CPA has partnered with the Berkeley Center for Green Chemistry in the development of their Public Library of Materials (PLuM) database,[42] with Healthy Building Network in the development of their Pharos database[43] and with the Wercs in the development of their GreenWERCS database[44] by working with these organizations to integrate the GreenScreen™ List Translator into their software tools.

The Wercs provides software tools and services to help advance the health and safety of humans and the environment with a focus on meeting and going beyond regulatory and sustainability compliance. Their Green-WERCS software, developed in partnership with the retail industry, provides

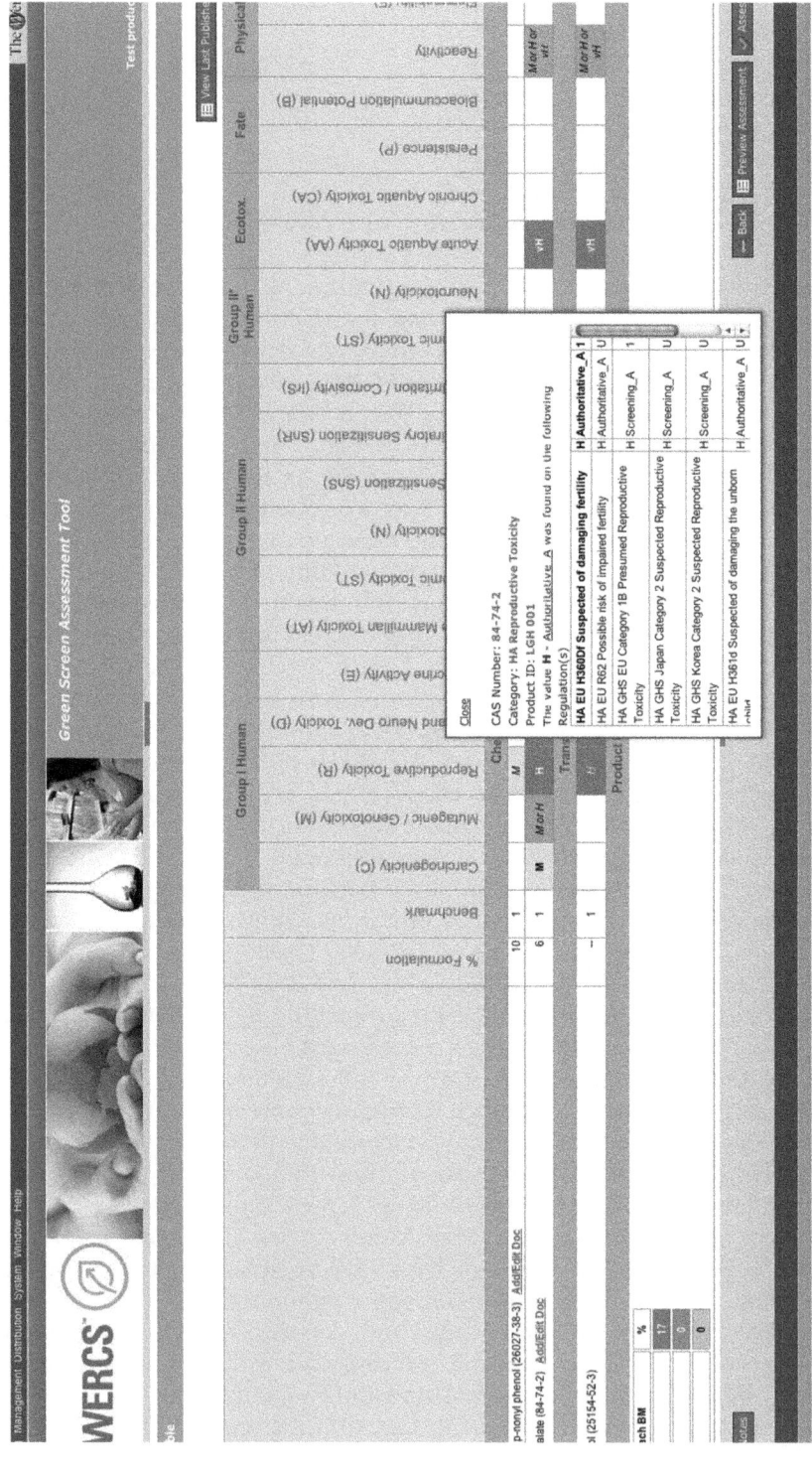

Figure 6.2 Screenshot from the GreenScreen™ List Translator in GreenWERCS. Results are reported in a GreenScreen™ hazard table and the Benchmark logic is applied. Any box can be highlighted to view regulations and hazards associated with each CAS No.

hazard screening and visual ranking based on a limited set of regulatory and authoritative lists to support retailers in identifying and selecting products with lower chemical hazard and that are less likely to generate hazardous waste.

The Wercs has integrated the GreenScreen™ List Translator, the hazard table and benchmarking approach into their GreenWERCS software solution. Use of GreenWERCS software with the GreenScreen™ List Translator can help identify hazard classifications based on over 850 source lists and sublists to assign a List Translator (LT) score of LT-1 (equivalent to Benchmark 1), LT-P1 (Possible Benchmark 1) or LT-U (Unspecified) (see Figure 6.2). It is not possible to classify higher Benchmark levels using the GreenScreen™ List Translator only. Typically, chemicals that are not hazardous are not identified on regulatory and other hazard lists. There are few sources of 'positive' or 'white' lists that identify chemicals based on low hazard overall or for a particular endpoint. Automation of the GreenScreen™ List Translator can replace many hours of manual searching. However, to do a full GreenScreen™ assessment still requires collecting data from the literature including test results, analogs and more and thus requires the input of an expert assessor.

Current goals of the GreenScreen™ Program involve the development of a verification program. The purpose of the verification program is to provide increased assurance of the comprehensiveness and accuracy of results from GreenScreen™ assessments by applying a secondary level of expert review. In addition, currently, public promotional claims based on GreenScreen™ results are not allowed and all reports generated by certified GreenScreen™ profilers are labeled 'Draft – Has Not Been Validated.' Promotional claims may be considered once the verification program is launched to assure sufficient integrity of results.

A number of organizations have expressed interest in using or creating a public repository of GreenScreen™ assessments in order to create efficiencies and to reduce costs. CPA is currently evaluating strategies, including partnerships with GreenBlue Institute, Interstate Chemicals Clearinghouse and Pharos, to create shared repositories of GreenScreen™ assessments. A reliable verification program will support the creation of public repositories.

6.4 Conclusion

The GreenScreen™ is a freely accessible, transparent, science-based method for CHA that supports businesses, governments and other organizations in the identification of chemicals of concern and in the assessment of safer alternatives in support of sustainable materials management. Although a comprehensive GreenScreen™ assessment requires expertise to perform, the results of a GreenScreen™ assessment can be used by experts and non-experts alike to make more informed decisions in the selection of chemicals and materials for activities ranging from procurement to product design and development. The hazard criteria and Benchmarks developed for the GreenScreen™ are based on scientific and regulatory precedents wherever feasible and are intended to harmonize with government and other authoritative initiatives such as the EU REACH legislation, the US EPA DfE and the Globally Harmonized System for Classification and Labeling.

So far, a handful of proprietary methods have been developed to evaluate and identify safer chemicals and materials and to help define the path to safer, healthier chemicals in product design. Notable examples include the Cradle to Cradle Design Protocol,[45] Greenlist™ developed by SC Johnson and Son, Inc.,[46] the Dye and Chemistry Protocol developed by Interface Fabrics[47] and the bluesign[®48] standard. To date, none of these methods fully discloses all the decision elements, including threshold values for hazard criteria, prioritization of hazard endpoints and life-cycle considerations.

The GreenScreen™ covers one aspect of sustainability in-depth, namely CHA. It can be used in conjunction with other sustainability indicators and tools that address aspects such as social equity and life-cycle impacts. Using results from complementary assessment frameworks provides a broader set of considerations for promoting sustainability.

The power of the CHA approach, particularly when comparing chemicals for similar functional uses, is that it allows one to compare and select safer alternatives. Risk assessment addresses the probability of harm and supports decisions about whether a chemical can be managed with adequate safety for a particular use. CHA, on the other hand, supports continual improvement by facilitating comparison of chemical options to determine which is safer. As more companies are faced with the need to phase out chemicals of concern used in their industry – whether prompted by regulations or internal initiatives – they are challenged to find alternatives and to prove that the alternatives are safer. From the perspective of the authors, the following quotation made in reference to the GreenScreen™ reflects the value of CHA:

> *The more you know about what you are putting into your products, the more likely you are to make better choices in product development.*
> *Jonathan Plisco, PolyOne, personal communication, 2011*

Acknowledgement

Ken Geiser of the University of Massachusetts Lowell is thanked for applying the phrases 'aversion chemistry,' 'safer chemistry' and 'green chemistry' to the three stages described in Section 6.2.3.

References

1. P. T. Anastas and J. Warner, *Green Chemistry Theory and Practice,* Oxford University Press, New York, 1999.
2. Public Law 101-508, 5 November 1990, Omnibus Budget Reconciliation Act of 1990 (Pollution Prevention Act of 1990).

3. European Agency for Safety and Health at Work, *Facts 34. Elimination and Substitution of Dangerous Substances*, 2003; http://osha.europa.eu/en/publications/factsheets/34 (last accessed June 2012).
4. US EPA, *Environmental Profiles of Chemical Flame-Retardant Alternatives for Low-Density Polyurethane*, Volume 1, 2005; http://www.epa.gov/dfe/pubs/flameret/ffr-alt.htm (last accessed February 2013).
5. US EPA, *Draft Report. Partnership to Evaluate Flame Retardants in Printed Circuit Boards*, 2012; http://www.epa.gov/dfe/pubs/projects/pcb/full_report_pcb_flame_retardants_report_draft_11_10_08_to_e.pdf (last accessed February 2013).
6. US EPA, *New Chemicals*, 2012; http://www.epa.gov/oppt/newchems/ (last accessed June 2012).
7. US EPA, *Existing Chemicals*, 2012; http://www.epa.gov/oppt/existingchemicals/pubs/ecactionpln.html (last accessed June 2012).
8. US EPA, *Environmental Profiles of Chemical Flame-Retardant Alternatives for Low-Density Polyurethane*, Volume 1, 2005; http://www.epa.gov/dfe/pubs/flameret/ffr-alt.htm (last accessed February 2013).
9. M. Rossi and L. Heine, *The Green Screen for Safer Chemicals v 1.0. Evaluating Flame Retardants for TV Enclosures*, 2007; http://www.clean-production.org/library/Green_Screen_Report.pdf (last accessed June 2012).
10. United Nations Economic Commission for Europe, *Globally Harmonized System of Classification and Labelling of Chemicals (GHS)*, Download latest version of GHS at http://www.unece.org/trans/danger/publi/ghs/ghs_welcome_e.html (last accessed February 2013).
11. Regulation (EC) No. 1907/2006 of the European Parliament and of the Council of 18 December 2006 Concerning the Registration, Evaluation, Authorization and Restriction of Chemicals (REACH), establishing a European Chemicals Agency. *Off. J. Eur. Union*, L396, 30 December.2006, 1–849.
12. US EPA, *Design for the Environment Program Alternatives Assessment Criteria for Hazard Evaluation, Version 2.0*, 2011; http://www.epa.gov/dfe/alternatives_assessment_criteria_for_hazard_eval.pdf (last accessed June 2012).
13. An intentionally added chemical in a product means a chemical in a product that serves an intended function in the product component: http://www.ecy.wa.gov/pubs/wac173334.pdf (last accessed June 2012). Any other chemical in the product is therefore an impurity (last accessed June 2012).
14. Clean Production Action, *Welcome to GreenScreen™ for Safer Chemicals v 1.2*, 2012; http://www.cleanproduction.org/Greenscreen.v1-2.php (last accessed June 2012).
15. Clean Production Action, *GreenScreen™ Reporting Template*, 2012; Download document at http://www.cleanproduction.org/Greenscreen.v1-2.php (last accessed June 2012).

16. Environment Canada Existing Substance Branch, *Guidance Manual for the Categorization of Organic and Inorganic Substances on Canada's Domestic Substances List,* Environment Canada, Gatineau, QC, 2003.

17. SRC, *FatePointers Search Module*, 2012; http://esc.syrres.com/fatepointer/search.asp (last accessed June 2012).

18. University of Minnesota Biocatalysis/Biodegradation Database, *UM-BBD Pathway Prediction System*, 2012; http://umbbd.msi.umn.edu/predict/ (last accessed June 2012).

19. US National Library of Medicine, *TOXNET Toxicology Data Network*, 2011; http://toxnet.nlm.nih.gov/cgi-bin/sis/htmlgen?HSDB (last accessed June 2012).

20. OECD, *ChemPortal. The Global Portal to Information on Chemical Substances, 2012*; http://www.echemportal.org/echemportal/index?pageID = 0&request_locale = en (last accessed June 2012).

21. OECD, *Manual for the Assessment of Chemicals*, 2012; http://www.oecd.org/document/7/0,3343,en_2649_34379_1947463_1_1_1_1,00.html (last accessed June 2012).

22. Environment Canada, *Canadian Environmental Protection Act (CEPA) Environmental Registry Domestic Substances Lists,* 2012; http://www.ec.gc.ca/CEPARegistry/subs_list/dsl/dslsearch.cfm (last accessed June 2012).

23. Regulation (EC) No. 1907/2006 of the European Parliament and of the Council of 18 December 2006 Concerning the Registration, Evaluation, Authorization and Restriction of Chemicals (REACH), establishing a European Chemicals Agency.

24. Regulation (EC) No. 1272/2008 of the European Parliament and of the Council of 16 December 2008 on classification, labelling and packaging of substances and mixtures, amending and repealing Directives 67/548/EEC and 1999/45/EC and amending Regulation (EC) No 1907/2006.

25. S. Webb, R. Tallman, R. Becker, K. Onuma and K. Igarashi, Risk perception: a chemical industry view of endocrine disruption in wildlife, *Pure Appl. Chem.*, 2003, **75**, 2575–2591.

26. Danish Centre on Endocrine Disrupters, *Evaluation of 22 SIN List 2.0 Substances According to the Danish Proposal on Criteria for Endocrine-Disruptors*, May 2012.

27. Clean Production Action, *GreenScreen™ for Safer Chemicals Version 1.2 Criteria,* 2012; http://www.cleanproduction.org/library/greenscreen-hazard-criteria-2012-03.pdf (last accessed June 2012).

28. International Agency for Research on Cancer, *IARC Monographs on the Evaluation of Carcinogenic Risks to Humans*, 2012; http://monographs.iarc.fr/ENG/Classification/index.php (last accessed June 2012).

29. US Department of Health and Human Services, Public Health Service, National Toxicology Program, *Report on Carcinogens*, 12th edn, 2011; http://ntp.niehs.nih.gov/go/roc (last accessed June 2012).

30. State of California, Environmental Protection Agency, Office of Environmental Health Hazard Assessment, *Safe Drinking Water and Toxic*

Enforcement Act of 1986. Chemicals Known to the State to Cause Cancer or Reproductive Toxicity, 2012; http://www.oehha.ca.gov/prop65/prop65_list/Newlist.html (last accessed June 2012).

31. European Commission, *CLP/GHS – Classification, Labelling and Packaging of Substances and Mixtures*, 2012, Regulation on the Classification, Labelling and Packaging of Substances and Mixtures (CLP), EC 1272/2008 and subsequent amendments. [Conversion of CMR list from 67-548-EEC (Categories 1–3) to GHS Categories (Category 1A, 1B, 2)]; http://ec.europa.eu/enterprise/sectors/chemicals/classification/index_en.htm (accessed June 2012).

32. US EPA, *Integrated Risk Information System (IRIS)*, 2012; http://www.epa.gov/iris/search_human.htm (last accessed June 2012).

33. Clean Production Action, *GreenScreen™ List Translator*, 2012; http://www.cleanproduction.org/library/greenscreen-translator-benchmark1-possible%20benchmark1.pdf (last accessed June 2012).

34. US EPA, New Chemicals, 2012; http://www.epa.gov/oppt/newchems/ (last accessed June 2012).

35. United States Government Accountability Office, *Chemical Regulation: Options Exist to Improve EPA's Ability to Assess Health Risks and Manage Its Chemical Review Program*, GAO-05-458, 2005; http://www.gao.gov/new.items/d05458.pdf (last accessed June 2012).

36. Department of Ecology, State of Washington, *The Quick Chemical Assessment Tool*, 2012; http://www.ecy.wa.gov/programs/hwtr/ChemAlternatives/QCAT.html (last accessed June 2012).

37. Clean Production Action, *Requirements for Becoming a Certified Green-Screen™ Profiler*, 2012; http://www.cleanproduction.org/library/requirements-for-becoming-a-greenscreen-chemical-profiler.pdf (last accessed June 2012).

38. US EPA Office of Pollution Prevention and Toxics, *Design for the Environment Program Alternatives Assessment Criteria for Hazard Evaluation, Version 2.0*, 2011; http://www.epa.gov/dfe/alternatives_assessment_criteria_for_hazard_eval.pdf (last accessed June 2012).

39. ToxServices, Inc. http://www.toxservices.com/cleanproduction.html (last accessed June 2012).

40. NSF International, *Clean Production Action GreenScreen™*, 2012; http://www.nsf.org/business/laboratory_services/green_screen/su_greenscreen_insert_web.pdf (last accessed June 2012).

41. Substitution Support Portal (SUBSPORT), *Alternatives Identification and Assessment Training*, 2011; http://www.subsport.eu/training (last accessed June 2012).

42. Berkeley Center for Green Chemistry, *PLuM Database*, 2012; http://bcgc.berkeley.edu/databases (last accessed June 2012).

43. Healthy Building Network, *Pharos*, 2012; http://www.pharosproject.net/ (last accessed June 2012).

44. The Wercs, *GreenWERCS*, 2012; http://www.thewercs.com/products-and-services/greenwercs (last accessed June 2012).

45. M. McDonough, M. Braungart, P. T. Anastas and J. B. Zimmerman, Applying the principles of green engineering to cradle-to-cradle design, *Environ. Sci. Technol.*, 2003, **37**, 434A–441A.
46. US EPA, *Presidential Green Chemistry Challenge. 2006 Award Recipients. Designing Greener Chemicals Award*, 2012; http://www.epa.gov/greenchemistry/pubs/pgcc/past.html (last accessed June 2012).
47. T. Greiner, M. Rossi, B. Thorpe and B. Kerr, *Healthy Business Strategies for Transforming the Toxic Chemical Economy*, Clean Production Action, Spring Brook, NY, 2006; http://www.cleanproduction.org/library/CPA-HealthyBusiness–1.pdf (last accessed June 2012).
48. Bluesign Technologies, *bluesign*® *Standard*, 2012; http://www.bluesign.com/index.php?id = 151 (last accessed June 2012).

Hewlett-Packard's Use of the GreenScreen™ for Safer Chemicals

H. A. HOLDER*, P. H. MAZURKIEWICZ, C. D. ROBERTSON AND
C. A. WRAY

ABSTRACT

The electronics industry is now being impacted by a rapidly growing number of chemical regulations and substance restrictions. In the light of this trend, there is a significant risk of facing multiple substitutions for the same application unless potential replacement technologies are properly assessed against environmental and human health criteria in advance of their widespread adoption. To ensure that replacements for restricted substances have lower impact to human health and the environment, it is essential to be able to determine the impacts of the alternatives compared with the original substances. In this work, we evaluated tools and methods for comparing restricted substances to their alternatives, including restricted substance list (RSL) screening, risk phrase or hazard statement screening, multi-criteria decision analysis, life-cycle analysis (LCA), risk assessment and hazard scoring schemes. This chapter describes how these tools were incorporated into an integrated alternatives assessment protocol for evaluating the environmental and human health impacts of replacements for restricted substances.

7.1 Introduction

In February 2003, a new EU Directive was adopted that broadly affected materials used in the electronics industry. *Directive 2002/95/EC on the*

---

*Corresponding author

Issues in Environmental Science and Technology, 36
Chemical Alternatives Assessments
Edited by R.E. Hester and R.M. Harrison
© The Royal Society of Chemistry 2013
Published by the Royal Society of Chemistry, www.rsc.org

Restriction of the Use of Certain Hazardous Substances in Electrical and Electronic Equipment (commonly referred to as RoHS) was part of an effort to reduce the inherent toxicity of electronics waste and to mitigate the effects of its disposal. The directive restricted the use of six substances in electronic products: lead (Pb), mercury (Hg), cadmium (Cd), hexavalent chromium [Cr(VI)] and their compounds and two types of brominated flame retardants, polybrominated biphenyls (PBBs) and polybrominated diphenyl ethers (PBDEs). With these restrictions in place, the electronics industry began replacing some of the most critical (and most widely studied) materials used in electronics.

Although the goal of reducing the toxicity of electronics waste was well intentioned, many people were concerned that the alternative substances had not been sufficiently evaluated to ensure that they would have less impact on human health and the environment. Assessments showed mixed results for the alternatives in comparison with the original materials.[1,2]

From a business perspective, it is undesirable to face future restrictions for the same application due to a poor choice of replacement materials. Chemical substitutions can be costly and the required changes can be disruptive to product releases. In the light of the trend towards more chemical regulation and substance restriction, there is a growing risk of multiple substitutions unless potential replacement technologies are properly assessed against environmental and human health criteria in advance of their widespread adoption.

7.2　Tools and Methods

To ensure that replacements for restricted substances have a lower impact on human health and the environment, it is essential to be able to determine the impacts of the alternatives compared with the original substances. As part of Hewlett-Packard (HP)'s global materials group, we reviewed several available methods and tools to help with that assessment, evaluating the quality and logic of the assessment methods and results and the ease of use, based on the following criteria:

Logic of the assessment method and quality of the results exemplified by:
- Use of objective and scientific criteria.
- Physically meaningful thresholds, indicative of actual substance behavior.
- Results that can stand up to peer review.
- Reproducible method, such that different assessors should come to similar conclusions about a substance.
- Transparency in both the method and the data.
- Balanced consideration of the potential impact to both human health and the environment.
- Alignment with criteria used by regulators and other stakeholders, in order to reduce multiple substitutions.

Ease of use:
- Results that are intuitive to use and applicable in different cultures and job functions.
- Assessments that can be completed quickly enough to keep pace with the speed of product development cycles.
- Affordable implementation.

The classes of tools and methods considered included:
- Restricted substance list (RSL) screening.
- Risk phrase or hazard statement screening.
- Multi-criteria approaches.
- Life-cycle analyses (LCAs).
- Risk assessment.
- Scoring schemes.

A brief discussion of each type of method follows. Many of the tools mentioned in this section are discussed in more detail elsewhere.[3,4]

7.2.1 Restricted Substance List (RSL) Screening

A rudimentary way to screen alternatives to restricted substances is to ensure that replacements are not restricted on other RSLs. For example, replacements for the brominated flame retardants (BFRs) eliminated under RoHS could be screened against the *ChemSec SIN (Substitute It Now!) List* or the *Joint Industry Guide (JIG) list*. The broader the RSL, the more protective is the screening.

While this approach may avoid obviously unsuitable replacements, RSLs are of limited value because they are inherently backward looking and might not reflect new, high-risk substances or applications. If a substance is being used in a new application, an RSL may not list the substance because of lower past potential exposure scenarios, even if it has a high-risk potential in the new application.

Although RSL screening is common, the broadest list screens cannot identify alternatives that are likely to be of interest to regulators in the future.

7.2.2 Risk Phrase or Hazard Statement Screening

Many screening approaches rely on assessing alternatives based on risk phrases (R-phrases) or hazard statements (H-statements) disclosed on Material Safety Data Sheets (MSDSs). R-phrases and the H-statements that are replacing them are standardized descriptions of the hazards of chemical substances and mixtures. These descriptions are part of the Globally Harmonized System of Classification and Labeling of Chemicals (GHS). Examples of R-phrase-based systems include the Column Model[5] and PRIO.[6]

Also, some ecolabels, including EPEAT[7] and TCO,[8] use R-phrases to disqualify certain replacements for PBDE flame retardants in large plastics parts in electronic products.

H- and R-phrase screening systems can be useful additions to RSL screening because they directly assess substances based on their inherent characteristics, which can anticipate potential future regulation. This is, of course, preferable to waiting for authoritative bodies to decide that certain substances are of too high a risk to remain in use.

Although H- and R-phrase systems are an improvement over RSL screening, they have some limitations. Important hazard topics of interest may not have assigned GHS categories. H- and R-phrases do not consider the transformation products of substances. In addition, these restriction systems do not necessarily ensure that the restricted traits are a good balance between human health and the environment.

Another concern is that H- and R-phrases are self-declared and not maintained in a central list. The GHS does not require testing for health hazards and the assignment of an H- or R-phrase is based on currently available data. This means that the absence of an H- or R-phrase for a substance on an MSDS might be a data gap rather than a negative result for that hazard in a standard test. Finally, any method relying on the disclosures found in an MSDS is at risk that some constituents of interest are not even listed, especially if they are not on RSLs.

Although H- and R-phrase screening systems can be useful, most do not meet the requirements for integrity of the underlying data or for balancing human health and environmental criteria.

7.2.3   Multi-criteria Decision Analyses

One of the most fundamental ways to compare alternatives in any context is multi-criteria decision analysis, which can range from simple comparison matrices to complex, analytical models.[9-13] Multi-criteria matrix methods involve evaluating options against a variety of criteria, which can be weighted to demonstrate more clearly tradeoffs and inform decisions about alternatives. Examples of multi-criteria approaches include P2OaSys[14] and the STPP Decision-Making Model.[15]

These methods are most useful when a unique, optimal solution is not apparent and the decision-maker must choose between alternatives. Multi-criteria methods are flexible; they can consider as many factors as desired. Because subjective judgments are required to define and weight the criteria, any matrix system must be transparent in terms of the criteria and assumptions. Without structured, standard factors and criteria, matrix methods can be used to justify virtually any replacement, negating the value of the assessment. For example, in a free-form, generic matrix assessment, cost could be weighted 100 times more than any other factor, justifying the use of the least expensive solution regardless of environmental factors.

Although matrix methods are useful, we did not want to create a new set of criteria. We wanted a forward-looking tool based on existing relevant, balanced and standardized criteria that aligned with criteria used by key regulators, ideally using the same thresholds and endpoints.

7.2.4 Life-cycle Analysis (LCA)

Life-cycle analysis (LCA) attempts to evaluate all impacts associated with a product through its entire life-cycle, from raw material extraction through eventual disposal or recycling. It is beyond the scope of this chapter to give a thorough treatment of LCA techniques and methodology.[16] This discussion will be limited to the utility of LCA for assessing replacements for restricted substances.

In theory, LCA should be a primary tool for such evaluations and indeed it has been applied in a few important cases, such as the replacements for SnPb solder.[2]

However, like RSLs, LCAs tend to be backward looking because data might not be available for new or less common substances. For example, an early investigation tried to use LCA to compare beryllium copper and phosphor bronze connector contacts. At that time, standard LCA libraries did not include a module for beryllium or beryllium copper, nor did the connector production modules reference beryllium. Although databases have improved and now contain information on beryllium, critical information was not available when it was needed for decision-making and important data might still be missing.

On a more philosophical note, it is important to recognize that the entire premise of LCA is to make trade-offs between different impacts. Hence it becomes theoretically acceptable to introduce a high-hazard substance to obtain other life-cycle benefits, a situation we find unacceptable. We have concluded that if a substance presents a high enough hazard, it should not be introduced even to gain other life-cycle benefits.

Our expert LCA practitioners agree that there are better tools than LCA for comparing chemical hazards and that trade-offs between life-cycle impacts should only be made after high-hazard options have been removed from consideration through more specialized screening. This view strongly influenced the development of our integrated assessment protocol, which includes a dedicated hazard screening step early in the process. Such an approach significantly reduces the risk that a trade-off introducing high hazard could occur in an LCA.

7.2.5 Risk Assessment

Another method that we considered was traditional risk assessment, a powerful, structured and well-defined method for evaluating the risk to human health and the environment from the actual or potential presence of pollutants. Risk assessment considers the source of chemicals, fate and transport mechanisms, dose and thresholds and exposure scenarios.

Risk assessments are highly detailed views of substances. They are indispensable for the development of effective environmental regulations. However, when it comes to deciding whether to accept a replacement for a restricted substance, we are reminded that:[17]

> *The primary purpose of a risk assessment is not to make or recommend any particular decisions; rather, it gives the risk manager information to consider along with other pertinent information. [Regulators and risk managers use] risk assessment as a key source of scientific information for making good, sound decisions about managing risks to human health and the environment. Examples of such decisions include deciding permissible release levels of toxic chemicals, granting permits for hazardous waste treatment operations and selecting methods for remediating Superfund sites.*

When seen in this light, risk assessment definitely does not qualify as a general-use alternatives assessment tool.

Risk assessments require expert insight to use and interpret and take significant time and resources to complete, which is consistent with the needs and skills of regulators and industrial hygienists. However, in the electronics industry, the people who make decisions about materials are electrical and mechanical engineers, chemists, industrial designers and technical staff with training in fields other than toxicology. The highly detailed information in risk assessments can be overwhelming to these decision-makers. Much of the information is irrelevant to them and, critically, the information is not in a useful format for comparing options. For example, the human health portion of a risk assessment on lead and its compounds is more than 800 pages long.[18]

Formal, traditional risk assessment provides valuable information and analysis for regulators and industrial hygienists, but it may not be the right tool for comparing the *relative* impacts of replacement and restricted substances. Nevertheless, there is a concern among some practitioners that shifting away from risk-based screening overemphasizes the inherent hazard of substances and does not account for exposure controls that are available to protect human health. While exposure controls are important, especially in industrial settings, a key reason to pursue hazard reduction is that perfect exposure control is not possible.

A basic premise of Green Chemistry[19] is that chemical risk is most effectively managed by reducing hazards because exposure controls can and do fail, products are used in unintended ways and end-of-life management of obsolete equipment is often problematic. Therefore, a dedicated hazard screening step in the alternatives assessment process to identify lower hazard options is consistent with the principles of Green Chemistry.

Another concern about shifting away from risk assessment as a primary assessment tool is that important exposure issues might be overlooked. However, in most cases of a straightforward substitution of one chemical for another in an application, the exposure potential for the chemical remains approximately the same. Lavoie *et al.*[20] noted that:

> *Where similar product and chemical use patterns are expected, exposure can be considered a constant. As a result, risk (defined as hazard multiplied by exposure) can be reduced through a reduction in chemical hazard.*

For screening out unsuitable replacements for restricted substances, hazard becomes an effective proxy for risk within the same application. This simplification allows us to consider more specialized tools.

7.2.6 Scoring Schemes, Including the GreenScreen for Safer Chemicals

There are many systems for scoring chemicals against criteria to enable comparisons to be made. Most scores are based on weighted criteria, which are then used to generate one or more indicators of the chemical's environmental and human health impacts. Standardized criteria are used to achieve reproducible and comparable results. The results from these systems are often easier to use than other analyses. Available systems with standardized criteria include the GreenScreen for Safer Chemicals[21] (GreenScreen) and the Cradle to Cradle® Design Protocol.[22]

The scoring system that best fit our needs and was most successful in the pilot was the GreenScreen. This open-source, comparative chemical hazard assessment tool was launched in 2007 by the non-governmental organization Clean Production Action. The tool provides a clear and transparent decision logic that assesses 17 globally harmonized endpoints for environmental fate, human health and environmental toxicity and then generates benchmark scores ranging from the most to the least hazardous.

The GreenScreen classification system aligns with the Globally Harmonized System of Classification and Labeling of Chemicals (GHS) and with the US Environmental Protection Agency (EPA) Design for Environment (DfE) program's chemical alternatives assessment approach.[20] This repeatable, scientifically based method can screen materials and provide a score that can be used in conjunction with other key factors, such as cost, performance, recycling and energy consumption, to assess alternatives.

The GreenScreen is a hazard-based tool, meaning that it focuses on the intrinsic properties of substances to assess their inherent hazards and does not explicitly address exposure or life-cycle issues. It is more specialized than some of the other tools discussed earlier, but has very useful features for material selection and product design.

Hazard assessments are significantly faster and easier to complete than LCAs or risk assessments because hazard endpoints are well defined and are based on directly observable intrinsic traits or the behavior of a substance. Also, hazard data are generally available now that new EU chemical regulations require more disclosure of test data (REACH).[23]

The integer scoring system is especially useful from a decision-making perspective. Once an expert assessor has evaluated a group of chemicals, the results can easily be translated into procurement guidance for people without toxicology or chemistry backgrounds.

The GreenScreen also considers the potential impacts of a chemical once released into the environment, including known and predicted degradates. Including degradation products is important, particularly for electronic products, where end-of-life management can pose environmental problems that often are not adequately assessed.

7.2.6.1 The GreenScreen Scoring System and Hazard Endpoints. The GreenScreen Benchmark scoring system employs structured decision logic to assign a single integer score to each chemical being assessed. This scheme incorporates national and international precedents to weigh and prioritize combinations of hazard endpoints.

The GreenScreen defines four Benchmark levels for substances (plus U for unspecified):

Benchmark 1 – *Avoid – Chemical of High Concern*
Benchmark 2 – *Use But Search for Safer Substitutes*
Benchmark 3 – *Use But Still Opportunity for Improvement*
Benchmark 4 – *Prefer – Safer Chemical*
Benchmark U – *Unspecified Due to Insufficient Data*

Each Benchmark includes a set of hazard criteria that a chemical, along with its known and predicted transformation products, must pass. The criteria for each Benchmark become progressively more demanding, with Benchmark 4 representing the most preferred chemicals. For example, to pass from Benchmark 1 to Benchmark 2, none of the criteria specified under Benchmark 1 can apply to a chemical (including its transformation products). Likewise, to advance from Benchmark 2 to Benchmark 3, none of the criteria specified under Benchmark 1 or 2 can apply to the chemical or its transformation products, and so on.

Benchmark 1 criteria
  (a) PBT = High P + High B + [very High T (Ecotoxicity or Group II Human) or High T (Group I or II* Human)]
  (b) vPvB = very High P + very High B
  (c) vPT = very High P + [very High T (Ecotoxicity or Group II Human) or High T (Group I or II* Human)]
  (d) vBT = very High B + [very High T (Ecotoxicity or Group II Human) or High T (Group I or II* Human)]
  (e) High T (Group I Human)
Benchmark 2 Criteria
  (a) Moderate P + Moderate B + Moderate T (Ecotoxicity or Group I, II or II* Human)
  (b) High P + High B

    (c) High P + Moderate T (Ecotoxicity or Group I, II or II* Human)
    (d) High B + Moderate T (Ecotoxicity or Group I, II or II* Human)
    (e) Moderate T (Group I Human)
    (f) Very High T (Ecotoxicity or Group II Human) or High T (Group II* Human)
    (g) High Flammability or High Reactivity

Benchmark 3 criteria
    (a) Moderate P or Moderate B
    (b) Moderate Ecotoxicity
    (c) Moderate T (Group II or II* Human)
    (d) Moderate Flammability or Moderate Reactivity

Benchmark 4 criteria
    (a) Low P* + Low B + Low T (Ecotoxicity, Group I, II and II* Human) + Low Physical Hazards (Flammability and Reactivity) + Low (additional ecotoxicity endpoints) + complete data set

Benchmark U
    (a) Insufficient data

The GreenScreen hazard endpoints are divided into the groups shown in Table 7.1.

In some cases, data gaps will exist for specific hazard endpoints. The GreenScreen allows the use of quantitative structure–activity relationships (QSARs) to fill data gaps until such time that actual empirical data are available.

Highly detailed information about these endpoints is available at the Clean Production Action website: www.cleanproduction.org.

7.2.6.1.1 Group I Human Health

Group I endpoints reflect priorities that are consistent with those found in national and international governmental regulations. These endpoints cover

Table 7.1 GreenScreen version 1.2 hazard endpoints and groups.

Environmental Fate	Environmental Health	Human Health Group I	Human Health Group II	Physical Hazards
Persistence	Acute Aquatic Toxicity	Carcinogenicity	Acute Mammalian Toxicity	Reactivity
Bioaccumulation	Chronic Aquatic Toxicity	Mutagenicity and Genotoxicity	Systemic Toxicity and Organ Effects	Flammability
		Reproductive Toxicity	Neurotoxicity	
		Developmental Toxicity	Respiratory and Skin Sensitization	
		Endocrine Activity	Skin and Eye Irritation	

Table 7.2   Thresholds for classifying hazard level for carcinogenicity in GreenScreen version 1.2.

Information source	High (H)	Moderate (M)	Low (L)
GHS Criteria and Guidance	GHS Category 1A (Known) or 1B (Presumed) for any route of exposure	GHS Category 2 (Suspected) for any route of exposure	Adequate data available and negative studies, no structural alerts and GHS not classified
EPA-C (1986)	Group A, B1 or B2	Group C	Group E
EPA-C (1996, 1999, 2005)	Known or Likely	–	Not Likely
EU CMR (1)	Category 1 or 2	Category 3	–
EU CMR (2)	Carc 1A or 1B	Carc 2	–
EU H-statements	H350 or H350i	H351	–
EU R-phrases	R45 or R49	R40	–
EU SVHC	Reason for inclusion: carcinogenic	–	–
IARC	Group 1 or 2A	Group 2B	Group 4
MAK	Carcinogenic Group 1 or 2	Carcinogenic Group 3, 4 or 5	–
NIOSH-C	Occupational cancer	–	–
NTP-RoC	Known or reasonably anticipated	–	–
Prop 65	Known to the State to cause cancer	–	–
EPA-C(1986)	Group D		
EPA-C (1999)	Suggestive evidence, but not sufficient to assess human carcinogenic potential		
EPA-C (2005)	Suggestive evidence of carcinogenic potential		
IARC	Group 3		

hazards that can lead to chronic or life-threatening effects or adverse impacts that can potentially be induced at low doses and transferred between generations.

It is beyond the scope of this chapter to review all the individual endpoints in this group, but the carcinogenicity thresholds are shown in Table 7.2 as an example.

7.2.6.1.2   *Group II and II* Human Health*

Group II and II* Human Health endpoints are differentiated in the Benchmark system because Group II endpoints have four hazard levels (vH, H, M and L) whereas Group II* endpoints have three (H, M and L). Systemic Toxicity/ Organ Effects and Neurotoxicity endpoints can belong in either Group II or

Group II*, depending on whether the data are generated from single exposure (acute) or repeated exposure (sub-chronic or chronic) studies.

7.2.6.2 GreenScreen Pilot Program. In the initial GreenScreen pilot, we applied version 1.0 assessment criteria to a variety of chemicals and mixtures to understand the tool and its possible uses better. More than 40 substances were assessed in the first year. The results were immediately useful to designers and procurement engineers selecting replacements for BFRs.

GreenScreen 1.0 was most successful at differentiating hazard profiles when it compared similar functional materials, such as flame retardants or plasticizers. It was well suited to analyzing organic compounds, but less so to inorganic compounds. However, in both cases, the results could meaningfully inform decision-making. Data gaps were a problem and some substances were rejected because they could not be scored. (GreenScreen 1.2, released in October 2011, addressed the issue of how to score substances when there are data gaps. The handling of inorganics was also improved in version 1.2 and is expected to evolve more in future revisions.)

Despite these limitations, the GreenScreen successfully differentiated the hazards of flame retardants, plasticizers and other chemicals of interest. It also provided comparative hazard information in a format that enabled better decisions to be made about substitute materials.

7.2.6.3 PVC-free Power Cord Program. The results from the first phase of the pilot were positive enough to support a larger, second trial where the GreenScreen was applied to power cord materials that were being changed from poly(vinyl chloride) (PVC) to thermoplastic elastomers (TPEs).

In this second phase of the pilot, the screening of materials was a mandatory part of the approval process, in addition to the existing technical requirements. Suppliers of PVC-free power cord materials were trained on GreenScreen 1.0 and both HP and its suppliers assessed the additives present above 1000 ppm in the candidate materials.

Since GreenScreen has no allowance for calculating a weighted score based on the amount of each constituent, the lowest score of the constituents was assigned to the overall material.

Despite initial challenges in arranging sufficient disclosure of the constituents, eventually all candidate materials were assessed. The results of the screening led to a number of materials advancing to regulatory testing and several are now approved for use. Since that time, a formal material specification for PVC-free power cord material has been released and an approved material list (AML) is now being used to manage these materials.

7.2.6.4 External Pilots. In addition to work done inside HP, we have collaborated with other groups piloting the GreenScreen on alternatives to BFRs[24] and plasticizers[25] and have found it to be a useful tool for differentiating between the environmental and human health impacts of potential replacement substances.

7.3   Integrated Alternatives Assessment Protocol

Through the pilot programs, the GreenScreen identified preferable options for replacing restricted substances. It was also helpful for communicating materials goals and criteria to suppliers. However, as a comparative chemical hazard screening tool, GreenScreen could not address the full range of life-cycle and exposure impacts. It made sense to build it into a system that would compensate for its limitations. LCA, risk assessment, multi-criteria decision analysis and even RSL screening are most valuable when integrated in such a manner that each tool operates as intended.

Based on the features and limitations of each tool, an integrated alternative assessment protocol (IAAP) was created and is shown in Figure 7.1. The IAAP shares many features with the BizNGO Chemical Alternatives Assessment Protocol, with which it was co-developed.[26] Both protocols build upon the work of the Lowell Center for Sustainable Production's Alternatives Assessment Framework,[27] the US EPA DfE program's approach to chemical alternatives assessment,[20] the United Nations Environment Programme, the Stockholm Convention on Persistent Organic Pollutants, the Persistent Organic Pollutants Review Committee's guidance document on substitution

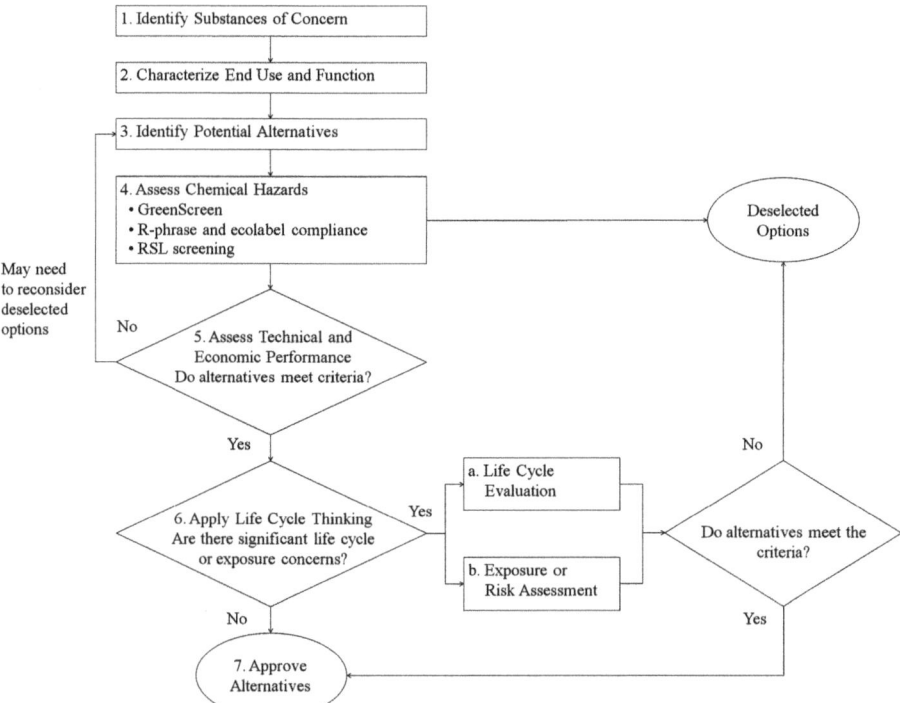

Figure 7.1   Integrated alternatives assessment protocol for evaluating replacements for restricted substances.

and alternatives,[28] the Massachusetts Toxics Use Reduction Institute's Five Chemicals Alternatives Assessment Study,[29] and Ökopol and Kooperations-stelle Hamburg's substitution report for the European Union.[30]

There are seven steps in the protocol, which begins with the identification of a substance of concern and concludes with at least one replacement selected.

7.3.1 Step 1 – Identify Substances of Concern

In the case of legally restricted substances, regulators or other authoritative bodies generally identify substances of concern through rigorous risk assessment and other analyses. This step is explicitly included in the protocol because market pressure, ecolabels or independent scientific findings could trigger a search for alternatives to certain substances even before regulations mandate a change.

7.3.2 Step 2 – Characterize Function and End Uses of the Substance

Once a substance has been selected for restriction, its function and uses must be characterized in order to find suitable alternatives. Functionality is a core element of the DfE's chemical alternatives assessment approach, where 'functional use' characterizes the purpose of the chemical and simplifies assessment.[20]

Information on a chemical's function, such as whether it is a flame retardant or plasticizer, is relatively easy to find. However, identifying where specific substances are used within assembled products can be difficult. Complex supply chains – and a lack of disclosure of chemical data along the supply chain – can make it hard to know which chemicals are in which components within products.

The primary source of such information is the supplier. Owing to existing substance restrictions, such as RoHS, there are mechanisms for obtaining this type of information from suppliers. Baseline information gathering is supplemented with expert knowledge about components and materials (to ensure that the correct suppliers are queried) and with analytical testing to validate reported information.

7.3.3 Step 3 – Identify Potential Alternatives

Owing to the complex nature of electronic products, restricted substances often impact materials and components far up the supply chain. For that reason, we usually work with suppliers to develop innovative solutions that meet our environmental, regulatory and quality requirements. Although some approaches aggressively screen options based on technical performance, at this point in the assessment process we prefer to keep as many options open as possible to give suppliers maximum flexibility and to ensure that we do not overlook potentially innovative solutions, including designing out the substance or function.

7.3.4   Step 4 – Assess Chemical Hazards

Chemical hazard assessment is a way to deselect alternatives that are of equivalent or greater hazard than the restricted substance. Hazard assessment takes place early in the process, preceding technical and economic assessments, because we do not want to evaluate alternatives unless they would have a lower impact on human health and the environment.

The GreenScreen is the central screening method for assessing the chemical hazards of the potential alternatives. However, RSL screening and R-phrase screening against ecolabel criteria are done simultaneously since many of the same information sources are used.

Unsuitable alternatives are set aside and may potentially be revisited if no viable technical solutions are found in Step 5.

7.3.5   Step 5 – Evaluate Technical and Economic Performance

Potential alternatives are next evaluated on technical and economic criteria before the time- and resource-intensive life-cycle or exposure assessments.

Alternatives must meet performance and functionality requirements of the original material or product, including compliance with all applicable legal requirements. If there are no existing specifications for a material or application, critical performance requirements may need to be determined at this point.

Correctly setting performance requirements can be challenging, but it is critical. The specifications should be neither over- nor under-engineered because suppliers might offer different solutions. Provided that the proposed replacements can meet the requirements, suppliers can choose the solution that best suits them. This approach encourages suppliers to find the most cost-effective and innovative solutions.

7.3.6   Step 6 – Apply Life-cycle Thinking

Life-cycle thinking is the process of evaluating the potential impacts on human health or the environment across the life-cycle stages of a chemical, material or product. The application of life-cycle thinking enables us to consider a range of potential human health and environmental impacts, including global warming, end-of-life material disposition and worker exposure. The depth of this assessment varies depending on the use and function of the original substance and the alternatives under consideration. The TURI Five Chemicals Alternatives Assessment Study[29] provides a good example of life-cycle thinking.

Based on the outcome of this process, two additional evaluations might be triggered:

(a) *LCA* – Alternatives that involve significant material or process changes may require LCA. These evaluations can vary widely in type and scope, ranging from carbon footprint calculations to full quantitative life-cycle analyses. In cases where a single alternative must be selected (as opposed to qualifying a range of solutions), life-cycle evaluations help differentiate between the alternatives.

(b) *Exposure* – If any alternatives have a significantly different use pattern or life-cycle than the original substance, exposure concerns may trigger additional assessments, such as an exposure assessment or even a full risk assessment.

7.3.7   Step 7 – Approve Alternatives

Step 7 represents a departure from many alternative assessment protocols because our goal is often not to select a single replacement, but rather to approve multiple acceptable replacements. As noted in Step 5, different suppliers may have different optimal solutions for their own circumstances. We try to give them the flexibility to choose replacement materials and technologies that meet their needs provided that they meet our requirements.

In certain cases, materials that are approved for use in products may be listed in a formal document, as is the case with the PVC-free power cord materials. Approved material lists act as "white lists" of replacements and facilitate wider adoption of screened materials.

7.4   Opportunities for Improvement

Although the GreenScreen and the integrated assessment approach have been very useful, there are opportunities to improve the methods themselves in addition to the infrastructure for creating and sharing assessments. There are also opportunities to promote the adoption of environmentally preferable replacements for restricted substances through the material criteria in ecolabels.

7.4.1   Methods

The GreenScreen method should be adjusted to handle inorganic chemicals better, to harmonize with the DfE Alternatives Assessment Criteria for Hazard Evaluation and to clarify the guidance on transformation products. Also, the integrated approach may benefit from standard triggers for each type of assessment and should harmonize with regulations requiring alternatives assessments.

7.4.1.1   Inorganic Chemicals in the GreenScreen.   Risk and hazard assessment of inorganic chemicals is a developing field.[31] The GreenScreen was primarily developed for organic substances and requires changes to assess the hazards of inorganic chemicals more accurately. Specifically, the GreenScreen should adjust its treatment of persistence and physical properties.

7.4.1.1.1   Persistence
As noted above, the GreenScreen did not always correctly assign hazard Benchmarks for inorganic substances in the pilot program. With organic

chemicals, persistence is closely tied to hazard and GreenScreen penalizes chemicals for persistence in the scoring logic. However, inorganic chemicals are often already in a mineralized form and can be considered highly persistent by currently accepted definitions of persistence. This persistence, in and of itself, does not always represent a hazard.

GreenScreen 1.2 makes a scoring exception for inorganic compounds that are highly persistent but have low hazard characteristics with respect to human health and aquatic toxicity. Chemicals that have high persistence where all other hazards are low may achieve a score of Benchmark 4. Although this change is an improvement, it is expected that future revisions will more accurately reflect the impacts of persistence in all scores for inorganic chemicals and more closely reflect actual hazard of inorganic chemicals.

7.4.1.1.2 Physical Properties

GreenScreen 1.2 also explicitly recognizes that the physical properties of inorganic chemicals are relevant to assessing their inherent hazard and toxicity. For example, water solubility can modify the classification of aquatic toxicity and particle size and shape can determine the potential to cause respiratory irritation. Version 1.2 introduced the requirement that particle size (*e.g.*, silica particles <10 μm), structure (*e.g.*, amorphous *versus* crystalline) and mobility (*e.g.*, water solubility) must be defined for inorganic substances and, in certain cases, assessed separately, as in the case of respirable crystalline silica.

7.4.1.2 Transformation Products.

The need for more guidance on transformation products was first identified in assessments of inorganic chemicals. GreenScreen 1.2 added guidance on how to identify and assess feasible and likely transformation products of parent substances, including both organic and inorganic chemicals. Feasible transformation pathways include biodegradation, hydrolysis, photolysis, oxidation and combustion. Any feasible transformation pathway may generate transformation products. Such products are considered relevant if they can reasonably be expected to be formed and if there is a real potential for human or environmental exposure to the transformation product.

Transformation products are considered feasible based on:

- the use of the parent chemical in a material or product (*e.g.*, the substance is exposed to the air when used, so oxidation is feasible) and
- the product's end-of-life management patterns (*e.g.*, the product is typically disposed of down the drain, so aquatic biodegradation is a feasible breakdown mechanism).

Transformation products with potential for extended exposure are also now considered. A transformation product is not considered relevant if expert judgment determines that it is either transient or unlikely to occur. For example, a chemical that is a transient intermediate during biodegradation may not be relevant.

A naturally occurring transformation product is not considered relevant if expert judgment determines that the product would result in exposure less than or equivalent to the normal levels found in the relevant environmental compartment(s). Relative exposure is particularly important when considering the transformation of metals relative to background metal levels.

These adjustments and refinements permit a more sophisticated assessment of inorganic substances. However, the process still requires expert judgment and additional guidance may be added over time.

7.4.1.3 GreenScreen Harmonization with the US EPA Design for the Environment Alternatives Assessment Criteria. The GreenScreen is an actively maintained tool and will evolve as underlying standards and criteria change. The DfE Alternatives Assessment Criteria for Hazard Evaluation[32] framework forms the core of GreenScreen. The DfE program released its updated criteria in August 2011 and it is expected that the next revision of the GreenScreen will be harmonized with the new criteria.

7.4.1.4 Integrated Alternative Assessment Approach. As more material types have been brought into this program, our approach has continued to evolve. We are developing clearer triggers and checklists for LCA and exposure assessments. We have also adjusted our approach such that it is harmonized with local jurisdictions and regulations where alternatives assessments will be required for restricted substances, such as in California.[33]

7.4.2 Infrastructure

One disadvantage of the GreenScreen is that, as a relatively new tool, there is less infrastructure for having assessments made and validated. This area could definitely be improved.

7.4.2.1 Assessments. Although there are some licensed GreenScreen profilers now, Clean Production Action is working with several partners to license more profilers in more locations to provide services to those who want (or need) to have third parties complete the initial assessments.

7.4.2.2 Validation. As an open source tool, GreenScreen is free, fully transparent and publicly accessible. It was primarily created for internal and business-to-business information sharing, rather than for promotional purposes. While preliminary assessments can still be useful within an organization, a validation program is needed so that high-confidence scores can be shared publicly across the supply chain. Any assessments of chemicals or materials that have not gone through an approved GreenScreen validation program must be considered drafts and are not eligible to make public or promotional claims using the GreenScreen name or marks.

Clean Production Action is currently developing a formal validation program for GreenScreen scores. In our experience with screens performed both internally and externally (including those done by licensed GreenScreen profilers), almost all assessments are corrected or significantly improved through peer review. We recommend that the validation program retain a level of critical review beyond simple checks for completeness.

7.4.2.3 Libraries or Repositories of Scores. In addition to licensing more profilers and developing a validation program, another evolving infrastructure area is the creation of libraries and repositories for scores. Both private and not-for-profit groups are developing complementary services and websites to help organizations share scores for chemicals and materials. The libraries will encourage the adoption of better materials by sharing information on known-good or 'white list' materials. Libraries will also reduce the redundancy from assessing the same common chemical multiple times.

7.4.3   Criteria for Preferred Material in Ecolabels

Using GreenScreen Benchmarks instead of H- and R-phrases for identifying preferred materials in ecolabels would encourage the sharing of information on, and adoption of, environmentally preferable replacements for substances of concern. If ecolabels also mandated fully validated Benchmark scores, there would be greater confidence that preferred materials were being identified based on data, rather than promoting alternatives with data gaps.

7.5   Conclusion

HP has investigated several tools for evaluating the human health and environmental impact of replacements for restricted substances in electronics. After successful pilot programs, we are structuring our integrated alternatives assessment approach around a dedicated hazard screening step early on in the assessment process, with complementary tools to assess other concerns. This approach meets our goals for quality, logic and ease of use. We can now give our suppliers much clearer guidance on our materials goals and criteria. Critically, this approach is also well aligned with regulators, which should reduce the likelihood of having to do multiple substitutions within the same application.

References

1. C. A. Ascencio and J. J. Madsen, *LCA Comparison of Alternative Soldering Techniques,* PRé Consultants, Amersfoort, 2005.
2. J. R. Geibig, *Solders in Electronics: a Life-Cycle Assessment,* EPA 744-R-05-001, US Environmental Protection Agency, Washington, DC, 2005.
3. S. Edwards, J. Tickner, Y. Torrie, M. Coffin and L. Kernan, *A Compendium of Methods and Tools for Chemical Hazard Assessment,* Lowell Center for Sustainable Production, University of Massachusetts Lowell, Lowell, MA, 2011.

4. S. Edwards, M. Rossi and P. Civie, *Alternatives Assessment for Toxics Use Reduction: a Survey of Methods and Tools*, Toxics Use Reduction Institute, University of Massachusetts, Lowell, Lowell, MA, 2005.
5. T. Smola, *The Column Model*, Institut für Arbeitsschutz der Deutschen Gesetzlichen Unfall-versicherung (IFA), Berlin, 2009.
6. KEMI, Swedish Chemicals Agency, *PRIO, a Tool for Risk Reduction of Chemicals*, 2006.
7. IEEE, *Standard for Environmental Assessment of Personal Computer Products*, P1680.1, IEEE, New York, 2010.
8. M. Söderberg, *TCO Certified Notebooks 4.0*, TCO Development, Stockholm, 2012.
9. Y. Hatamura (ed.), *Decision-Making in Engineering Design*, Springer, London, 2006.
10. R. L. Keeney and H. Raiffa, *Decisions with Multiple Objectives: Preferences and Value Tradeoffs*, Wiley, New York, 1976.
11. J. N. Siddall, *Analytical Decision-Making in Engineering Design*, Prentice-Hall, Englewood Cliffs, NJ, 1972.
12. W. Edwards, R. F. Miles Jr. and D. von Winterfeldt, *Advances in Decision Analysis: from Foundations to Applications*, Cambridge University Press, Cambridge, 2007.
13. J. X. Wang, *What Every Engineer Should Know About Decision Making Under Uncertainty*, Marcel Dekker, New York, 2002.
14. Toxics Use Reduction Institute, *Pollution Prevention Options Analysis System (P2OASys)*, University of Massachusetts Lowell, Lowell, MA, 2005.
15. T. F. Malloy, P. J. Sinsheimer, A. Blake and I. Linkov, *Developing Regulatory Alternatives Analysis Methodologies for the California Green Chemistry Initiative*, UCLA Sustainable Technology and Policy Program, University of California Los Angeles, Los Angeles, CA, 2011.
16. Scientific Applications International Corporation (SAIC), *Life Cycle Assessment: Principles and Practice*, US Environmental Protection Agency, National Risk Management Research Laboratory, Office of Research and Development, Cincinnati, OH, 2006.
17. K. Dearfield, *An Examination of EPA Risk Assessment Principles and Practices*, EPA/100/B-04/001, US Environmental Protection Agency, Washington, DC, 2004.
18. C. Boreiko and R. Battersby, *Voluntary Risk Assessment Report on Lead and Some Inorganic Lead Compounds*, LDAI Lead Risk Assessment Working Group, Lead Development Association International, London, 2008.
19. P. T. Anastas and J. C. Warner, *Green Chemistry: Theory and Practice*, Oxford University Press, Oxford, 2000.
20. E. T. Lavoie, L. G. Heine, H. Holder, M. S. Rossi, R. E. Lee II, E. A. Connor, M. A. Vrabel, D. M. DiFiore and C. L. Davies, *Environ. Sci. Technol.*, 2010, **44**, 9244–9249.
21. M. Rossi and L. Heine, *The Green Screen for Safer Chemicals: Evaluating Flame Retardants for TV Enclosures*, Clean Production Action, Somerville, MA, 2007.

22. W. McDonough, M. Braungart, P. Anastas and J. Zimmerman, *Environ. Sci. Technol.*, 2003, **37**, 434A–441A.
23. European Commission, *EC Regulation on Registration, Evaluation, Authorisation and Restriction of Chemicals (REACH)*, No. 1907/2006, European Commission, Brussels, 2007.
24. A. Beard, M. Klimes and U. Wietschorke, *Non-halogenated phosphorus, inorganic and nitrogen flame retardants for electronics: update on market situation, drivers and trends*, Electronics Goes Green 2012 + (EGG), Brussels, Belgium, 2012.
25. Green Chemistry and Commerce Council (GC3), *Assessment of Alternate Plasticizers*, 2011.
26. M. Rossi, C. Peele and B. Thorpe, *BizNGO Chemical Alternatives Assessment Protocol*, Business-NGO Working Group, Clean Production Action, Somerville, MA, 2011.
27. M. Rossi, J. Tickner and K. Geiser, *Alternatives Assessment Framework of the Lowell Center for Sustainable Production*, University of Massachusetts Lowell, Lowell, MA, 2006.
28. UNEP, Stockholm Convention on Persistent Organic Pollutants, *Report of the Persistent Organic Pollutants Review Committee on the work of its fifth meeting, Addendum: General guidance on consideration related to alternatives and substitutes for persistent organic pollutants and candidate chemicals (UNEP/POPS/POPRC.5/10/Add.1)*, Stockholm Convention, Châtelaine, 2009.
29. M. Ellenbecker, *Five Chemicals Alternatives Assessment Study*, Toxics Use Reduction Institute, University of Massachusetts Lowell, Lowell, MA, 2006.
30. J. Lohse, M. Wirts, A. Ahrens, K. Heitmann, S. Lundie, L. Lissner and A. Wagner, *Substitution of Hazardous Chemicals in Products and Processes*, Report compiled for the Directorate General Environment, Nuclear Safety and Civil Protection of the Commission of the European Communities, Hamburg, 2003.
31. P. M. Chapman, *Hum. Ecol. Risk Assess.*, 2010, **14**, 5–40.
32. US EPA, *Design for the Environment Program Alternatives Assessment Criteria for Hazard Evaluation*, US Environmental Protection Agency, Washington, DC, 2011.
33. M. Feuer, *Assembly Bill No. 1879*, California State Assembly, Sacramento, CA, 2008.

DSM's Sustainability Journey Towards a Proactive Ingredient Policy for Gaining Effectiveness in the Design of Better Products

THOMAS A. J. WEGMAN, FREDRIC PETIT, ANNETTE WILSCHUT, THEO JONGELING AND GAELLE M. NICOLLE*

ABSTRACT

Many opportunities can be leveraged through the development of highly performing, sustainable, eco-friendly products. Being compliant with local regulations is a prerequisite for responsible business; however, the industry could contribute much more to sustainable benefits by focusing beyond the minimum compliance level. For instance, material solutions can be developed with no substance of potential environmental concern throughout their life-cycle and still achieve high performances. The path to more sustainable products and the challenges encountered are explained in this chapter.

8.1 General Introduction: Sustainability as Business Growth Driver Opportunity and Global Trends

Societal issues posed by global shifts (demographic shifts, urbanization, high-growth economies, usage of resources, impact of new technology), climate change, energy, health and wellness unleash an imperative to act.

*Corresponding author

Issues in Environmental Science and Technology, 36
Chemical Alternatives Assessments
Edited by R.E. Hester and R.M. Harrison
© The Royal Society of Chemistry 2013
Published by the Royal Society of Chemistry, www.rsc.org

Vision 2050 developed by the World Business Council for Sustainable Development studied and described this imperative.[1] It has also been widely recognized in global discussion platforms such as the World Economic Forum and Rio$^+$20 that the engagement of governments, institutions, organizations and industries will be crucial in providing better alternatives and solutions to those issues at an acceptable speed and scale. This is a massive challenge which requires traction to catalyze the change. The driving force for industry remains business – sustainable business which can sustain its own profit for the long term. Contributing to solving these global issues can be the driving force for creating new business opportunities, which, if followed by the mainstream industry, will create sufficient traction to respond to the global issues with the right speed and scale. It requires, of course, a new mindset for better product design by focusing innovation on improving environmental and social impact.

8.1.1   Ecological Benefits: Transparent Assessment with Life-Cycle Assessment (LCA)

'Sustainable development' means 'meeting the needs of the present generation without compromising the ability of future generations to meet their own needs.' This is the widely accepted definition that the Brundtland Commission published in 1987.[2] By simultaneously pursuing social responsibility, environmental quality and economic performance, values on the three dimensions of People, Planet and Profit (Triple P focus) can be created. Brand owners and consumers show increasing interest in sustainability. Increasing the sustainability performance of their products, services and solutions over time and enabling both upstream and downstream stakeholders of the value chain to do the same can be a clear differentiator for industry. Clear added value can be created by designing high or even higher performing products with less environmental impact. Ecological benefit can be created at any stage of the life-cycle – from the raw material, manufacturing, use, to potential re-use and end-of-life disposal. The yardstick commonly used to measure the ecological impact of products and to bring transparency in industry claims is the so-called life-cycle assessment (LCA). LCA identifies the material, energy and waste flows associated with a product over its life-cycle to determine environmental impacts or potential improvements. The environmental impact, or eco-footprint, is assessed in a number of key areas, including resource depletion, global warming potential, acidification and eutrophication and human and eco-toxicity.

LCA helps us to:

- understand eco-efficiency of products and solutions
- quantify environmental impact of DSM's solutions *versus* alternative solutions
- steer product/process development in the most sustainable direction.

Social impact measurement tools still need to be further developed to enable industry to bring transparency to the social impact of their products.

8.1.2 Importance of Global Platforms and Partnerships

One prerequisite (but also an additional challenge) for quality LCA is the right data input. The life-cycle can be complex. It is a real challenge to understand the product value chain from the sourcing of the raw materials, manufacturing, use, reprocessing, recycle/re-use or end-of-life disposal. The focus on better alternatives for any stage of the life of a product has to be integrated in the product design and innovation process. To obtain the required insight for all dimensions of the value chain that it can leverage on, involvement of key stakeholders in the value chain of the products is crucial. Partnership with customers and also other upstream or downstream stakeholders in the value chain increases data insight for better product design. Active participation in platforms such as the World Business Council for Sustainable Development (WBCSD) or The Sustainability Consortium is also important in term of alignment and discussion with those stakeholders.

Achievements once made and tangibly measured have to be further transparently communicated. This is also part of gaining the customer's trust and credibility within the sector or cross-sectors. Without such credibility, companies may offer better products on the market that may not be recognized as such. Providing transparency on sustainable achievement is also key to achieving traction of sustainable consumption and clear differences in response to global trends. For instance, the Global Reporting Initiative (GRI) has developed Sustainability Reporting Guidelines that strive to increase the transparency and accountability of economic, environmental and social performance.[3] The GRI was established in 1997 in partnership with the United Nations' Environment Programme. It is an international, multi-stakeholder and independent institution whose mission is to develop and disseminate globally applicable Sustainability Reporting Guidelines. These Guidelines are used voluntarily by organizations for reporting the economic, environmental and social dimensions of their activities, products and services. Ernst&Young confirmed that DSM has applied the GRI G3 guidelines to an A+ level for the integrated annual report 2011.

8.1.3 The Challenges Ahead for a Life Sciences and Materials Sciences Company

To achieve its sustainability goals, DSM focuses on lowering the eco-footprint of its solutions across the value chain: by increasing the bio-based product portfolio and proactively working on better ingredient alternatives to limit the use, release and formation of substances of hazardous concern in the product value chain. Looking at the broad range of the DSM product portfolio

– from chemical intermediates to pharmaceutical and nutritional finer products
– these focuses result in various challenges for the company.

8.1.3.1 Moving Towards a Bio-based Economy. DSM has a strong belief
that bio-based economy can be part of the solution towards a Sustainable
Society. One typical challenge related to a bio-based economy concerns the
use of biomass. Biomass is renewable organic material, *e.g.*, agricultural
crops or residue streams, wood, grasses or municipal waste. In the context of
biomass as a feedstock for energy or products, it is often referred to as plant-
based material. Whereas fossil feedstocks are finite, biomass is renewable – it
can be grown on the land. Biomass has been hailed as a viable alternative to
fossil feedstocks, for mitigating greenhouse gas emissions and providing a
major opportunity for rural development increasing prosperity on a global
scale. Biomass as feedstock can be used for the production of power and
heat and as a raw material for the production of bio-based products such as
fuels, chemicals and materials. One can distinguish between first-, second-
and third-generation renewable feedstocks.

First-generation feedstocks comprise plant oil, starch and sugar, which can
also be used in the food chain. Second-generation feedstocks comprise non-
edible plant material and residues from forestry, agriculture and industry and
energy crops that do not directly compete with food crops. Third-generation
feedstocks comprise the direct use of CO_2 and can be based on, for example,
algae, which do not require fertile agricultural land.

Biomass – a source for food, feed, energy and bio-based products – raises
questions and (public) concern. Although industry would like to do everything in the correct way from the start, unfortunately it does not yet have all
the answers to these questions and concerns. One concern centers on the use
of crops for food or feed production. DSM believes that it is not food versus
fuel but food and fuel and bio-based products enabled by increasingly
efficient methods of agricultural production, especially in developing areas.
The cultivation of yet uncultivated arable land for the production of
food and feed crops, urbanization and industrialization may lead to direct
land use change. If existing arable land is used for industrial purposes, *e.g.*,
the production of biofuels instead of food, this will likely cause indirect
land use change (ILUC) owing to the necessity to replace land for food
and feed production. Although studies reveal that ILUC factors vary
greatly,[4] it is important to take ILUC effects into account. DSM acknowledges that the subject of land use change is of great complexity and that the
scientific work on the modeling of the impact of land use change for biofuel
does not yet provide clear and unambiguous answers. However, DSM
strongly believes that jeopardizing an endangered habitat for cultivation of
feedstocks for the bio-based economy should be avoided. For that, measures
are needed to improve agricultural management systems to increase agricultural
productivity, carefully balance water use, protect endangered habitats and
secure biodiversity.

Other concerns include depletion of carbon and minerals in the top soil, loss of biodiversity, the use of genetically modified seeds, crops or microorganisms and the use of fertilizers, insecticides and water. In addition to the benefits of biomass for bio-based products and materials, one needs to acknowledge these concerns, albeit on the basis of facts. Responsible sourcing and use of sustainable biomass are key in meeting the needs for food, energy and materials of the present generation without compromising the ability of future generations to meet their own needs.

Two bio-based product examples from DSM's Materials Sciences cluster are EcoPaXX™, a 70% bio-based polyamide with a carbon neutral footprint (cradle to gate) that offers exceptional performance even in the most demanding environments and the resin Palapreg® ECO. This resin has 55% bio-renewable content (the highest bio-based content in resins on the market) and excellent properties while simultaneously supporting sustainability in the value chain.

8.1.3.2   From Driven by Compliance Towards a Proactive Ingredient Policy. DSM recognizes that the end-of-life phase of a product is generally left to the responsibility of the end-user and normally becomes subject to a waste management system. Although specific ingredients may bring benefits or properties in the manufacturing and/or in the use phase, they may cause problems in the end-of-life phase or there could be limited recycling options. Being compliant with regulations is the minimum in taking responsibility, but there is a great opportunity to be proactive in ingredient selection. The general concern on certain ingredients, although the health and environmental risk might be controlled, should be taken into account when moving from compliance towards a more proactive ingredient policy.

In the past, DSM focused merely on its own products and markets as legal requirements would lead to the removal of certain ingredients. DSM is now clearly moving beyond this reactive mode to a company that is far more proactive in the way in which products and markets are approached. Examples of products that offer an alternative solution to products with ingredients under discussion because of their hazardous concerns are described further in this chapter. DSM strives to bring sustainable solutions to the market with clear advantages compared with competing products.

Engagement or partnership with other organizations including non-government organizations (NGOs) and their active participation in global discussion platforms reveal that there are more benefits to provide to people around the globe than are actually covered by existing regulations. One clear benefit could be provided by the proactive design of material solutions with the absence of, or absence of formation of, substances of potential concern in their life-cycle. The remainder of this chapter highlights the potential solutions in DSM's Materials Sciences cluster and what means are required to develop those better products, while providing the reader with insight into the benefits that can be created across the value chain.

8.2   A Brighter Future with Composites

8.2.1   Leading-edge Performance

Products and product solutions based on composites have become increasingly popular over recent decades. Composites are material systems that consist of a combination of resin (typically providing ductility, chemical resistance and durability) with a fibrous reinforcement (providing strength and stiffness). Typical resin systems can be unsaturated polyesters, vinyl esters, epoxies, polyurethanes and hybrid systems. The most commonly used fibers are based on glass, but increasingly carbon fibers, UHDPE fibers (*e.g.*, Dyneema® fibers), aramid fibers (*e.g.*, Kevlar® fibers) and in some cases natural fibers are also being used.

The benefits associated with the use of composites include durability (through their great resistance to a variety of chemicals), mechanical strength and stiffness, low weight (through their inherently low density), design freedom and excellent fatigue resistance.

Typical application examples of composites include:

- Body panels and exterior parts for trucks and cars. Benefits over steel include lower fuel consumption and CO_2 emissions through the light weight, personalized shapes and superior a esthetics.
- Piping for water distribution, sewage and cooling water supply. Here the key advantage drivers are strength, resistance to corrosion (resulting in low maintenance and continued process operation) and easy installation associated with the lower weight.
- Windmill blades. The aerodynamic design and weight are optimized, so the windmill can achieve high efficiencies. In many cases a 1% difference in efficiency over a 15 year period is paying for the investment of the windmill. The excellent fatigue resistance ensures reliable operation with low maintenance.
- Building and infrastructure components, such as profiles, shaped roof sections and façade panels. Windows based on composites profiles are slimmer than their aluminium equivalents, letting through more light while insulating the heat better. The increasing popularity of so-called 'BLOB' architecture (fully utilizing the freedom of design for making buildings and roofs) is a trend that is expected to gain in importance, driven by the desire of architects to create unique shapes and home functionality. Good examples include the Yitzhak Rabin Center in Tel Aviv, Israel, and the recently reopened Gemeentelijk Museum in Amsterdam, The Netherlands.

Global trends for increasing quality of life, personal health and wellness and convenience are driving the introduction of new consumer products and services. Meanwhile, commodities such as clean drinking water, continued power supply and availability of transportation are considered to be essential for an increasingly growing world population.

Bearing in mind the scarce resources available for the world's population, this means that material solutions are required that can provide better functionality at a reduced cost and that these materials should be processed using parts and components with a limited environmental footprint. Composite pipes are instrumental for continued operation of power plant cooling water systems, ensuring that consumers have uninterrupted access to power. For sea water desalination plants, the composites avoid line breakdown as typically steel would be corroded. Tanks and vessels manufactured in composites need little maintenance and do not affect the quality of the products contained and processed.

Composites are highly appreciated as a material solution for car body designs, where consumers are looking to express their personality through unique shapes.

8.2.2 Lower Eco-footprint

Composite materials are being considered as a viable and environmentally friendly alternative to traditional materials such as steel, concrete and aluminium. Apart from the functional benefits described above, solutions based on composite materials tend to have a significantly lower eco-footprint. Good examples include windows based on pultruded composite profiles and storage tanks manufactured through filament winding.

A very common expression of an LCA result is the carbon footprint. Although it is an easy and well-accepted way of comparing material solutions, the carbon footprint alone provides no insight into other environmental impacts of a product as climate change is only one of the impact categories that can be investigated. The eco-footprint is a more balanced perspective on a material's impact on the environment. The challenge is therefore to improve all parameters that impact the eco-footprint, including energy consumption, waste generation, emission of greenhouse gases and the smart choice of raw materials by using LCAs as a key decision-making tool.

Process-wise, DSM has in-house expertise and a solid track record in performing LCAs. DSM is actively involved in the development of international standards and guidelines for LCA. Consequently the LCAs that we do are performed in line with ISO 14040 series standards. We make use of material databases such as Eco-invent (V 2.2) and use the well-accepted SimaPro 7.3 calculation software with the following two methods:

- The IPCC GWP 100a method assesses the carbon footprint: emissions of greenhouse gases expressed as kilograms of CO_2 equivalent.
- The ReCiPe method assesses the eco-footprint, including all environmental impacts (including human health, natural resources and eco-system quality).

Using standard software, internationally accepted calculation methods and well-maintained material databases helps to make the right life-cycle assessments, leading to better insights and decisions.

8.2.2.1   Windows Made from Composite Profiles.   DSM has performed an
LCA on the use of composite profiles in windows, including the manufacture
of raw materials and the finalized window. The objective was to understand

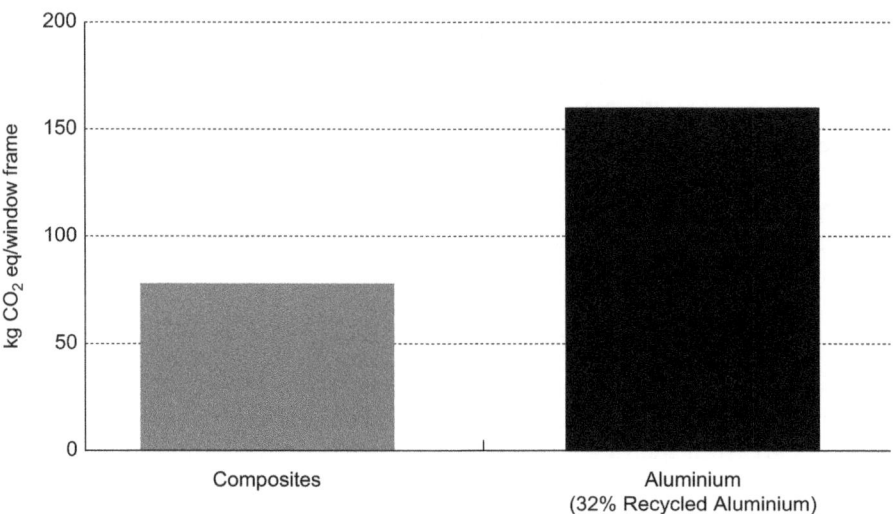

Figure 8.1   Reduced carbon footprint of windows made from composites.

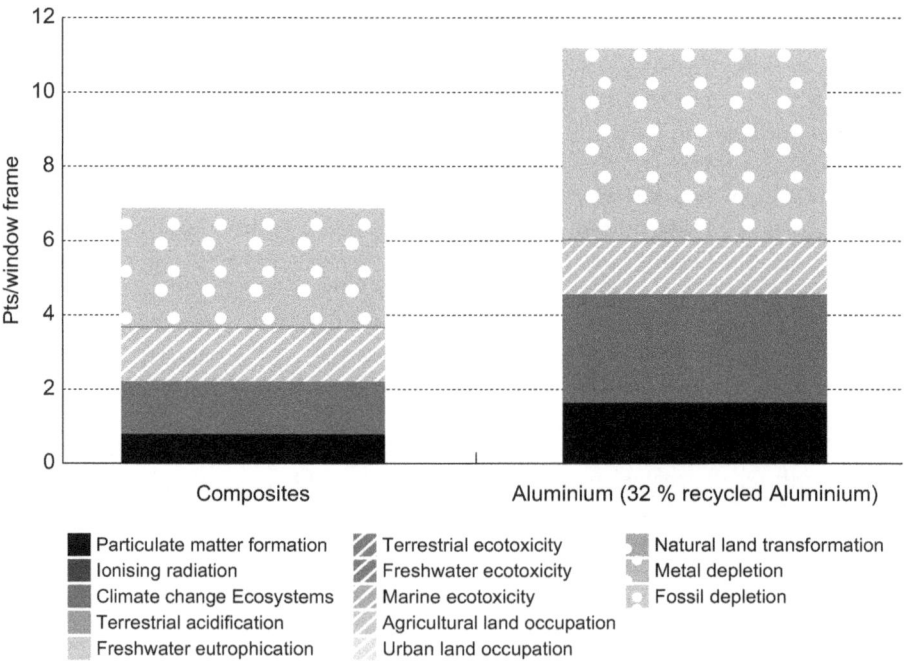

Figure 8.2   Eco-footprint comparison for window frames.

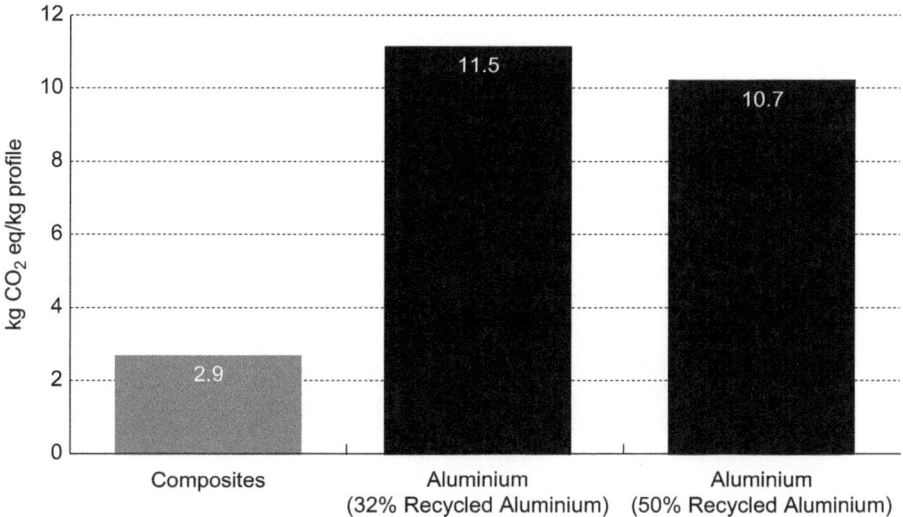

Figure 8.3   Reduced carbon footprint of composite profile *versus* aluminium profile.

the key environmental impacts compared with aluminium, another commonly used material solution for making windows.

It is clear from the analysis depicted in Figure 8.1 that the composite solution has a lower carbon footprint than the equivalent solution in aluminium.

This is consistent with the analysis shown in Figure 8.2, performed to evaluate the eco-footprint of window frames made of composite or aluminium. Again, the composite solution is advantaged.

8.2.2.2   Sensitivity Analysis.   One of the key assumptions of this LCA was the recycled content of the aluminium. Although today the consumption of aluminium significantly outweighs demand, in the future larger quantities of recycled aluminium may become available.

Interestingly, the picture does not change significantly with increasing recycled aluminium content, as shown in Figure 8.3.

8.2.2.3   External Certification of LCA Results.   In order to provide transparency in the LCA calculations and obtain an objective third party opinion on the quality of the assessment, the accredited company CE Delft has reviewed and certified the calculations for composite profiles depicted in Figure 8.4.

8.2.2.4   Composites Are Fully Recyclable.   It was demonstrated many years ago that composites are fully recyclable. Complete material recycling through co-processing in cement manufacture is a proven and feasible technology, with composite regrind being an interesting alternative fuel that can reduce the LCA footprint in cement manufacturing.[5,6]

Through co-processing, composites regrinds are replacing traditional fuels and raw materials (*e.g.*, coal, oil, limestone, clay), providing energy-efficient

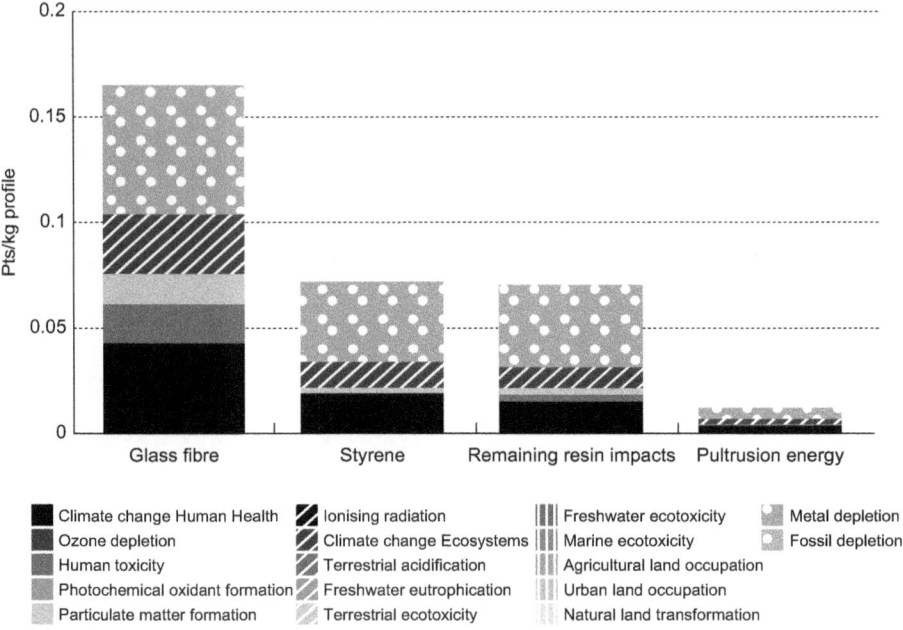

Figure 8.4   Eco-footprint breakdown – 1 kg GRP profile.

alternative fuels and raw materials for quality cement manufacturing. Composites are a valuable source of materials for new materials and are compliant with the Waste Framework Directive (WFD) and End-of-Life Vehicle Directive (ELV).

8.2.3   Elimination of Substances of Hazardous Concern

Composite systems typically consist of a resin matrix and a combination of fibrous reinforcement and filler. Materials used have to provide the right level of performance and functionality while being capable of proper processing. Most commonly used resin systems include unsaturated polyester resins (UPRs), vinyl ester resins (VEs) and epoxy resins.

The resins used for making composite components are complex chemical systems, containing numerous ingredients that play a role in achieving the right resin stability and shelf life, curing, processing and performance in the final application. Fine-tuning of such a chemical system is a complex process, which may require a considerable amount of time and resources. DSM has a lot of expertise in optimizing resin composition so that simultaneously attention is paid to 'people, planet and profit.'

In line with its commitment made through the industry-wide initiative 'Responsible Care,' DSM has been eliminating hazardous substances from its resin formulations for many years. Such a process can start when new insights on material toxicity become available. For instance, through the information published by the REACH (the European regulation for Registration, Evaluation, Authorization and Restriction of Chemicals that came into force on 1 June 2007) consortia or by the European Authorities [*e.g.*, ECHA (European CHemicals Agency)] or when there is a change in legislation on the use of certain substances.

In recent years, DSM has changed its philosophy with regard to the use of hazardous substances. Instead of reacting to changing legislation, DSM has been proactively introducing greener alternatives for substances that are expected to become reclassified or for substances that are less preferred by end customers and consumers because of their anticipated health and safety risk. There are several good examples of where DSM is investing in major development programs to lead the composites industry towards using more sustainable material solutions. These include the introduction of cobalt-free curing systems, the development of alternatives to styrene and the introduction of good manufacturing practice (GMP) for the manufacture of products used in contact with food and portable water.

8.2.4 BluCure™ Cobalt-free Curing Systems

Anticipating the increasing environmental pressure on cobalt, DSM has been working for many years on the development of cobalt-free resin curing systems. The company has successfully scaled up the technology and filed a broad range of patents covering many types of accelerator systems and materials.

In 2012, DSM and AkzoNobel announced together the introduction of the BluCure™ brand, in order to give composite component manufacturers an easy way to recognize that the products they use are 100% cobalt-free and their customers a way to recognize that their composite parts are produced from 100% cobalt-free products. The BluCure™ brand and seal stand for 100% cobalt-free and true sustainability. BluCure™ products are either cobalt-free accelerators or cobalt-free pre-accelerated resins and meet strict criteria:

- 100% cobalt-free
- easy-to-use in composite manufacturing processes
- not labeled as mutagenic, carcinogenic or toxic to human beings
- do not contain chemical components listed in the European Substances of Very High Concern (SVHC) List (*i.e.*, no CMR 1A, CMR 1B or PBT ingredients).

The abbreviations CMR and PBT refer to carcinogenic, mutagenic, reprotoxic and to persistent bio-accumulative toxic, respectively, effects either on human health and safety or on the environment. The annotations 1A and 1B indicate a high level of risk with regard to CMR performance.

Cobalt octoate is the main component used today in accelerators for curing unsaturated polyester and vinyl ester resins, either incorporated in the resin (so-called pre-accelerated resin) or added separately to a non-accelerated resin. Pre-accelerated resins only require an organic peroxide to cure. For that reason, they are easier to handle and have fewer parameters to adjust in the curing process. Using a non-accelerated resin with a separate accelerator will provide more flexibility to fine-tune curing. This may be required depending on the transformation process and desired end properties.

In Europe, it is discussed whether cobalt octoate should be considered as a CMR 1B substance or not, following the CMR2 reclassification of cobalt octoate by the Cobalt REACH Consortium in August 2012.[18] A CMR 1B classification might be possible in line with the current classification of cobalt(II) compounds.[7]

In case cobalt octoate is classified as CMR 1B, the exposure of workers, consumers and the environment with the substance involved will have to be minimized. Chemicals containing more than 0.1% of a CMR 1B have to be classified and labeled accordingly. For the specific cobalt salts this limit may even become a factor lower, *i.e.*, 0.01% (calculated for the cobalt metal content). For manufacturers of composite parts, this means that the process for obtaining operational permits is significantly more complicated if they continue to work with these substances.

The solution is to make a cobalt-free product, although this is not straightforward. Over the years, the resins and peroxides have been optimized to work well together with cobalt. DSM and AkzoNobel have strived to let the accelerators work with existing resins and peroxides, but no solution was found that works for all combinations. Hence multiple solutions were developed which can match different resin types, peroxides and transformation processes.

When DSM started to develop cobalt-free resins, clear objectives were defined at the beginning of the project.[8] These included that no CMR/SVHC listed ingredients could be used (*i.e.*, there must be a definite improvement in terms of sustainability), the resins must be cured with standard peroxides, resin handling should be similar to handling of conventional resins and performance of the cured components would not be compromised. Compounds based on copper, manganese and iron were selected as the best alternatives to cobalt, for reasons of performance, cost and toxicity. Adding the pure metal ion did not provide proper resin curing, but the addition of auxiliary components (including ligands) greatly enhanced system performance.[8]

As was done for the cobalt-containing formulations in the past, each formulation has to be optimized to give the best performance with the new cure systems. In terms of overall mechanical, thermal and chemical resistance properties, cobalt-free laminates based on BluCure™ technology have similar performance characteristics to their cobalt-containing equivalents.[8] For some accelerators and resins, it has been found that the new systems can even give a better cure than the cobalt-containing reference, resulting in a faster buildup of

Barcol hardness and reduction of residual styrene. In some applications, this may eliminate the need for post-cure.

BluCure™ is the first cobalt-free curing technology in the world for ambient temperature cure. Through sub-licensing it is immediately and easily accessible to all composite component and resin manufacturers around the world. BluCure™ technology offers opportunities for outstanding performance and sustainable end-user value.

8.2.5 Alternatives to Styrene Reactive Diluents

In unsaturated polyester and vinyl ester resin systems, the predominantly used reactive diluent is styrene. In June 2011, the US Department of Health listed styrene as a 'reasonably anticipated human carcinogen,' based on an assessment of publicly available toxicity data. Meanwhile, the US Environmental Protection Agency (EPA) has not given styrene a formal carcinogen classification or listed it as a carcinogen on the Integrated Risk Information System (IRIS) database, which is a widely used reference for a chemical's potential carcinogenicity.[9]

The US Occupational Safety and Health Administration (OSHA) regulates styrene on the basis of avoidance of narcosis in the workplace. The current OSHA permissible exposure limit (PEL) for styrene is 100 ppm, meaning that a worker should not be exposed to more than an average of 100 ppm of styrene in air during a regular workday without respiratory protection. However, under an agreement with OSHA and for added worker protection, Styrene Information & Research Center members voluntarily adhere to a styrene PEL of 50 ppm.[10]

In Europe, discussions on carcinogenicity concluded in 2007 that styrene is not classified for human carcinogenicity. In a voting of the Technical Committee for Classification and Labelling (TC C&L) under the European Chemical Bureau (ECB), there was no majority for a classification of styrene as CMR 1B Carcinogenic. Meanwhile, in December 2012, the European Committee for Risk Assessment (RAC) proposed to classify styrene as CMR2 (Reprotoxic). The final decision for harmonised classification and labelling will be taken by the European Commission.[19]

In 2010, the Styrene REACH consortium filed their dossier and proposed a reclassification of styrene that resulted in additional Hazard Statements. However, classification on CMR was not proposed. In addition, the Styrene REACH Consortium proposed a DNEL (Derived No Effect Level) for worker inhalation exposure of 20 ppm as an 8 h time-weighted average in their REACH dossier. In some markets it is not trivial to be REACH compliant and to achieve Safe Use conditions. As there are EU Member States that have an occupational exposure limit (OEL) of >20 ppm (such as the UK, 100 ppm), being REACH compliant will be even more difficult.

The composites industry has been minimizing styrene emissions by using resins with additives that reduce styrene evaporation, using low-styrene resins and ensuring air quality at the working spot (*e.g.*, with air ventilation systems). Resins with a reduced content of styrene can provide temporary relief and will help reduce styrene emissions to some extent. The discussion about the hazardous properties of styrene has been ongoing for decades and still has to reach a conclusion. So we need to be realistic about the pace of change. However, once a viable styrene-free technology becomes available, the likelihood that customers will want to implement such a sustainable solution will definitely increase. It is fair to say that the US NTP listing and the new proposed CMR2 reclassification are very important milestones towards creating an environment where more efforts will need to be undertaken to develop cleaner, more sustainable materials.

Styrene-free products have been on the radar for municipal authorities for a number of years due to the environmental downside of using styrene and epoxies in pipe relining. There has also been a lot of interest in applications such as in the marine, automotive, food and construction industries. Here, closed mould systems have been developed to protect workers' health and safety (*e.g.*, vacuum infusion techniques, hot cure SMC/BMC processes). Meanwhile, for traditional processing techniques such as hand lay up and spray up, we see that customers are looking for solutions to minimize and preferably eliminate styrene emissions. DSM is actively working on the development of such resin systems. Interest in these solutions has understandably increased markedly in the aftermath of the US decision to list styrene as a potential carcinogen.

DSM identified styrene as a potentially harmful substance years ago, which is why they began investing in styrene-free alternatives and is why, today, they have a range of styrene-free products on the market. These products have been attracting a lot of interest from customers who in the past were not aware that alternatives to styrene existed.

However, for many applications, styrene-free is not an option at present, so it is crucial that companies, while supporting existing industry initiatives such as safe use guidelines, also continue to develop closed mould applications to eliminate any risk to human health.

In the styrene-free resin systems currently offered by DSM, alternative reactive diluents are used, typically based on methacrylates. Apart from the lower (or in some cases close-to-zero) emissions, these can help in creating unique system performance while being safe to use and not suspect in terms of human toxicity. A key challenge in such a development is to maintain advantages associated with the robustness in processing offered by styrene-containing resin systems and to find a solution that can come close in terms of cost.

Styrene-free resins will have slightly different characteristics to styrenated resins, simply because the chemical composition is different. In some cases, the performance is better (*e.g.*, the chemical resistance of Atlac® Premium 600 resin to organic solvents compared with Atlac® 430 resin) and sometimes it is slightly worse.

Selecting the right methacrylates in combination with optimization of resin backbone chemistry has been shown to be the key for achieving the right system performance. It is clear that material systems used to repair industrial storage tanks will require a different technical solution to a system to repair drinking water pipelines. Also, in some markets and applications, avoiding liabilities and reducing product risk can outweigh the higher initial cost of styrene-free resin systems.

Depending on the specific end-use requirements and processing needs, DSM is proposing alternative resins that meet performance requirements at the lowest possible cost. One important aspect here is that most glass-fiber types have been optimized for use with styrenated resins. The new styrene-free resins may in some cases require adjustment of fiber sizing and binder.

The discussion about whether styrene is a carcinogen has been ongoing for decades and still has to reach a conclusion, so we need to be realistic about the pace of change. However, once a technical solution becomes available, the likelihood that regulators take action will increase.

8.2.6 Introduction of Good Manufacturing Practice (GMP)

Pipes, vessels and storage tanks operating in direct contact with foodstuffs or drinking water require compliance with specific European Food and Drinking Water industrial standards and regulations. This assures customers, operators and consumers that the raw materials used in the resin composition have been assessed by the European Food Safety Authority and the European Council for their effects on human health and have been approved for use in food contact articles.

DSM takes its responsibility to introduce more environmentally sustainable products and solutions very seriously. While enabling food processing and drinking water operations to run more smoothly through the durability and resistance to chemicals of vinyl ester and unsaturated polyester resins, DSM supports the industry in creating products that are safer to use and consume.

Identifying suitable food contact raw materials as defined by European legislation was the first step in this project. In line with DSM's policy on eliminating (where possible) harmful raw materials from the workplace, the available literature was scrutinized for developments which may affect the toxicological status and the future suitability of raw materials for food contact and water applications.

Many syntheses were carried out using these raw materials to optimize the resin properties, achieving the best physical and chemical properties. In parallel, the appropriate procedures and technologies were developed for ensuring that the production of the resin was in line with GMP common in the food processing industries. Since 2011, DSM has introduced many resins that are made according to GMP and European Framework Regulation 1935/2004 and that are compliant with the 'positive list' (list of allowed materials) of the European Food Contact Directive 2002/72/EC and European Commission Regulation 1895/2005.

GMP is the internationally accepted quality control system used in the nutritional, food and pharmaceutical industries that sets standards and guidelines for the safe production, testing and use of a product. GMP ensures that materials and parts are made consistently at a high quality level and that all the right precautions are taken for avoiding cross-contamination.

In line with European Framework Regulation 1935/2004 or drinking water contact regulations (which tend to differ per country), the components are to be evaluated in a leaching test (dependent on the specific country and the end-use application). It is the responsibility of the component manufacturer to obtain the applicable test certificates.

As per the recommendations of Plastics Europe, the production of food contact 'materials' (*i.e.*, both base materials and finished components) shall be done according to GMP. In particular, in a factory that produces both food contact materials and non-food contact materials, there is a risk that cross-contamination can harm the quality of the food contact materials.

Companies supplying components for contact with foodstuffs and potable water (for instance tanks, pipes and joints) also have to prove that their products are made according to GMP regulation 2023/2006 (Good Manufacturing Practice) or equivalent. In particular, if in a factory both food contact materials and non-food contact materials are produced there is a risk that cross contamination can harm the quality of the food contact materials. DSM can help customers by means of auditing their plants and propose process improvements (manufacturing procedures, dedicated equipment, cleaning and traceability).

Certifying that the ingredients of GMP resins are compliant with the 'positive list' simplifies the approval process for DSM's direct customers. Thanks to this resin, obtaining the approval for the final components which are manufactured from this resin is easier, quicker and can be achieved with reduced testing efforts.

With the growing global population and the desire for improved health, the market for food and drinking water will increase significantly over the next decades. More than 20% of the world will be affected by water scarcity in the next 20 years. The supply of clean water and management of waste water will become more and more crucial. Furthermore, as health, safety and environmental regulatory and legislative measures are likely to become more stringent, the demand for composite-based storage facilities such as tanks and pipes is set to grow.

Both the European Council and the US government [through the US Food and Drug Administration (FDA)] has put in place regulations to ensure consumer health and safety. The FDA requires manufacturers of components and materials to state that materials in contact with foodstuffs are based on ingredients from a positive list, but does not actually verify the composition.

In Europe, this compliance is verified and certified by independent bodies. Also, components are subjected to specific leaching tests, ensuring that any leached out substances are well below generally accepted levels.

For composite components in contact with foodstuffs, there is a clear legislative requirement to have GMP in place. For components in contact with drinking water, this is less defined: many local regulations in individual EU countries require testing and certification of components. By using resins that have been made in line with GMP, manufacturers of components for drinking water contact will obtain faster certification of their parts and can better control their production process, as the resin system simply does not contain hazardous components that may leach out.

8.3 Facing the Technical, Performance and Cost Challenges: the Introduction of Halogen-free Alternatives

DSM was one of the first chemical companies to offer a range of halogen-free products that can be widely used in electronics or other electrical applications. DSM overcame technical, performance and cost challenges to produce its new bromine- and chlorine-free high-temperature plastics. These new products can be used not only as PVC replacements for electronic wires and cables but also for many other electronic applications such as in connectors and in electric motor components.

DSM's long history of and commitment to product stewardship has guided its journey in developing new bromine- and chlorine-free materials for electronic products. DSM's Bright Science Brighter Living approach to sustainable product and process design for engineering plastics includes four key tenets:

- reducing the use of substances of hazardous concern
- improving eco-efficiency throughout the value chain
- promoting recycling with the ultimate goal to create closed loops
- developing bio-based polymers for critical technical parts.

By keeping abreast of market trends, DSM became one of the first companies to recognize the value of developing solutions to replace bromine and chlorine in electronic connectors and cables. Over the last 2 decades, the growing demand for bromine- and chlorine-free products justified the investment required to develop a range of new halogen-free products, including polyamides (6, 46, 4T and 4.10) and polyesters poly(butylene terephthalate) (PBT) and poly(ethylene terephthalate) (PET) and thermoplastic copolyester elastomer (TPC)].

Until recently, the (electronics) industry generally considered brominated flame retardants (BFRs) and PVC plastic to have an ideal performance/safety balance. BFRs are very effective in reducing flame retardancy and at the same time keep the mechanical properties of the engineering plastics at an acceptable level. However, the environmental and toxicological performance of this class of flame retardants and the polybrominated biphenyls (PBBs) and poly-brominated diphenyl ethers (PBDEs) particularly has led to more and more negative attention. It even resulted in a specific RoHS directive banning penta-BDE and octa-BDE in 2004. In addition, publications about dioxin and furan formation as a result of the inappropriate incineration of municipal waste containing halogenated waste or during uncontrolled and informal recycling processes of end-of-life electronics equipment have led to a growing concern that these materials can have risks to human health and to the environment.

Additionally, it can be stated that BFRs are generally used in combination with antimony trioxide (Sb_2O_3), acting as a synergist for maximum flame retardant activity. Although various studies are still in progress, it can be stated that from the toxicity point of view Sb_2O_3 is also a high-priority candidate for replacement.

Furthermore, some serious fires in private buildings (*e.g.*, Düsseldorf airport on 12 April 1996) or in electronic equipment installations have shown that the burning of PVC or of plastics with BFRs leads to severe smoke formation and the evolution of toxic and corrosive gases. DSM recognized these concerns in the 1990s and actively developed alternative products since then, although the real 'boost' occurred in the years after the turn of the century when several original equipment manufacturers (OEMs) brought it to the company's attention.

By working together with partners throughout its entire value chain, including OEMs and suppliers, DSM developed and now produces new bromine- and chlorine-free engineering plastics that meet high technical and environmental performance standards. Most of the alternative solutions nowadays contain phosphorus- or nitrogen-based chemicals, and it appears to be a real challenge to combine high-temperature resistance (for use in engineering plastics in the electronics sector) with flame-retardant efficiency, while maintaining other physical and mechanical properties of the end product at an acceptable level. These solutions enforce the competitive advantage for the emerging market demand for BFR- and PVC-free products in the electronics sector. DSM was among the first chemical companies to offer a complete portfolio of engineering plastics that are free of these substances.[11] An extensive overview of flame retardant developments and their markets can be found in an SRI consulting report.[12]

Two key bromine- and chlorine-free DSM products with desirable qualities for electronic connectors and cables are:

- *Arnitel XG*[13] – A high-performing thermoplastic copolyester that contains no BFRs, PVC, halogens or plasticizers. The product has been successfully commercialized for PVC replacement for use in electronic wires and cables.
- *Stanyl ForTii*[14] – A fully halogen-free polyamide material that can be used for various types of electronic connectors. Stanyl ForTii has the optimal balance of (physical) properties desired in high-temperature polyamides: high strength and stiffness, high melting temperature and high glass transition temperature. The material retains its good mechanical and thermal performance in all stages of its life-cycle, ranging from the production and end-use phase, up to and including recycling processes at OEMs.

Retooling or specialized equipment is not required to use these new plastics to produce connectors and cable products. This significantly reduces the costs for electronics manufacturers using these products. DSM is producing its halogen-free plastic resins in high volume to meet the increasing demand projected as more customers move away from the use of BFRs, PVC and other halogen-containing materials.

8.3.1 Overcoming Technical Challenges

When DSM began its quest to develop halogen-free versions of the high-temperature plastics used in (electronics) connectors and cable insulation, the viability of such formulations was in question owing to reliability issues, such as brittleness, blooming and corrosion. The company formed a large multi-disciplinary team to conduct its own in-house research and development effort to find better solutions. The company's material scientists and engineering teams credit some of their success in solving many of the reliability issues to the working relationships that they established with other manufacturers in the large and diverse Electrical and Electronics supply chain who were also grappling with some of the same challenges in their efforts to remove bromine and chlorine from their products.[15]

These efforts included large OEM clients who were attempting to convert complete product lines, and also 'Tier 1' connector and cable manufacturers who needed viable engineering plastics. These companies collaborated to set up a feedback system whereby customers could report on the performance characteristics of new compounds. The information gleaned through this system allowed DSM's engineering teams quickly to address problems and incorporate changes into new versions of their products. The company also worked closely with suppliers to identify environmentally preferable flame retardants. DSM's engineering teams conducted both internal and external safety, health and environment studies to ensure that the new compounds met high environmental standards.

In addition to overcoming the technical, performance and cost challenges that previously inhibited commercialization of new bromine- and chlorine-free high-temperature plastics, DSM also helped facilitate the development of new flame retardancy standards. For the past decade, electronics suppliers and manufacturers only used plastic materials that conformed to the Underwriters Laboratories UL94-V0 flammability standard. This blanket approach to fire safety did not provide an incentive for innovative designs. In some cases, it even encouraged the use of flame retardants in applications where the risk of fire was low.

DSM developed green design strategies based on a new fire safety standard (IEC 62368) being proposed by the International Electrotechnical Commission (IEC). The new standard would allow designers to address fire-safety by either preventing ignition (distancing the placement of flammable materials and heat sources) or controlling the spread of fire (using flame retardants and or fire barriers).

8.3.2 Moving Forward

Bio-based plastic polymers , like the 70% bio-based polyamide 4.10 and the up to 50% biobased TPC-elastomer, that avoid or reduce the use of fossil fuels have been developed. They can also improve the recyclability and eco-efficiency of its engineering plastics.

To further reduce the use of substances of hazardous concern, DSM has engaged with various global initiatives. To innovate and improve further their product development process on Cradle to Cradle®, DSM works together with MBDC and EPEA. Moreover, DSM has filed various cases with ChemSec's SubsPort.[16] The use and support of development tools such as the Green-Screen™[17] by Clean Production Action also strengthen the proactive ingredient policy development.

8.4 Conclusion

The focus on new alternatives is often under discussion in the scientific community but not yet part of any legal requirement. Still, tangibly better resins and plastics can be designed with high functional performance, lower ecological impact and less impact on human health. Life-cycle assessment from cradle to grave has been used to benchmark the ecological benefits of the alternatives against other current solutions. For a company such as DSM, expanding their focus on proactive ingredient policy across the entire organization will help effectively provide truly healthful products throughout their life-cycle. Taken to a more holistic level, the effort in further developing a proactive ingredient policy is also an important milestone towards a circular economy, since it limits the entry of substances of concern into the biosphere.

References

1. World Business Council for Sustainable Development, *Vision 2050, the New Agenda for Business,* WBCSD, Geneva, 2010.
2. World Commission on Environment and Development (1987). *Our Common Future,* Oxford University Press, Oxford, 1987.
3. Global Reporting Initiative, *G3.1 Guidelines – an update and completion of G3,* GRI, Amsterdam, 2011.
4. T. Searchinger, R. Heimlich, R. A. Houghton, F. Dong, A. Elobeid, J. Fabiosa, S. Tokgoz, D. Hayes and T.-H. Yu, *Science,* 2008, **319**, 1238–1240.
5. J.-P. Degré and E. Schmidl, Co-processing of fiber reinforced plastics, presented European Alliance for SMC/BMC Sustainability Day 'Light Composite Materials for a Sustainable Future', 26–27 January, 2012, Brussels.
6. J. Lempke, Composites are recyclable, presented at European Alliance for SMC/BMC Sustainability Day 'Light Composite Materials for a Sustainable Future,' 26–27 January 2012, Brussels.
7. Third Annex XIV Recommendation of European Chemical Agency (ECHA), 20 December 2011. Details can be found on http://echa.europa.eu/candidate-list-table.

8. Th. Wegman and W. Posthumus, Cobalt free! And you?, Technical Presentation at JEC 2012, Paris, March 28, 2013.

9. Styrene Information and Research Center, *U.S. EPA Regulation of Styrene*; http://www.styrene.org/regulatory/epa_regulation.html (last accessed 10 January 2013).

10. Styrene Information and Research Center, *OSHA Regulation of Styrene*; http://www.styrene.org/regulatory/osha_regulation.html (last accessed 10 January 2013).

11. T. P. Sidiki and R. D. Hilty, Design for low-halogen green electronics, green design in electronics, presented at the IPC APEC Expo 2009, 29 March–2 April 2009, Las Vegas, NV.

12. H. Janshekar, H. Chinn, W. Yang and Y. Ishikawa, Flame Retardants, SRI Consulting, SRI International, Menlo Park, CA, 2011.

13. DSM, *Welcome to Arnitel®*; www.arnitel.com.

14. DSM, *Welcome to Stanyl® ForTii™*; www.fortii.com.

15. M. Degenhardt, *Kunstst. Int.*, 2012, **4**, 61–64.

16. DSM, *SUBSPORT: Moving Towards Safer Alternatives*; www.subsport.eu/case-stories?search = DSM&sektor = 0&Function = 0&prozess = 0&produkt = 0&type = case_studies.

17. Clean Production Action, *GreenScreen™ for Safer Chemicals*; www.cleanproduction.org/Greenscreen.php.

18. Environmental classification Cobalt bis (2-Hexyldecanoate) (205-250-6), received through Cobalt REACH Consortium, http://www.cobaltreachconsortium.org/, October 2012.

19. RAC adopts seventeen scientific opinions, European Chemicals Agency web site, http://echa.europa.eu/en/web/guest/view-article/-/journal_content/c89bdb13-09e9-497c-8e73-ddae13a842c8, December 7, 2012.

US Environmental Protection Agency's Design for the Environment (DfE) Alternatives Assessment Program

CLIVE DAVIES,* MELANIE ADAMS, EMILY CONNOR,
ELIZABETH SOMMER, CAROLINE BAIER-ANDERSON,
EMMA LAVOIE, LAURA ROMANO AND DAVID DIFIORE

ABSTRACT

Design for the Environment (DfE) Chemical Alternatives Assessments (CAA) are an approach to chemical substitution used to evaluate chemicals targeted by the US Environmental Protection Agency (EPA) for action. CAAs have helped stakeholders consider the health and environmental profiles of chemicals along with the more traditional factors of cost and performance as they choose alternatives. Recent improvements to the CAA methodology and criteria for differentiating chemicals have enhanced transparency and made the methodology implementable by parties outside the EPA. DfE is working to harmonize methodologies for CAA in the USA and plans to begin working with other countries on harmonization. Harmonization would have a number of benefits, introducing efficiencies by reducing the need for redundant assessments, and making it possible to have a central data repository where those conducting CAAs and those choosing safer chemicals can go for information.

*Corresponding author

Issues in Environmental Science and Technology, 36
Chemical Alternatives Assessments
Edited by R.E. Hester and R.M. Harrison
© The Royal Society of Chemistry 2013
Published by the Royal Society of Chemistry, www.rsc.org

9.1 Introduction

The US Environmental Protection Agency (EPA) Design for the Environment (DfE) program has developed a methodology and hazard criteria for conducting Chemical Alternatives Assessments (CAAs) for chemicals that the EPA has targeted for action. DfE CAAs are intended to encourage substitution to safer alternatives. In some cases, a demonstration that safer alternatives are available may also complement regulatory activities. DfE carries out its CAAs in partnership with a broad range of stakeholders who can use information on safer alternatives, in combination with data on cost, performance and other factors, to choose the safest possible, high-functioning alternatives.

In addition to CAAs, the DfE program conducts life-cycle assessments (LCAs) and encourages the use of safer chemicals in consumer and industrial products, such as cleaners, by allowing the use of its logo on those that meet the program's stringent criteria. Through these programs, DfE furthers the EPA goal of reducing risk to people and the environment by promoting the use of green chemistry, safer chemical alternatives, and materials that pose fewer life-cycle impacts. DfE focuses on industries that combine the potential for chemical risk reduction with a strong motivation to make lasting, positive changes. DfE approaches rely heavily on the special expertise and tools of the EPA Office of Pollution Prevention and Toxics (OPPT), developed for implementation of the Toxic Substances Control Act (TSCA). DfE also relies on the technical expertise, experience and perspectives of its stakeholders to ensure that partnerships provide an outcome useful to decision-makers.

In DfE partnerships, robust representation from relevant stakeholders ensures transparency, full consideration of environmental issues and an understanding of performance and cost that help in finding lasting solutions. DfE partners are drawn from the full breadth of product supply chains. Raw material suppliers, chemical and product manufacturing industry representatives and recyclers are included. Environmental, consumer, and worker advocates, in addition to other US government agencies, local government and foreign governments are also represented.

Richard Denison, Senior Scientist, at the Environmental Defense Fund (EDF), believes that much can be accomplished when transparency is a guiding partnership principle:

> *From a public interest perspective, one of the key benefits of DfE is its emphasis on transparency, both in the development of criteria and in program implementation. Transparency is so important because it helps level the playing field, allowing groups of various sizes and resources – from the supply chain and the NGO [non-governmental organizations] community – to understand and participate as effective stakeholders in the program.*[1]

As public awareness of the potential effects of exposure to some chemicals has grown, DfE approaches have become more valuable to the program's stakeholders.[2] Companies such as Walmart, Home Depot and Staples are implementing programs to encourage the use of safer chemicals in products, and leading product and chemical manufacturers are designing more sustainable products to meet retailer demands and enhance sustainability programs in their own companies.[3–5]

DfE takes a broad perspective when considering risk management challenges. The program focuses on reducing risk by encouraging use of safer chemicals; this is especially appropriate for the relatively small number of chemicals that the EPA has targeted for action. Even with this focus, DfE recognizes the usefulness of other approaches. Conducting a CAA may be the appropriate role for EPA and DfE, and some industries may be able to use information from DfE's report as the basis for decision-making. Especially in the case that clearly safer alternatives are not available, some industries may need to supplement the DfE information with risk assessment or other tools to ensure they are choosing the right alternative for their uses.

Application of life-cycle thinking can be useful in tailoring robust risk management approaches and understanding the context for chemical substitution. The DfE program and its collaborators have recognized that broadening substitution analysis to include other relevant life-cycle impacts can improve the understanding of real-world trade-offs and reduce the likelihood of unintended consequences in chemical substitution.

9.2   Selecting an Approach for Chemical Substitution

The DfE program considers hazard assessment an important tool in chemical substitution. Hazard assessment must be a part, and usually the first step, of the process. However, DfE also recognizes that hazard assessment may not always be the primary consideration in selecting alternatives and that the science or innovations for providing safer alternatives may not yet be available. Figure 9.1 illustrates one method for choosing risk mitigation approaches.

Figure 9.1 gives a general sense of the logic that DfE applies as it chooses risk management approaches. It shows that where a chemical poses an acute hazard, is likely to impact people or the environment and safer alternatives are not available, best practices for use of the chemical may be the most effective risk management strategy.[6] Auto refinishing using diisocyanate-based chemistry is a good candidate for best practices.[7] Figure 9.1 also shows that if exposure to potentially hazardous chemicals is not a critical focus and life-cycle factors such as raw material extraction or energy use during product manufacture are of primary concern, then tools such as a LCA may be best for guiding environmental improvement.[8] LCA is an excellent tool for identifying the most significant impacts of a product manufacturing process and can point to areas ripe for improvement and innovation.[9]

In cases of chemical substitution, and in particular in cases of substitution away from chemicals that the EPA has targeted for action, the DfE alternatives

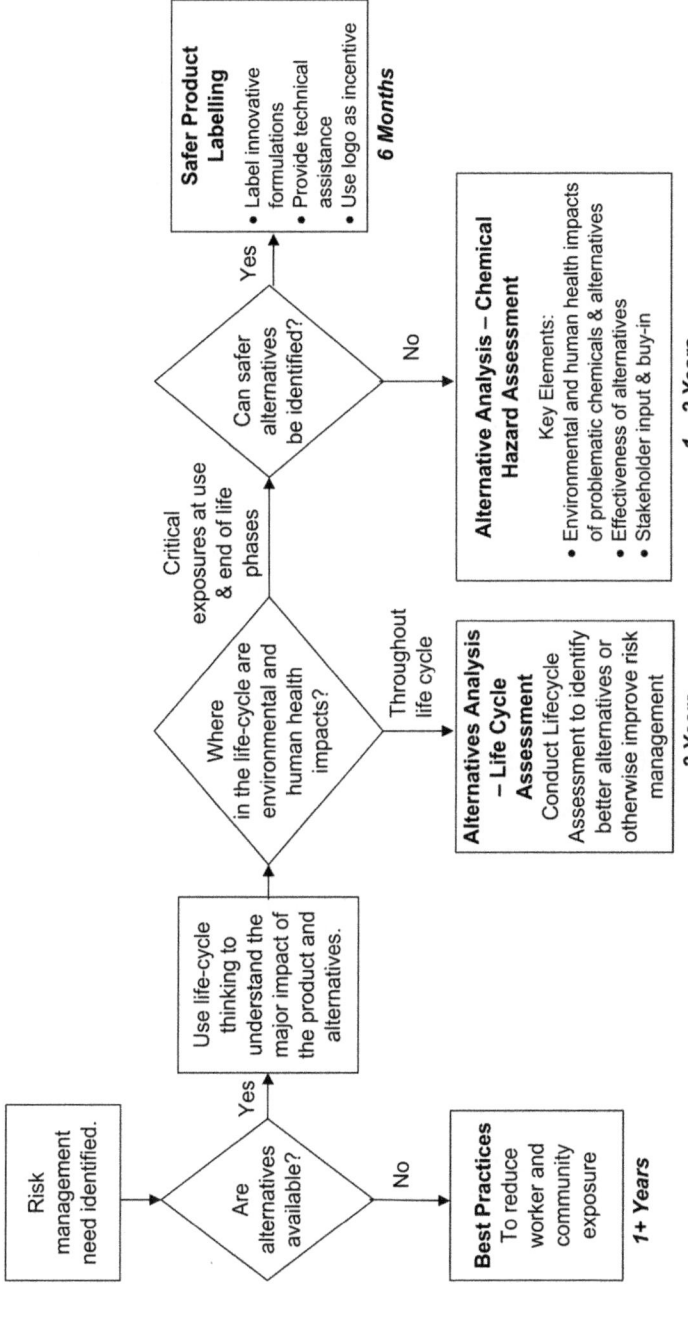

Figure 9.1 Decision logic for DfE approaches.

assessment methodology may be a useful tool. Once alternatives assessment is chosen as the methodology, life-cycle thinking is needed again for focusing the alternatives assessment on hazards at critical exposure points. The CAAs discussed starting in Section 9.3 illustrate this point.

Figure 9.1 also shows that providing information through product labeling can encourage the use of safer chemicals.[10,11] This is particularly the case where a range of functional safer alternatives are available to be substituted for a chemical of potential concern. A good example is the broad range of highly functional alternatives available as substitutes for the problematic nonylphenol ethoxylate (NPE) class of surfactants. By providing information on product content and the availability of products containing safer surfactants, DfE product manufacturing and retailing partners were able to reduce greatly, if not virtually eliminate, the use of NPEs in US consumer products.[11] Section 9.6 provides more information on this approach.

Hazard evaluation in the context of a DfE CAA has proven an important complement to EPA regulatory scoping for several chemicals chosen for action. The following case studies give descriptions of why a DfE alternatives assessment was the appropriate analytical tool, how the DfE methodology has evolved based on our experience in conducting assessments, and the value that assessments offer in uncovering the range of available alternatives. The examples also explain the comprehensive hazard profiles that are the heart of the assessments and how they can guide beneficial substitution, be built upon for multi-component comparisons or point to the need for further innovation in safer chemistry.

9.3 PentaBDE Case Study

In the early 2000s, flame-retardant chemicals, such as pentabromodiphenyl ether (pentaBDE). were found widely in the environment and in human serum and breast milk, and scientists observed that these chemicals appeared to behave similarly to the polychlorinated biphenyl family in terms of their fate in the environment and presence in humans.[12–14] Stakeholder concern was high, especially in the USA, where pentaBDE breast milk levels were found to be higher than in other countries.[14] The presence of pentaBDE in people appeared to correspond to the use of pentaBDE as a flame retardant in polyurethane furniture foam. PentaBDE production levels were ~19 million lb (8.6 million kg) per year.[15]

In response to stakeholder concerns among non-government organizations (NGOs), but also in the furniture industry and throughout the furniture manufacturing value chain, the EPA convened DfE's Furniture Flame Retardancy Partnership.[12–14] Stakeholders in the partnership agreed that fire safety was a critical need, but also agreed that it should be achieved in a way that minimizes risk to human health and the environment.

Shortly after the partnership was formed, the sole manufacturer of pentaBDE in the USA announced plans to phase out production voluntarily at the end of 2004. The EPA began work to backstop the voluntary phase-out with regulatory action that would not allow renewed production without prior notification – an action that essentially bans future manufacture in the USA. These events added real-time urgency to DfE's investigation and identification of available flame-retardant alternatives that could protect lives and property. Furniture foam manufacturers selected alternatives in the next few months, and the window of opportunity for providing data to inform substitution decisions – and minimize the likelihood of unintended consequences –lasted only a few months.

9.3.1 How DfE Conducted CAA for PentaBDE

9.3.1.1 Partnership. DfE convened a diverse group of stakeholders, including members of the furniture industry, chemical manufacturers, environmental groups, fire safety advocates and government representatives. Drop-in chemical alternatives for pentaBDE were the focus of the group's activities because of the need for cost-effective fire safety. The group considered alternative, and generally more expensive, ways to achieve fire safety, such as barrier fabrics and inherently flame-resistant materials, but this was a secondary focus.

All participants desired a rapid transition to alternatives, but concern was high, notably in the furniture industry, that alternatives could share the concerns for toxicity, persistence and bio-accumulation associated with pentaBDE. The EPA, in partnership with stakeholders, developed a method for conducting a hazard-based CAA so that information on the potential concerns associated with the alternatives would be available and easy to understand.

Under the CAA methodology, partners provided information on potential alternative flame-retardant formulations, and demonstrated feasibility and cost-effectiveness while the EPA reviewed and predicted health and environmental hazard information. This approach led to evaluation of the health and environmental profiles of 14 commercially available alternative flame-retardant formulations. The project focused on the potential for exposure at the most significant stages of the life-cycle, including the use and disposal phases.

9.3.1.2 Methods of Chemical Review. The CAA for furniture foam was intended to provide the best information available in the literature or that could be modeled. The methodology for evaluating alternative flame retardants included a screening-level assessment of the chemicals that focused on potential hazards and exposure routes, in addition to the potential for the chemicals to bioaccumulate and persist in the environment. The assessments considered information on the alternatives from five sources and used a weight-of-evidence approach for assigning hazard values. These sources were publicly available measured (experimental) data; measured data contained in

EPA-held confidential business information; structure–activity relationship (SAR)-based estimations from the EPA's Pollution Prevention Framework and Sustainable Futures predictive methods; professional judgment of EPA staff based on experimental data for chemical analogs; and unpublished experimental studies supplied by the chemical manufacturers.

9.3.1.3   Presentation of Data.   The final partnership report summarized the toxicological information for alternatives to pentaBDE. Volume 1 contained the purpose and scope of the assessment, a description of the general characteristics of flame retardants, a general overview of exposure pathways and routes for flame retardants used in flexible polyurethane foam, and the results of the assessments of 14 formulations of flame-retardant products most likely to replace pentaBDE. Volume 2 contained the detailed chemical hazard reviews.

After reviewing early report drafts, stakeholders felt that the information presented in these two volumes was not adequately transparent to readers who did not have chemistry and toxicology in their backgrounds. They explained that understanding how the effects seen at certain dose levels corresponded to concern for hazard for a given endpoint was challenging. Using this information to draw comparisons between chemical alternatives was even more difficult. With the stakeholders, DfE developed a three-level scheme for data presentation that better met the needs of all of the audiences that would use the report, maximize transparency and aid in decision-making. A summary of the three levels is given below.

1. A top-level table with a qualitative summary of the hazard assessment for each chemical, presented in the context of the flame-retardant formulation that would be used in furniture foam. The table includes caveats and footnotes, with additional relevant information, such as ongoing testing. Readers used this information to compare and choose alternatives. Table 9.1 is the first page of the DfE report's Table 4-1 and shows the potential hazard concern (High, Moderate, Low) and its basis (measured data or SAR analysis) for each key human health and environmental endpoint, and indicates where there is a potential for exposure based on physical-chemical properties.
2. Quantitative summaries of the toxicity and exposure data. The sources for this data are in Section 9.4.5 – Step 5.
3. Detailed hazard data reviews with a summary of the availability and adequacy of data and a full data review by endpoint, including references.

The report confirmed that available alternatives were safer than pentaBDE, and enabled decision-makers to consider health and environmental impacts in choosing alternatives. The report indicated that the alternatives had a lower level of concern for persistence, bio-accumulation and toxicity compared with

Table 9.1 Screening level toxicology and exposure summary.

L = Low hazard concern   N = No
M¹ = Moderate hazard concern   Y = Yes
H = High hazard concern   P = Yes for pure chemical
*Ongoing studies may result in a change in this endpoint
▲Persistent degradation products expected
L, *M¹*, or *H* = Italicized letter indicates that endpoint was assigned using estimated values and professional judgment (Structure Activity Relationships)

Company	Chemical	% In Formulation²	Cancer Hazard	Skin Sensitiser	Reproductive	Developmental	Neurological	Systemic	Genotoxicity	Acute	Chronic	Persistence	Bioaccumulation	Worker Inhalation	Worker Dermal	Worker Ingestion	Gen. Pop. Inhalation	Gen. Pop. Dermal	Gen. Pop. Ingestion	Gen. Pop. Aquatic	Reactive or Additive?
Albemarle	SAYTEX RX-8500																				
	Proprietary D Reactive brominated flame retardant		*L*	M	*L*	*L*	M	M	*L*	M	M	*L*▲	*L*	N	Y	Y	N	N	Y	Y	Reactive
	Proprietary B Aryl phosphate		*L*	*L*	M*	M*	M	M*	*L*	H	H	*L*	*M*	N	Y	Y	N	Y	N	N	Additive
	Triphenyl Phosphate CAS # 115-86-6		*L*	*L*	*L*	*L*	*L*	M	*L*	H	H	*L*	*L*	Y	Y	Y	Y	Y	Y	Y	Additive
Ameribrom	FR513																				
	Tribromoneopentyl Alcohol CAS # 36483-57-5		*M*	L	*M*	*M*	*M*	M	M	M	M	*L*	*L*	Y	Y	Y	N	N	Y	Y	Reactive
Supresta³	AC073																				
	Triphenyl Phosphate CAS # 115-86-6	38-48%	*L*	*L*	*L*	*L*	*L*	M	*L*	H	H	*L*	*L*	Y	Y	Y	Y	Y	Y	Y	Additive
	Proprietary J Aryl phosphate	40-46%	*L*	*L*	*L*	*L*	*L*	M	M*	*L*	H	*L*	*L*	Y	Y	Y	Y	Y	Y	Y	Additive
	Proprietary K Aryl phosphate	12-18%	*L*	*L*	*L*	*L*	*L*	*M*	*L*	*L*	*L*	*L*	*L*	P	Y	Y	N	Y	N	N	Additive
	Proprietary L Aryl phosphate	1-3%	*L*	*L*	*L*	*L*	*L*	*M*	*L*	*L*	*L*	*L*	*L*	P	Y	Y	N	Y	N	N	Additive

¹ The moderate designation captures a broad range of concerns for hazard.
² Chemical concentrations are listed in descending order; only chemicals with concentrations greater than one percent in the formulation were evaluated.
³ Supresta was acquired by Israel Chemicals Ltd. (ICL) in August 2007.

pentaBDE; however, each alternative was of concern for at least one of the key hazard endpoints, and exposure to the alternatives, *via* at least some routes, was possible.

9.3.1.4 Outcomes. The assessment provided a framework for choosing and designing safer alternatives, by highlighting structural alerts and distinguishing characteristics across an inclusive list of human health and environmental hazard endpoints. The report did not name a 'favored' alternative to pentaBDE or identify better or worse alternatives. Instead, the report provided data that empowered the furniture industry value chain, informed through conversation with relevant stakeholders, to weigh human health and environmental impacts with more traditional decision factors as they chose alternatives.

The furniture industry supply chain made the move to alternative flame-retardant formulations with the benefit of the best information on potential health and environmental effects that could be obtained. Without the EPA's involvement in making information on the health and environmental effects available and understandable, the choice of alternatives might have been based on cost and effectiveness with only minimal consideration of the potential for unintended consequences.

In addition to being used as a decision-making tool by the foam industry and others in the period leading up to the voluntary phase-out of pentaBDE, the assessment was also useful in accomplishing a related EPA regulatory action. By showing that safer alternatives were available, the DfE CAA eased the development and finalization of a TSCA Significant New Use Rule (SNUR) that required notification of the EPA before any future US manufacture of pentaBDE more straightforward.

9.3.2 Limitations of Original Methodology and Criteria

The methodology for conducting the alternatives assessment and the criteria for differentiating concern for hazard that DfE used to conduct the CAA for pentaBDE were important advances in alternatives assessment. The partnership made use of the expertise and tools that the EPA had developed through almost 30 years of implementing TSCA. It also involved a broad range of stakeholders that represented the supply chain from chemical, foam and furniture manufacture through product disposal and also included the NGO community. DfE and the stakeholders collectively developed a format for presenting the project's most important data that communicated the outcome, in one table, to the full range of involved stakeholders.

Even with these advances, the DfE program saw substantial room for improvement. US states, leaders in industry and NGOs indicated a desire to use the DfE approach, but found it difficult to reproduce the tools, expertise and approaches used to assess chemicals under TSCA. The methodology used by DfE for CAAs needed to be documented and the criteria for hazard evaluation also needed to be made transparent.

The risk assessment tools developed to implement TSCA, although appropriate for regulation, have limitations for use in CAA. TSCA assessment methodologies are intended for case-by-case application and are internal to EPA. DfE CAAs are a tool intended to be applied in a range of chemical substitution circumstances. This work is done in a multi-stakeholder environment, where transparency and reproducibility of results are critical. The developing community of practice around CAA required criteria that could be implemented outside the EPA. Existing tools did not meet these needs.

Stakeholder feedback confirmed that better documented criteria could provide the transparency and reproducibility that would make CAA a more effective tool for the EPA and make it possible for other organizations to conduct such assessments. A factor contributing to the need were the Chemical Action Plans that specified alternatives assessments in risk management actions; some of these were intended as complements to regulatory actions and would see a high degree of scrutiny.[16]

No one authoritative source contained the information needed to evaluate alternative chemicals and assign values of High, Moderate and Low for concern for hazard. In fact, the classification schemes that existed were not complete or transparent and in some cases conflicted. The program considered the full range of sources, using existing TSCA risk assessment tools as a starting point and reviewing and extracting information from other EPA classification tools, such as the High Production Volume (HPV) Challenge Program Methodology for Risk-Based Prioritization, the Office of Pesticide Programs' Acute Toxicity Categories and international tools, such as those under the Registration, Evaluation, Authorization and Restriction of Chemicals (REACH) and the Globally Harmonized System (GHS) for classification of chemicals. For each human health and environmental endpoint, the program developed a rationale for differentiating among chemicals. In many cases, DfE relied on precedents from TSCA tools, especially when these precedents were harmonized with other systems. In other cases, DfE chose tools such as GHS as the basis for distinguishing chemicals as having Low, Moderate and High concern for hazard.

Existing precedents for some endpoints were not useful for distinguishing among chemicals; for example, EPA's PBT criteria are intended to help the EPA identify chemicals that are problematic (*i.e.*, persistent, bioaccumulative and toxic).[17] However, the criteria were not intended to help identify chemicals that are safer, *e.g.*, chemicals that degrade rapidly, have low bio-accumulation potential and are minimally toxic. In such cases, DfE turned to the latest peer-reviewed science to help modify existing criteria so that they could differentiate both High hazard chemicals as well as Low hazard chemicals.

The outcome was the Alternatives Assessment Criteria for Hazard Evaluation, referenced in Step 5 of the methodology outlined in Section 9.4.5. The Criteria were published in the *Federal Register* and on the DfE website, giving the public the opportunity to comment and recommend changes. DfE incorporated these stakeholder and public comments into the final version published in August 2011, which is available on the DfE website.[18]

9.4 New CAA Methodology and Criteria: Steps to Conducting a CAA

This section presents the seven broad steps of the DfE methodology for conducting CAAs (Figure 9.2). The criteria for evaluating and differentiating chemicals are embedded in this methodology.

9.4.1 Step 1: Determine Feasibility

Since the program began conducting alternatives assessments in 2004, the demand for these partnerships has outstripped the program's resources. To prioritize needs and understand the potential benefits of an alternatives assessment, DfE considers whether alternatives: are commercially available and cost-effective, have the potential for an improved health and environmental profile and are likely to result in lasting change. Stakeholder interest is also a key consideration. CAAs to find safer alternatives to problematic chemicals have been initiated by industry or consumer request and by policy or regulatory drivers. Chemicals prioritized by the EPA for consideration under TSCA such as Existing Chemical Action Plans and Work Plan Chemicals are generally the primary current source for identifying chemical candidates for risk management.[19] The EPA chooses from among these chemicals and will generally conduct alternatives assessments where they are likely to result in the greatest benefit for health and the environment.

9.4.2 Step 2: Collect Information on Chemical Alternatives

Before engaging stakeholders in a formal partnership, the EPA gathers information to understand better the chemical that is the focus of the alternatives assessment and the alternatives that may take its place. DfE considers the availability of data for characterizing the possible alternatives, the chemical manufacturing process, the range of functional uses that the chemical serves and the feedstock or contaminants and residuals from the production process. DfE also considers the work of other organizations in exploring alternatives for the problematic chemical, similar chemicals and functional uses. Based on analysis of this information and preliminary stakeholder consultation, DfE develops a proposed project scope and an approach for the alternatives assessment.

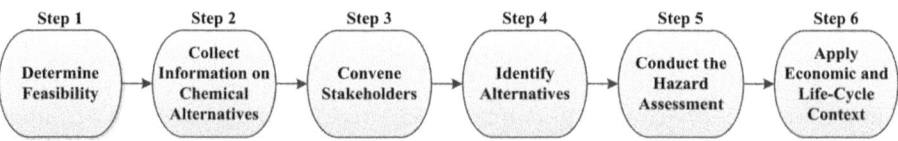

Figure 9.2 Steps to conducting a CAA.

9.4.3   Step 3: Convene Stakeholders

Stakeholder participation is critical to the success of DfE projects. DfE uses input from diverse perspectives to inform the project scope, identify alternatives and facilitate manufacturer and user adoption of safer chemicals. Stakeholders help to design the methodology, monitor its implementation and use the outcome to move to safer chemicals. Stakeholders are drawn from the entire supply chain and all life-cycle stages of the chemical that is the focus of the alternatives assessment. Involvement throughout the project helps to ensure that stakeholders contribute to, understand and support the outcome, enhancing credibility and promoting the adoption of the safer alternatives. Typical stakeholders include: chemical manufacturers, product manufacturers, NGOs (including environmental and worker advocates), US and international government agencies, academics, retailers, consumers and waste and recycling companies. Innovators designing and synthesizing new chemicals – purposely designed to avoid the environmental and health concerns associated with other chemical available for the same functional use (*e.g.*, flame retardants) – are critical members of the group. Likewise, technology innovators, who find ways to deliver the needed function in a new way, such as substituting electronic receipts (e-receipts) for thermal receipt paper containing unconjugated bisphenol A (BPA), are invaluable in the process.

9.4.4   Step 4: Identify Alternatives

Through literature review and discussion with stakeholders, DfE identifies alternatives that are likely to be effective replacements for chemicals that are the focus of the alternatives assessment. The alternative chemicals included are ideally drop-in replacements that are functional with minimum disruption to the manufacturing process. Often, these focus chemicals have multiple functionalities and the replacements are effective in only a subset of those applications and use of the alternatives may require adjustments to other chemicals in product formulations. In some cases, the alternatives included in the report may be 'research and development' chemicals. While functionality on an industrial scale may be theoretical, the program still evaluates such chemicals; they may be important safer alternatives in the future or their toxicological profiles – as shown in the report – may inform green chemists as they design new chemicals. For some focus chemicals, the partnership may choose to limit the size of the field of alternatives considered for cost-effectiveness of the evaluation or for other reasons. One mechanism for culling the number of alternatives is to require viability demonstrations by chemical and product manufacturers.

9.4.5   Step 5: Conduct the Hazard Assessment

Central to the alternatives assessment is the evaluation of a robust set of hazard endpoints. Based on the best data that are available from the literature or can be modeled, DfE assigns a descriptor of hazard concern level of High, Moderate or Low and in some cases Very High or Very Low for each alternative across a range of endpoints. These endpoints include acute and repeated dose toxicity, carcinogenicity and mutagenicity, reproductive and developmental toxicity, neurotoxicity, sensitization and irritation, acute and chronic aquatic toxicity and persistence and bio-accumulation. In addition, the program provides a qualitative description of potential endocrine activity.

Sources of information for assessment of hazard include:

1. Publicly available measured (experimental) data obtained from a literature review.
2. Measured data contained in confidential business information received by the EPA under TSCA regulations.
3. SAR-based estimations from the EPA's Pollution Prevention Framework and Sustainable Futures predictive methods.
4. Professional judgment of the EPA science experts with decades of chemical review experience, often predicated on experimental data for chemical analogs.
5. Confidential empirical data in experimental studies supplied by the chemical manufacturers.

In evaluating the hazard of chemicals and alternatives, DfE uses EPA data sources and expertise to inform the comparison of alternatives. When measured data are not available or adequate for an endpoint, DfE assigns a hazard concern level based on SAR and professional judgment to cover all endpoints and complete the hazard profile. The program uses data from estimation models when available, such as EPI Suite for fate, bio-accumulation and aquatic toxicity or OncoLogic for structural alerts for potential carcinogenicity.[20,21] DfE convenes a workgroup of experts in chemistry, toxicology and environmental fate to review the hazard profiles.

The Alternatives Assessment Criteria for Hazard Evaluation are the basis for designating levels of hazard concern. These criteria address a robust range of endpoints based on as complete an understanding of their human health and environmental characteristics as possible. DfE also designates a level of confidence associated with hazard calls and notes decisions based on little or conflicting evidence. The most instructive endpoints are those that reveal differences in toxicity among alternatives; these endpoints make it possible to distinguish the inherently safer chemicals from the less safe.

Most industrial chemicals are not fully characterized for hazard based on experimental data. As stated above, DfE filled data gaps with analogs, SAR and professional judgment. Estimated hazard levels are the best available basis for decision-making, but the confidence associated with estimates is lower than for hazard calls based on experimental data. Longer term stewardship for chemicals requires the development of more robust data sets for this class of chemicals. This is especially the case for chemicals with higher production. Estimated values for hazard could be used in the design of a plan to develop testing data so that a more robust characterization of the alternatives could lead to better choices of alternatives in the future.

9.4.6   Step 6: Apply Economic and Life-cycle Context

DfE makes its most important contributions to alternatives assessment in Steps 1–5. The program can contribute information, for Steps 6 and 7, but often the most robust expertise on these steps resides with DfE's stakeholder base. Stakeholders contribute heavily to report sections on economic and life-cycle context. With an understanding that stakeholder input is critical to successful implementation of Step 6 and that the program's investment in providing information should respond to stakeholder need, DfE offers this information as context for the hazard assessment and the work of the partnership. The information may include descriptions of manufacturing processes, use patterns and life-cycle stages that may pose special exposure concerns. The report may contain information on potential economic impacts associated with the selection of alternatives and may also contain information on alternative technologies that could result in safer chemicals, manufacturing processes and practices.

9.4.7   Step 7: Apply the Results in Decision-making for Safer Chemical Substitutes

Although DfE CAAs do not specify a favored alternative, when the CAA outcome is combined with information on cost, performance and other factors, CAAs can help stakeholders make informed decisions as they choose alternative chemicals. CAAs may show that safer, highly functioning alternatives are available, leading to the development and availability of environmentally preferable product formulations. Availability of safer, viable alternatives can also support the case for phase-out of a chemical. Alternatively, a demonstration that safer alternatives are not available might strengthen arguments for exposure controls and implementation of best practices. Where CAAs show that viable alternatives are not available for certain uses of a chemical targeted for action by EPA, innovators may seize the opportunity to develop safer chemicals or alternative and inherently safer processes.

Stakeholders involved in alternatives assessments, especially those choosing safer chemicals, have developed and implemented complementary approaches to guide chemical selection. Clean Production Action created the Green-Screen™ for Safer Chemicals (GreenScreen™) to assist manufacturers in selecting safer substitutes.[22] This tool is broadly intended to help move society toward the use of more sustainable chemical-intensive products. The Green-Screen™ focuses on hazard reduction and assigns a value for each chemical, based on hazard determinations from the CAA. The resulting data provide those choosing chemicals with a calibrated tool to weigh the health and environmental trade-offs of alternatives. Other similar tools are coming into use.

Washington State has developed the Quick Chemical Assessment Tool (QCAT) and Hewlett-Packard developed the Materials Choice Platform. The German Column Model and US National Institute for Occupational Safety and Health (NIOSH) Hazard banding approach are also available.[23–25]

The next sections give two examples of how the updated DfE CAA criteria are being used to explore safer alternatives to chemicals that the EPA has targeted for action.

9.5 Application of New CAA Criteria

9.5.1 BPA Alternatives in Thermal Paper Partnership

9.5.1.1 Problem Statement. Public attention and concern around exposure to BPA and its potential effects continues in the USA.[26] BPA is an HPV chemical with a US volume estimated at 2.4 billion lb (1.09 billion kg) in 2007 and an estimated value of almost US$2 billion (€1.54 billion).[27] A cost-effective chemical with many useful applications, BPA is used widely in manufacturing and consumer products and the public is routinely exposed to it. The US Centers for Disease Control and Prevention (CDC) have found that 95% of a sample size of 294 Americans had detectable levels of BPA in their urine.[28] BPA is also found in breast milk; in a study of 23 healthy women, all breast milk samples registered positive for BPA.[29]

Because BPA is a reproductive, developmental and systemic toxicant in animal studies and interacts with estrogen receptors, there are questions about its potential impact, particularly on children's health and ecosystems. Several government entities [*e.g.*, the European Union (EU), Health and Environment Canada, Japan's National Institute of Advanced Industrial Science and Technology (AIST) and the US Food and Drug Administration (FDA)] have published reports examining potential human health and environmental hazards associated with BPA exposure. Additional research is under way, particularly concerning whether BPA may cause effects at low concentrations.[27,30]

BPA is used in a variety of applications. While most human exposure to BPA is believed to come from food and beverage packaging made from polycarbonate plastic and epoxy resins, less than 5% of the BPA produced is used in food contact applications.[27] In the USA, food-related uses fall under the jurisdiction of the FDA rather than the EPA. Among their many applications, BPA-based materials are used in automotive and other transportation equipment, optical media such as DVDs, electrical and electronics equipment, construction, linings inside drinking water pipes, thermal paper and foundry casting.

The EPA is examining the use of BPA in thermal paper. In March 2010, the EPA released an Action Plan on BPA based on its screening-level review of hazard and exposure information. EPA's Action Plan calls for DfE to conduct a CAA of BPA in thermal paper coatings for applications including point-of-sale receipts, airline tickets, event and cinema tickets and labels.

When used as a developer in thermal paper, BPA is present as 'free' (*i.e.*, monomeric or unconjugated) BPA, which may be more available for exposure than BPA polymerized into a resin or plastic.[27] Upon handling, BPA in thermal paper can be transferred to the skin and there is some concern that residues on the hands could be ingested through incidental hand-to-mouth contact.[31] Some studies suggest that dermal absorption may contribute a small fraction of the overall human exposure.[31,32] In one US study, pregnant women who worked as cashiers, who presumably had frequent contact with thermal paper used in cash register receipts, had the highest urinary BPA concentrations compared with pregnant women in other occupations.[33] European data indicate that the use of BPA in paper may also contribute to the presence of BPA in the stream of recycled paper and in landfills.[34]

9.5.1.2 Scope and Approach. DfE initiated the BPA Alternatives in Thermal Paper Partnership in July 2010 to implement part of the EPA's Action Plan for BPA. The Partnership includes a diverse array of stakeholders including thermal paper manufacturers, thermal paper converters, chemical manufacturers, point-of-sale equipment manufacturers, retailers, trade associations, NGOs, green chemistry and technical experts and international governmental organizations. These partners helped DfE identify and evaluate potential alternatives to BPA in thermal paper and ensured that the points of view of all stakeholders are considered.

The scope of this DfE CAA is evaluating the potential hazards associated with thermal paper developers that are likely to be viable alternatives to BPA. Thermal paper systems include a developer along with dyes and sensitizers that optimize the chemical interaction. A change in the developer may require adjustments to the system, including the selection and amounts of other chemicals.

In partnership with stakeholders, EPA identified 19 alternatives to BPA in thermal paper. These alternatives were selected because they have the potential to be functional substitutes to BPA based on their physico-chemical properties and/or because they are already in commercial use. DfE conducted a hazard assessment on these 19 alternatives and the draft Partnership Report was published in 2012. A final version of the Partnership Report will be released in 2013.

9.5.1.3 Findings. In choosing to conduct this DfE CAA, the EPA planned to improve understanding of the potential environmental and human health impacts of BPA and alternative developers in thermal paper throughout their life-cycles. Unfortunately, the hazard evaluation of BPA and its 19 alternatives shows that no alternatives are clearly safer or preferable to BPA. Of the 19 alternatives, eight are structurally similar to BPA and have similar hazard profiles. Five of the chemicals are structurally similar to bisphenol S, a chemical commonly substituted for BPA in thermal paper. The hazard profile of bisphenol S is similar to that of BPA. Other chemicals evaluated through the partnership have various chemical structures, but all are associated with trade-offs. For example, some are slow to degrade in the environment, may bioaccumulate or are associated with aquatic toxicity. The hazard profiles for many of the chemicals in this report were not well characterized by experimental data. Concern for hazard was estimated for many endpoints based on structurally similar chemicals.

These findings stress the need for further toxicological testing of alternatives, and also out-of-the-box thinking to encourage the development and use of safer alternatives. One idea for encouraging the development of safer alternatives is a green chemistry challenge to develop safer chemical substitutes for use in thermal paper. Another idea that is gaining traction in the retail community is the use of e-receipts. E-receipts are being offered by Apple, Nordstrom, Whole Foods and other major retailers. They are either e-mailed directly to consumers or uploaded to a password-protected website. E-receipts allow consumers to keep searchable records of purchases on their computer. E-receipts could reduce paper waste and also limit exposure to BPA and other chemicals, making them an approach to be considered in alternatives discussions.

9.5.2 Flame Retardant Alternatives to DecaBDE Partnership

9.5.2.1 Problem Statement. Similarly to pentaBDE, the EPA is concerned that other polybrominated diphenyl ethers (PBDEs) are persistent, bioaccumulative and toxic to both humans and the environment. In December 2009, the EPA released the PBDEs Action Plan that summarizes hazard, exposure and use information for three commercial PBDE mixtures, including decabromodiphenyl ether (decaBDE). DecaBDE has been studied for ecotoxicity in mammals, birds, fish and invertebrates. In some cases, current levels of decaBDE exposure for wildlife may be at or near adverse effect levels. Human exposure to decaBDE can occur through the diet, at the workplace and in the home. According to the EPA's 2010 exposure assessment of PBDEs, individuals in occupations that would lead to higher exposures to specific congeners have higher concentrations of PBDE congeners in their blood than the general public.[35] The primary route of consumer exposure to decaBDE is through the ingestion of house dust or, for infants, ingestion of breast milk, followed by food and water ingestion and dermal absorption.[35–37]

In December 2009, the largest commercial producers and suppliers of decaBDE in the USA agreed to phase out the use of the chemical by 2012, similarly to the voluntary phase out of pentaBDE. The use of decaBDE has also been restricted in particular electrical and electronic equipment under the EU Restriction of Hazardous Substances Directive (RoHS), with some exemptions.[38,39] The US states of Maine, Maryland, Oregon, Vermont and Washington have imposed restrictions on the manufacture and/or use of dec-aBDE in certain applications.[40–44] Additional states have proposed legislation restricting the manufacture and/or use of decaBDE.

DecaBDE is effective in meeting fire safety standards for plastics and textiles that are used for the manufacture of consumer electronics, appliances, wire and cable insulation, building materials (flooring, wall coverings and roofing), seating, electronics and paneling for cars, buses and airplanes and storage and distribution products including plastic shipping pallets. Few potential alternatives to decaBDE are 'drop-in' replacements (those that require negligible process changes). The use of alternatives may necessitate additional changes in product formulation or movement to different classes of polymers.

9.5.2.2 Scope and Approach. Under the EPA's Action Plan for PBDEs, DfE initiated the Flame Retardant Alternatives to DecaBDE Partnership in October 2010. DfE has engaged with representatives from industrial, academic, governmental and non-governmental organizations to identify functional and viable alternatives, evaluate their human health and environmental profiles and inform decision-making in organizations choosing alternatives to decaBDE.

In partnership with stakeholders, DfE identified 32 alternatives to decaBDE that fit the scope of this project: to identify potentially functional, viable alternatives for use in the identified polyolefins, styrenics, engineering thermoplastics, thermosets, elastomers or waterborne emulsions and coatings and to evaluate their potential human health and environmental hazards. The alternatives evaluated in the report were not chosen based on environmental preferability but on their functionality and viability. The content of the report will help businesses choose environmentally preferable alternatives. As companies that have been using decaBDE in their products prepare for the phase out, the information generated by this DfE CAA is intended to reduce the potential for the unintended consequences that could result if functional but poorly understood alternatives are chosen.

9.5.2.3 Findings. The alternatives available to replace decaBDE come from three general classes of chemistry. These include brominated and chlorinated compounds, phosphorus- and nitrogen-based compounds and combinations of the two and inorganic compounds. Some of the alternatives are discrete molecules and some are polymeric. Only a few of the alternatives evaluated in the report are recent introductions to the market.

According to the report's evaluation, large polymers, with molecular weights much greater than 1000 Da, were generally associated with lower concern for hazard than discrete chemicals because the larger chemicals generally cannot be absorbed or easily metabolized. Smaller polymers and discrete chemicals, with molecular weights generally less than 1000 Da, were associated with some concern for hazard. At face value, these summary statements appear to suggest that polymeric flame retardants are better – but this would be an oversimplification.

All but two of the 33 chemicals in the evaluation are expected to persist in the environment; thus, DfE does not consider persistence to be a distinguishing characteristic for flame retardants. To be effective, chemical flame retardants must be stable by design to retain their flame-retardant properties.

Even though it is not distinguishing for this class of chemical, persistence has implications for health and the environment that must be evaluated. Persistent chemicals will eventually degrade in certain environmental conditions over months, years or much longer; degradation to lower brominated congeners is one of the concerns associated with decaBDE.[45] Persistence can be important in understanding the health and environmental profile of a chemical. The extreme example is that persistent compounds also associated with toxicity and bioaccumulative potential are of the highest possible concern. Even with these potential concerns, persistent compounds such as high molecular weight polymers may be one way to move toward the green end of the spectrum of continuous improvement. High molecular weight polymers that perform reliably through the life of a durable product, are not bioavailable and are therefore low in toxicity, even after disposal, may be a good choice from the available alternatives.

Some of the flame-retardant options evaluated in the report include inorganic compounds that are persistent or recalcitrant. The persistence of these compounds, usually metal oxides and polyphosphates (with molecular weights of $\sim 100\,000$ Da), should be considered pragmatically in the context of the metal and its oxidation state, the dissociation products (if any), the likely environmental form after release and environmental transport. The ultimate degradation is only to the element (*e.g.*, aluminium or phosphorus) and cannot be considered the same way in which organic compounds may degrade through mineralization to relatively inert carbon dioxide and water.

Similarly to most industrial chemicals, few of the alternatives to decaBDE were fully characterized for hazard based on experimental data. DfE filled data gaps with analogs, SAR and professional judgment according to the methodology outlined in Section 9.4. Since successful replacements for decaBDE are likely to see very high production volumes, using estimated values for hazard to design a plan to develop testing data to generate a more robust characterization of the alternatives could lead to better choices of alternatives in the future.

Environmental fate is a characteristic that should receive special attention for flame retardants likely to replace decaBDE. An understanding of the fate of decaBDE show that higher molecular weight flame retardants do not always behave as predicted.[46-49] An aggressive program to develop data for this endpoint for newer chemistries that are likely to find high-volume uses would help reduce the likelihood of unintended consequences.

An important point in considering alternative flame retardants is that, for some groups, concern over halogenated flame-retardant compounds is driven by the potential for the generation of combustion byproducts such as dibenzodioxins and dibenzofurans during combustion.[50] These may be generated by combustion, either through accidental fires or through uncontrolled 'recycling' practices, common in developing countries. As a result, regardless of how positive the health and environmental profile of a flame retardant in this class is, some will voice concern.

9.6 Safer Product Labeling Program Case Study

9.6.1 Functional Use Concept Advances Greener Chemistry

The core components of the DfE Safer Product Labeling Program (SPLP) are the assessment of chemical ingredients and identification of greener alternatives for chemical-based products. In labeling safer products, DfE works within product sectors and carefully evaluates every product ingredient, only allowing use of the safest possible chemicals as measured against the program's stringent human and environmental health criteria. In essence, when the DfE SPLP reviews a product, it conducts a series of small, hazard-focused alternative assessments, one for each ingredient evaluated as part of a functional use class of potential alternatives.

The functional use concept serves as an organizing principle and ensures appropriate chemical and product-level comparisons. DfE applies functional use thinking in two ways: (1) to select product classes for potential labeling (such as laundry detergents or fire-fighting foams) – the purpose a product serves and how it is used correlating with its chemical constituents and whole-product characteristics; and (2) to group chemical ingredients based on the properties they impart (such as surfactancy or solvency), which collectively give a product its distinct performance attributes.[51]

A focus on functional use helps drive the transition to greener chemistry. When evaluating product ingredients, DfE considers each chemical within its functional use class (*e.g.*, is it a surfactant or anti-foaming agent?) and applies evaluation criteria based on an array of hazard endpoints tailored to the component class whenever possible. Based on its hazard profile, a chemical is positioned along a continuum of improvement or preferability – from red to yellow to green – to identify the functional alternatives that pose the lowest human and environmental concerns. Since use patterns are similar within product classes and the chemicals along the continuum can often substitute for one another, the exposure component of the risk equation can be considered a constant and need not be evaluated. As a result, DfE can reduce risk – and drive greener chemistry – by focusing the attention of formulators on the selection of ingredients from the green end of the hazard spectrum.

It is worth noting that in its evaluations, DfE incorporates key elements of exposure that are intrinsic to a chemical, such as the potential for absorption, persistence and bio-accumulation. Other exposure-related factors, such as physical form, that do not vary within product classes are less relevant to the assessment, although the method of application, especially spraying, will heighten attention on ingredients that may pose respiratory or allergenic concerns.

9.6.2 Ingredient-class Criteria Define 'Safer' Chemicals

Working within the functional use framework, DfE has developed sets of criteria to define safer chemistries by ingredient class under the umbrella heading of 'Criteria for Safer Chemical Ingredients.'[52] Criteria currently exist for surfactants, solvents, chelating or sequestering agents and fragrances. The DfE master criteria serve as the safer chemical measure for all other component classes; adapting the master criteria to reflect a class's inherent functional characteristics (such as the need for certain polymers to resist degradation) or its key toxicological markers is an area of continuing advancement and improvement at DfE.

Similarly to the CAA criteria, the labeling program's safer chemical criteria cover a wide range of human and environmental health endpoints. Although the CAA and safer ingredient criteria are similar in many ways – in structure, hazard endpoints and profiling methodology – there is one key distinction: the safer ingredient criteria add a bright-line threshold to the hazard profile for each toxicological endpoint. This line demarcates a chemical's pass/fail status and eligibility for use in a DfE-labeled product. A designation of 'Low' or 'Very Low' hazard in a CAA equates to a 'pass' or 'acceptable' status in the labeling program. To qualify for use in a labeled product, a chemical must pass on all endpoints and be considered a 'low concern' ingredient. DfE-labeled products only contain low concern ingredients.

Characterizing ingredients based on a full range of toxicological and fate endpoints and comparing their characteristics in a same-functionality context make it possible to identify the key characteristics that distinguish a safer from a less safe member of the class. These distinguishing endpoints not only allow DfE to streamline the hazard evaluation process, but, by basing testing requirements on these endpoints, make it easier in a data-poor environment for DfE to determine if new chemistries will enter the pool of safer alternatives.

To define the low concern end of the spectrum, DfE uses toxicological thresholds established by respected health and environmental protection authorities, including the United Nations GHS for the Classification and Labeling of Hazardous Substances and the EPA's Office of Chemical Safety and Pollution Prevention's chemical management and prioritization programs. DfE calls on the expertise of EPA's chemistry, toxicology and environmental fate experts to review and interpret data and apply them to the threshold criteria.

By taking an ingredient-by-ingredient rather than whole-product approach to evaluating potential hazards, the DfE SPLP is able to review formulations thoroughly and make discriminating calls on potential concerns, and also easily identify safer substitutes. For example, a surfactant that is acutely toxic to aquatic organisms and environmentally persistent can appear to pose a low acute concern when blended with other less toxic and less persistent surfactants. It is likely that, even in a dilute state, the surfactant would still pose more difficult to detect chronic concerns. Water, typically the largest percentage ingredient even in concentrates, can mask the effects of a hazardous chemical if subject only to whole-product testing.

DfE's Criteria for Safer Chemical Ingredients and its Standard for Safer Cleaning Products have made the labeling program more transparent and encouraged chemical manufacturers to innovate.[53] After DfE established criteria for safer surfactants in cleaning products, chemical manufacturers synthesized chemicals to meet the criteria.[54] These new surfactants biodegrade rapidly and completely, reducing the potential toxicity to aquatic organisms. Many surfactants and other component-class ingredients that meet the DfE criteria are listed in CleanGredients, a database of safer chemicals administered by the NGO GreenBlue®.[55]

9.6.3 Partnership Process Fosters Teamwork in Safer Formulation

The product manufacturer plays a key role in evaluating the cost-effectiveness and technological feasibility of alternative chemistries. Once the product manufacturer is able to formulate a high-functioning product that uses the safest possible ingredients in each functional class, DfE offers the manufacturer the opportunity to partner with the EPA through a signed partnership agreement.

In the agreement, the company commits to, most notably, manufacture qualifying products with the same ingredients as revealed to the EPA in its statement of formula; submit to annual surveillance audits, including triennial site audits; demonstrate product performance based on appropriate industry standards; disclose publicly all ingredient information by CAS number and chemical name (trade secret ingredients must be listed by chemical-descriptive name); and follow the conditions for use of the DfE label. The agreement is valid for 3 years, during which time any changes to the formulation must be made in consultation with the EPA. At the end of 3 years, the product manufacturer must meet any new advances in green chemistry before the partnership can be renewed and use of the DfE label can be continued. If safer chemicals have been identified during the partnership period, the manufacturer must replace any chemicals that are no longer the safest in their class.

As part of its partnership service, DfE offers advice on safer alternatives for ingredients that do not meet its safety criteria. This consultation service, unique among ecolabels, helps formulating companies learn about safer chemistry and earn the DfE label (Figure 9.3), thereby distinguishing their products in the marketplace. Growing numbers of consumers are focusing on sustainability and are looking for advice on how to buy safer products for their families and to help protect the environment. The DfE label, backed by the EPA's scientific expertise and years of experience in evaluating chemicals, provides just the assurance that many consumers seek. Companies that have invested in safer chemistry and earned the label have an entry to an expanding marketplace for sustainable products. This marketplace includes major retailers, such as

Figure 9.3 DfE label for DfE-approved products.

Walmart and Home Depot, who have given special status to DfE-labeled products, and government purchasers who are increasingly specifying the DfE label in their purchasing requirements.

The six-step process described below presents the key details common to the development of most new-sector partnerships.

- *Step 1:* DfE opens product sectors to partnership largely based on the extent to which the sector uses chemicals of potential concern, opportunities exist for safer formulation and companies have developed or are actively working on innovative formulations.
- *Step 2:* Based on product type, DfE develops a knowledge base of the available alternatives for each functional component class used in the product, along with information on human health and environmental characteristics. For cleaning products, the main functional classes are typically surfactants, solvents, chelating agents, builders, colorants, fragrances and preservatives.
- *Step 3:* DfE convenes stakeholders to refine its understanding of the sector and to help develop or modify DfE component-class criteria which must be met to earn the label.[52] Stakeholders help by clarifying the functionality of ingredients, ensuring that the full range of possible alternatives is considered, describing limitations to the use of certain chemicals and optimal chemical combinations and providing information on how products formulated with safer ingredients perform. Manufacturers also share economic information on ingredient and product pricing. Cost-effectiveness is critical in ensuring that safer products are marketable and available to all consumers at a wide range of retailers.
- *Step 4:* Working through qualified third parties (hired by product manufacturers), DfE receives available information on each chemical in a candidate formulation and compiles it into hazard profiles. As in the flame retardant and other CAA assessments, the use of SAR estimation methods and professional judgment are important in filling data gaps and completing hazard profiles. Such profiles describe the toxicological characteristics of the ingredient and whether it satisfies DfE's transparent pass/fail component-class and labeling criteria. The DfE Criteria for Safer Chemical Ingredients include data requirements. In some cases, data gaps may be filled through estimation models based on physico-chemical properties or suitable analogs; in others, particularly where the endpoint is key in distinguishing among alternatives, measured data may be required.
- *Step 5:* The labeling program provides product manufacturers with a chemical assessment of each ingredient in their product formulation based on DfE's safer chemical criteria and its pass or fail status.
- *Step 6:* To become a DfE partner, a manufacturer must use the safest ingredients from each functional group, in addition to meeting other product-level requirements, such as pH and performance, as described in

the DfE Standard.[53] Product manufacturers must also sign a Partnership Agreement with the EPA/DfE that formalizes their commitment to make the recognized product(s) as described to the EPA, with improvements over time as science and innovation advance. Once the agreement is in place, partners may differentiate their products in the marketplace through use of the DfE label. An audit program verifies a partner's product ingredients, good manufacturing practices and other partnership requirements.

9.6.4 Development of Safer Alternatives

For some critical functionalities, a challenge encountered by DfE in facilitating the move away from chemicals targeted for action by EPA is the lack of safer, cost-effective alternatives. Recent improvements in this area, however, show a move along the continuum to safer chemicals. A number of alternatives for flame retardants such as decaBDE and hexabromocyclododecane (HBCD) were developed by chemical manufacturers with the intention of being safer alternatives. Also, hundreds of high-functioning, safer surfactants are now on the market and many can serve as replacements for NPEs. Initial indications are that the chemical manufacturers may have mitigated concerns with these new flame-retardant chemicals, but confirmation with experimental data will be needed. In the case of the surfactants, manufacturers have developed the data required by DfE criteria to show that hundreds of safer chemicals are now available.[55,56]

A helpful developing trend for many classes of chemicals is ensuring that chemicals do not persist after the end of their useful life. Design for degradation may not be appropriate for all classes of chemicals; flame retardants, for example, must be persistent in plastics and other flammable media to be effective. But for many other classes, concern for chemical fate and toxicity can be mitigated at the chemical design stage.[57] Green chemistry research for safer substitutes has often been a search for means to lower toxicity, but biodegradability can be an equally important part of chemical design. When a chemical is slow to degrade, it remains available to biota to exert toxic effects, not all of which may be known or predictable at the outset. The cornerstone of waste water treatment, microbial degradation is fundamental to the treatment of many waste streams. Responsible molecular design that incorporates degradability can increase both treatability of generated wastes and safety of materials that ultimately enter the environment.

Surfactants for detergents and cleaners offer an excellent example of chemicals that can be designed for biodegradation without sacrificing efficacy.[57] The classic detergent surfactant is made up of a hydrophilic head and hydrophobic tail. The surfactant tail attaches to soil and its head pulls the soil into the wash water. Since most high-performing surfactants are toxic to aquatic organisms, rapid degradation after soil removal is very desirable from an environmental health perspective. Many surfactant manufacturers, especially those interested in offering customers DfE-eligible chemicals, have designed surfactants with microbial degradation in mind. These surfactants degrade rapidly in waste water treatment systems and meet DfE's safer surfactant criteria. More than 200 safer surfactants are listed in the CleanGredients database and can be used in DfE-labeled products.

Another approach for encouraging the development of safer alternatives is Green Chemistry Challenges, perhaps structured similarly to an 'X Prize.'[58] Such challenges could be developed as part of a CAA. Especially where safer alternatives are not found, a Green Chemistry Challenge could identify performance needs, distinguishing toxicological characteristics and the types of chemical structures that had been developed to date but were associated with undesirable effects. The challenge could also describe the business case for investment in safer alternatives, such as the size of the market for a safer alternative.

9.7 What's Next

9.7.1 The Community of Practice is Growing

Alternatives assessments are proving to be a valuable approach and are being conducted by a broadening range of practitioners. Government in the USA is using CAAs to inform stakeholder decisions as they choose alternatives to high-profile chemicals, such as plasticizers and flame retardants. Governments elsewhere, notably under REACH, are encouraging alternatives assessment for high-priority chemicals. US states are using alternatives assessments to influence the move to safer chemicals and NGOs are developing tools to make the approach more broadly available. Leaders in industry are using the NGO tools, and tools they have developed themselves, to choose chemicals for consumer goods, such as personal computers and sporting goods.

In the USA, the EPA is applying the DfE approach for conducting CAAs in a range of contexts – as a complement to regulations and as a tool to encourage movement to safer chemicals in the absence of a regulatory driver. A CAA cannot replace regulation, but, as shown through the penta- and decaBDE examples, the EPA has found that this approach can demonstrate that safer, high-functioning alternatives are available and help justify a regulatory action to prevent future manufacture of a problematic chemical.

Alternatives assessment generally takes less time and resources than regulation and can operate well in a non-regulatory context. As a result, the method can be a useful tool for substitution challenges that would be difficult to address or not suitable for regulatory action. CAAs can help leaders in industry move away from chemicals of emerging concern to meet the demands of outside groups and consumers. The safer products differentiated by the SPLP show both the value of government involvement in this application of alternatives assessment and how effective the application can be in chemical substitution.

US states such as California, Washington, Minnesota, Connecticut and Maine are assessing alternatives to priority chemicals as they take actions that might limit their use.[59-63] These comparisons often consider hazard to be a primary factor in determining preferability and implementation of state policy to limit priority chemical use is often contingent on a demonstration that safer alternatives are available.

In the EU, alternatives assessments will be used by applicants for authorization under REACH for substances of very high concern (SVHCs). Applicants must replace SVHCs with a suitable alternative, but if alternatives are not available, applicants are required to conduct an alternatives assessment that accounts for risk reduction, functionality and cost-effectiveness.[64]

Leaders in industry such as Hewlett-Packard and Nike are assessing the chemicals in their products, using alternatives assessment to choose the safest, high-functioning alternatives available. Walmart, Home Depot, Staples and others are encouraging their suppliers to develop safer and more environmentally friendly products and highlighting the safer products in their stores. Methods of differentiating the products vary from focusing on a particular attribute of a product, such as recycled content and energy efficiency, to using sophisticated evaluation tools such as LCA and alternatives assessment.

The alternatives assessments conducted by DfE and the safer products distinguished by the program can be a useful addition to retailer sustainability programs. Roger McFadden, Vice President and Senior Scientist for Staples, speaks to the value of alternatives assessments in the retail market when he states:

> *The single most important element of the AAs DfE conducts, from a retailer's perspective, is having the scientific data on alternatives expertly analyzed and interpreted and presented as thorough and objective chemical profiles and comparative hazard assessments. Without this information, retailers, formulators and other chemical users are at a great disadvantage in trying to verify claims made by chemical suppliers and product manufacturers that the chemical alternatives they make and use have safer profiles than the problematic chemicals targeted for replacement. Quality alternatives assessments play a critical role in informing and strengthening company efforts to drive the chemical supply chain and marketplace to safer, more sustainable chemistries.[65]*

Interestingly, McFadden's comments point out a weak link in the current methodologies for CAA. The level of technical expertise needed to conduct a CAA can be a significant investment for a product manufacturer or retailer. If CAAs are to have a greater impact on chemical substitution, it will be important in going forward to make the methodology more accessible and implementable by a broader range of the potential practitioners.

One important way to increase capacity is to build out the scientific foundation and technical support elements for CAAs. Through training and continuing education, it will be possible to disseminate information more broadly and effectively on methodologies. Establishing a more formalized community of practice around alternatives assessment may be an important step in strengthening the informational links for conducting CAAs. Ready access to easy-to-understand information on the health and environmental profiles of chemicals is a challenging gap that must be addressed to make CAA implementable by the broadest possible audience. In the future, an information support system may be available to bridge this gap through web-based tools such as the ToxCast™ database or the Chemsec SUBSPORT.[66,67]

9.7.2 Promoting Harmony in CAAs

The DfE CAA methodology could be a useful basis for developing a standard tool for hazard-based alternatives assessment. The methodology is flexible and has been applied for difficult substitution challenges, in addition to more routine safer chemical comparisons. CAAs have been used to address the major substitution challenges associated with alternative flame retardants, but can also be used in addressing issues where a lower level of resources is available, as is the case for choosing safer surfactants for cleaning products.

National and even international harmonization of CAA methodologies would be beneficial because it would enable other practitioners to use existing resources on chemical assessments, allow sharing of data by governments and across industry and provide a greater degree of certainty for industry as the market is moving to safer chemicals.

In the USA, the EPA is collaborating with other arms of the US government, US states, businesses and NGOs to harmonize CAA methodologies and make data on chemicals more accessible and easy to interpret and apply. For example, EPA collaborates with Clean Production Action and its cross-industry group called the BizNGO Working Group on Safer Chemicals and Sustainable Materials.[68] Both Clean Production Action and members of the BizNGO use DfE's methodology as the foundation for their own alternatives assessments. Clean Production Action and DfE actively collaborate to ensure harmonization.

The EPA provides technical assistance to the States of California and Washington and the Technical Alternatives Assessment Guidance Team (TAAG). The TAAG team is a collaboration of state agencies who seek to develop guidance for industry on the use of alternatives assessment.[69]

The EPA also welcomes opportunities to work with private sector teams, such as the members of the Joint Roadmap to Zero Discharge of Hazard Chemicals. These companies in the apparel and footwear industry (Nike, Adidas, Puma, H&M, C&A and Li Ning) are working towards zero discharge of hazardous chemicals for all products by 2020.[70] One of the steps towards this goal is the development of and agreement on a tool to assess the hazards of chemicals used in apparel and footwear manufacturing. Once the assessment is completed, the group will create a chemical inventory categorized by intrinsic hazard and establish a sector-wide list of hazardous chemicals. The group has sought EPA guidance on cost-effective hazard assessment approaches to evaluate the thousands of chemicals used by this industry. In a related effort, the EPA is also working with the Outdoor Industry Association Sustainability Working Group to develop an approach to manage and assess chemicals in their products. Green chemistry is at the core of this work and the group was recently recognized by the US White House for their leadership in this area.[71]

The EPA hopes to broaden its collaboration in the USA and internationally to jointly develop a standardized but flexible approach for conducting alternatives assessment. The Organization for Economic Co-operation and Development (OECD) could be an excellent forum for this work. As part of an Ad Hoc Workgroup on Substitution, the EPA may be able to survey OECD countries on the methods and approaches they use now. From this survey, the EPA might develop a resource document and consult with other countries to craft a joint approach. Substantial work is being conducted by other countries and the EPA looks forward to collaboration with others to develop a robust approach useful to all countries.[72]

References

1. R. Denison, personal communication to C. Davies, 2012.
2. A. Igrejas, *HuffPost Green*, 2012; http://www.huffingtonpost.com/andy-igrejas/new-york-times-wealthy-ne_b_1354411.html (last accessed 7 February 2013).
3. Walmart, *2011 Global Responsibility Report: Sustainable Product Index*, 2011; http://walmartstores.com/sites/ResponsibilityReport/2011/environment_products_SustainableProductsIndex.aspx (last accessed 26 March 2012).
4. The Home Depot, *Eco Options*; http://www.ecooptions.homedepot.com (last accessed 7 February 2013).
5. Staples, *Staples Soul: Environment*, 2011; http://www.staples.com/sbd/cre/marketing/staples_soul/environment.html#id_e2 (last accessed 26 March 2012).
6. US EPA, *Design for the Environment: Best Practices*, 2012; http://www.epa.gov/dfe/best_practices.html (last accessed 28 March 2012).
7. US EPA, *Design for the Environment: Automotive Refinishing Partnership*, 2012; http://www.epa.gov/dfe/pubs/projects/auto/index.htm (last accessed 26 March 2012).
8. J. R. Geibig and M. L. Socolof, *Solders in Electronics: a Life-Cycle Assessment,* US EPA, Washington, DC, 2005.
9. US EPA, *Lithium-ion Batteries and Nanotechnology Partnership: Assessing Life-Cycle Impacts of Lithim-ion Batteries*, 2010; http://www.epa.gov/dfe/pubs/projects/lbnp/lithium-ion_nanotechnology-factsheet.pdf, 2010 (last accessed 18 April 2012).
10. US EPA, *What Does the DfE Label Mean?*, 2012; http://www.epa.gov/dfe/pubs/projects/formulat/label.htm; (last accessed 12 February 2013).
11. US EPA, *Design for the Environment: Safer Detergents Stewardship Initiative (SDSI)*, 2012; http://www.epa.gov/dfe/pubs/projects/formulat/sdsi.htm (last accessed 26 March 2012).

12. Agency for Toxic Substances and Disease Registry (ATSDR), *Toxicological Profile for Polybrominated Biphenyls and Polybrominated Diphenyl Ethers*, US Department of Health and Human Services, Atlanta, GA, 2004.

13. P. Eriksson, E. Jakobsson and A. Fredriksson, *Environ. Health Perspect.*, 2001, **109**, 903–908.

14. A. Schecter, M. Pavuk, O. Päpke, J. J. Ryan, L. Birnbaum and R. Rosen, *Environ. Health Perspect.*, 2003, **111**, 1723–1729.

15. J. Peltola and L. Ylä-Monone, *Pentabromodiphenyl ether as a global POP*, Nordic Chemical Group, Reykjavik, 2000.

16. US EPA, *Existing Chemical Action Plans*, 2012; http://www.epa.gov/oppt/existingchemicals/pubs/ecactionpln.html (last accessed 26 March 2012).

17. US EPA, *PBT Profiler*, 2011; http://www.epa.gov/oppt/sf/tools/pbtprofiler.htm (last accessed 28 March 2012).

18. US EPA, *Design for the Environment Program Alternatives Assessment Criteria for Hazard Evaluation*, US EPA, Washington, DC, 2011.

19. US EPA, *Existing Chemicals*, 2012; http://www.epa.gov/oppt/existingchemicals/ (last accessed 26 March 2012).

20. US EPA, *Estimation Program Interface (EPI) Suite*, 2011; http://www.epa.gov/oppt/exposure/pubs/episuite.htm (last accessed 18 April 2012).

21. Y.-T. Woo and D. Y. Lai, in *Predictive Toxicology*, ed. C. Helma, Marcel Dekker, New York, 2005, pp. 385–413.

22. M. Rossi and L. Heine, *GreenScreen™ for Safer Chemicals*, 2011; http://www.cleanproduction.org/Greenscreen.php (last accessed 18 April 2012).

23. OSHA, *Facts on Aligning the Hazard Communication Standard to the GHS*, 2012; http://www.osha.gov/as/opa/facts-hcs-ghs.html (last accessed 3 April 2012).

24. NIOSH, *Workplace Safety and Health Topics: Control Banding*, 2012; http://www.cdc.gov/niosh/topics/ctrlbanding/ (last accessed 3 April 2012).

25. IFA, *The GHS Column Model: an Aid to Substitute Assessment*, IFA, Berlin, 2011.

26. D. ElBoghdady, *Washington Post*, 2012; http://www.washingtonpost.com/business/economy/congressman-moves-to-ban-bpa-from-food-packaging/2012/03/16/gIQAXNNAHS_story.html (last accessed 7 February 2013).

27. US EPA, *Bisphenol A Action Plan*, US EPA, Washington, DC, 2010.

28. A. M. Calafat, X. Ye, L.-Y. Wong, J. A. Reidy and L. L. Needham, *Environ. Health Perspect.*, 2008, **116**, 39–44.

29. Y. Sun, M. Irie, M. Kishikawa, M. Wada, N. Kuroda and K. Nakashima, *Biomed. Chromatogr.*, 2004, **18**, 501–507.

30. L. S. Birnbaum, *Environ. Health Perspect.*, 2012, **120**, a143–a144.

31. D. Zalko, C. Jacques, H. Duplan, S. Bruel and E. Perdu, *Chemosphere*, 2011, **82**, 424–430.

32. S. Biedermann, P. Tschudin and K. Grob, *Anal. Bioanal. Chem.*, 2010, **398**, 571–576.

33. J. M. Braun, A. E. Kalkbrenner, A. M. Calafat, J. T. Bernert, X. Ye, M. J. Silva, D. B. Barr, S. Sathyanarayana and B. P. Lanphear, *Environ. Health Perspect.*, 2011, **119**, 131–137.

34. Joint Research Centre–Institute for Health and Consumer Protection (JRC–IHCP), *European Union Risk Assessment Report, 4,4′-Isopropylidenediphenol (Bisphenol-A)*, JRC, Ispra, 2010.
35. US EPA, *An Exposure Assessment of Polybrominated Diphenyl Ethers*, National Technical Information Service, Washington, DC, 2010.
36. C. Petito Boyce, S. N. Sax, D. G. Dodge, M. C. Pollock and J. E. Goodman, *J. Environ. Protect. Sci.*, 2009, **3**, 75–96.
37. M. Lorber, *J. Exposure Sci. Environ. Epidemiol.*, 2008, **18**, 2–19.
38. Council of the European Union, European Parliament and the Council of the European Union, *Off. J. Eur. Union*, 2003, Directive 2002/95/EC.
39. Council of the European Union, European Parliament and the Council of the European Union, *Off. J. Eur. Union*, 2011, Directive 2011/65/EU.
40. State of Maine, *Title 38 §1609: Restrictions on Sale and Distribution of Brominated Flame Retardants*, Maine State Legislature, Augusta, ME, 2010.
41. State of Maryland, *Senate Bill 556: Chapter 320, An Act Concerning Environment – Decabrominated Diphenyl Ether – Prohibitions*, Maryland General Assembly, Annapolis, MD, 2010.
42. State of Oregon, *Senate Bill 596, A Bill for an Act Relating to Decabrominated Diphenyl Ether*, Oregon Legislative Assembly, Salem, OR, 2009.
43. State of Vermont, *H.444: An Act Relating to Health Care Reform*, Vermont State Legislature, Montpelier, VT, 2009.
44. State of Washington, *House Bill (HB) 1488/Senate Bill (SB) 5515: An Act Relating to Brominated Flame Retardants,* Washington State Legislature, Olympia, WA, 2006.
45. US EPA, *Polybrominated Diphenyl Ethers (PBDEs) Action Plan*, US EPA, Washington, DC, 2009.
46. A. P. Vonderheide, K. E. Mueller, J. Meija and G. L. Welsh, *Sci. Total Environ.*, 2008, **400**, 425–436.
47. S. Shaw and K. Kannan, *Rev. Environ. Health*, 2009, **24**, 157–229.
48. Environment Canada, *Ecological Screening Assessment Report on Polybrominated Diphenyl Ethers (PBDEs)*, 2006.
49. Environment Canada, *Draft: State of the Science Report on the Bioaccumulation and Transformation of Decabromodiphenyl Ether*, 2009; http://www.ec.gc.ca/ceparegistry/subs_list/decaBDE/SR_TOC.cfm (last accessed 18 April 2012).
50. K. Brigden and I. Labunska, *Mixed Halogenated Dioxins and Furans: a Technical Background Document*, Greenpeace Research Laboratories Technical Note 03/2009, Greenpeace Research Laboratories, Exeter, 2009.
51. National Pollution Prevention and Toxics Advisory Committee (NPPTAC), *Recommendation to the EPA Administrator and Deputy Administrator on Incorporating the Functional Use Approach into OPPT Activities,* US EPA, Washington, DC, 2006.

52. US EPA, *DfE's Standard and Criteria for Safer Chemical Ingredients*, 2012; http://www.epa.gov/dfe/pubs/projects/gfcp/index.htm (last accessed 26 March 2012).

53. US EPA, *DfE's Standard for Safer Cleaning Products*, 2012; http://www.epa.gov/dfe/pubs/projects/gfcp/index.htm#Standard (last accessed 26 March 2012).

54. US EPA, *DfE Criteria for Surfactants*, 2012; http://www.epa.gov/dfe/pubs/projects/gfcp/index.htm#Surfactants (last accessed 26 March 2012).

55. CleanGredients, *CleanGredients® Home*, 2012; http://www.cleangredients.org/home (last accessed 26 March 2012).

56. US EPA, *DfE Alternatives Assessment for Nonylphenol Ethoxylates, September 26, 2011 Draft*, US EPA, Washington, DC, 2011.

57. R. S. Boething, E. Sommer and D. DiFiore, *Chem. Rev.*, 2007, **107**, 2207–2227.

58. X PRIZE Foundation, *X PRIZE Foundation: Revolution Through Competition*, 2012; http://www.xprize.org/ (last accessed 03 April 2012).

59. California Environmental Protection Agency, *Green Chemistry*, 2012; http://www.dtsc.ca.gov/PollutionPrevention/GreenChemistryInitiative/index.cfm#Green_Chemistry_Initiative_Documents_and_Information (last accessed 13 April 2012).

60. Washington State Department of Ecology, *Children's Safe Products Act: Steps Toward Safer Chemical Policy*, 2011, http://www.ecy.wa.gov/programs/swfa/cspa/ (last accessed 13 April 2012).

61. Minnesota Department of Health, *Toxic Free Kids Act: Chemicals of High Concern and Priority Chemicals*, http://www.health.state.mn.us/divs/eh/hazardous/topics/toxfreekids/index.html, 2011 (last accessed 13 April 2012).

62. Connecticut General Assembly, *Bill Status: Substitute for Raised H.B. No. 5650*, 2008; http://www.cga.ct.gov/asp/cgabillstatus/cgabillstatus.asp?selBillType = Bill&bill_num = 5650&which_year = 2008 (last accessed 13 April 2012).

63. Maine Department of Environmental Protection, *Safer Chemicals in Children's Products*, 2011; http://www.maine.gov/dep/safechem/ (last accessed 13 April 2012).

64. European Chemicals Agency (ECHA), *Guidance on the Preparation of an Application for Authorization*, ECHA, Helsinki, 2011.

65. R. McFadden, personal communication to D. DiFiore, 2012.

66. US EPA, *ToxCast™: Screening Chemicals to Predict Toxicity Faster and Better*, 2011; http://www.epa.gov/ncct/toxcast/ (last accessed 26 March 2012).

67. SUBSPORT, *SUBSPORT: Moving Towards Safer Alternatives*, http://www.subsport.eu/, 2012 (last accessed 12 April 2012).

68. BizNGO, *Business NGO Workgroup*, http://www.bizngo.org/, 2008 (last accessed 12 April 2012).

69. Washington State Department of Ecology, *Pollution Prevention: Assessing the Safety of Chemical Alternatives*, 2012; http://www.ecy.wa.gov/programs/hwtr/ChemAlternatives/index.html (last accessed 09 April 2012).

70. Nike, *Nike Roadmap Toward Zero Discharge of Hazardous Chemicals*, 2011; http://nikeinc.com/news/nike-roadmap-toward-zero-discharge-of-hazardous-chemicals (last accessed 12 April 2012).

71. F. Hugelmeyer, *Champions of Change*, 2012; http://www.whitehouse.gov/blog/2012/04/12/creating-sustainable-practices-change-way-world-does-business (last accessed 8 February 2013).

72. OECD, *The OECD Environment, Health and Safety Programme: Achievements, Strengths and Opportunities*, OECD, Paris, 2011.

NGO Initiatives in the EU – Identifying Substances of Very High Concern (SVHCs) and Driving Safer Chemical Substitutes in Response to REACH

JERKER J. LIGTHART

ABSTRACT

Tens of thousands of man-made chemicals have been developed over the last few decades. Many are used in everyday products, from toys, textiles and paints to shampoo, electronic equipment and building materials. The problem is that we know little about their short- and long-term effects on our health and on the environment. But we do know enough to understand that many of the chemicals in widespread use today are hazardous. However, in order to control the use of hazardous substances properly, one first needs to identify which substances should actually be prioritised, then different strategies can be used to promote the substitution of hazardous chemicals with safer alternatives. REACH, the new European Chemicals regulation, provides a legal framework both for the identification of hazardous substances and for substitution. This chapter outline show ChemSec, as an environmental non-governmental organisation (NGO), has been influencing the phase out of hazardous chemicals together with key stakeholders such as policy makers, progressive companies and financial investors. ChemSec has 10 years of experience in encouraging the development of chemicals management by offering an open dialogue, and also concrete tools with an emphasis on identifying hazardous substances through the SIN List and more recently promoting the use of safer alternatives via SUBSPORT. The chapter also give a

Issues in Environmental Science and Technology, 36
Chemical Alternatives Assessments
Edited by R.E. Hester and R.M. Harrison
© The Royal Society of Chemistry 2013
Published by the Royal Society of Chemistry, www.rsc.org

broad overview of the steps that NGOs take to influence and encourage companies and regulators both to strengthen chemicals regulation and to push them to perform better on all levels.

10.1 ChemSec Background

ChemSec, the International Chemical Secretariat, is a non-profit organisation based in Gothenburg, Sweden, working towards a world free from hazardous chemicals. ChemSec was founded in 2002 by four leading Swedish environmental organisations: Friends of the Earth, WWF, Swedish Society for Nature Conservation and Nature and Youth.

ChemSec is primarily focused on monitoring, improving and promoting European Union (EU) chemicals regulation, but it also follows other global chemicals issues such as SAICM (Strategic Approach to International Chemicals Management) and the Stockholm Convention.[1,2] By being a highly focused, single-issue organisation, its work and activities are committed to improving the management of industrial chemicals used in everyday products. The ChemSec staff are a highly dedicated team of chemists, political scientists, business experts and communicators providing a unique mix of professionals with one common goal.

ChemSec operates through support from a broad spectrum of society. One of the main contributors is the Swedish Government, but ChemSec also receives financial support from charitable foundations and other non-governmental organisations (NGOs). However, ChemSec has never accepted financial support from companies or organisations with a financial interest in chemicals management, making it fully independent from the organisations it tries to influence, which is of paramount importance from a credibility point of view.

ChemSec strives to promote collaboration and an open dialogue between all parties interested in phasing out hazardous chemicals, in order to bridge the gap between decision-makers, industry, NGOs and scientists. At the same time, ChemSec ensures that current scientific knowledge is made available and understandable during regulatory processes. It offers expertise and guidance on chemical management policies in order to influence the development of progressive chemicals legislation. A perfect example of such collaboration is the SIN List of hazardous chemicals.

To achieve the goal of a toxic-free environment, ChemSec is actively working to:

- Achieve broader acceptance in society for the key principles of Precaution, Substitution, Polluter Pays and Right to Know for consumers and society.
- Promote legislation that builds upon these principles, while also ensuring that regulations are respected and enforced.
- Encourage business stakeholders to support and apply these principles – by showing that substituting hazardous chemicals with safer alternatives is both technically and economically feasible.

- Help financial investors to include chemicals when assessing a company's environmental performance and the financial risks of producing chemicals that are likely to face future regulation.
- Serve as an open and dynamic forum for all parties working for more effective control on chemicals.

10.2 Overview of NGO Initiatives – Putting REACH into Practice and Working with Companies

In June 2007, the European Union's new framework policy on industrial chemicals, REACH,[3] came into force. REACH stands for registration, evaluation, authorisation and restriction of chemicals. This is the first time that chemicals regulations have required commercially available chemicals to be assessed with regard to their intrinsic properties and to have that information published for anyone to access. It is unique in its high ambition to protect both human health and the environment with its declared intention to phase out the most hazardous substances and replace them with safer alternatives. However, since it treads on new regulatory territory, it will also require a longer and steeper learning curve before it truly lives up to its own ambitions. NGOs and many other stakeholders are striving to make sure that all the possibilities of REACH are made use of and this section highlights a few areas where special attention has been invested by NGOs.

10.2.1 The SIN List

Developed by ChemSec in cooperation with a number of major NGOs, this list of hazardous chemicals is based on the criteria for Substances of Very High Concern as laid down in REACH, the EU chemicals regulation. The aim of the project is to provide a platform of concrete examples of the most hazardous substances in order to speed up implementation of chemical policies in the EU and encourage businesses to substitute hazardous substances used in products with safer alternatives ahead of regulation. ChemSec works actively to increase the awareness and use of the SIN List, in addition to keeping it updated. It has proven to be a huge success with numerous companies and regulators referencing the SIN List both in a regulatory context and in proactive chemicals management and work on toxic use reduction among companies.

10.2.2 Substitution Support Portal – SUBSPORT

ChemSec is one of four partners in the development of a free-of-charge, multilingual, Internet-based Substitution Support Portal (SUBSPORT) that helps companies in fulfilling the substitution requirements of EU legislation and is a resource for stakeholders such as authorities, environmental organisations, trade unions and scientific institutions. SUBSPORT aims to become the leading database for substitution worldwide. The online resource provides extensive information on different regulations addressing substitution, how chemicals should be substituted, and a database made up of substitution examples – 'case stories' – from companies and organisations. This database will continue to grow and also serve as an interactive discussion forum. In addition to the web resource, the project has developed materials to provide training in identifying and assessing alternatives to hazardous chemicals.

The SIN List and SUBSPORT are discussed in more detail in the following sections.

10.2.3   Transparency

The chemical manufacturing industry has for a long time been shrouded in secrecy and protected from public scrutiny with regard to which hazardous chemicals they produce and how they are used. This situation has been the default due to the protection of so-called confidential business information (CBI) that has never been questioned or even asked for in the past. However, with REACH, an increased focus on supply chain communication as a means of phasing out hazardous substances, and legislative demands for transparency in certain cases, for instance, chemical safety reports (CSRs), has forced the industry to be more transparent.[†] Even if the tools for disclosing chemical and environmental information are available in the legislation, there has been little action from the authorities actually to disseminate this information. Transparency and access to substance information are a key factor when companies decide to substitute a specific substance or not, especially among retailers and downstream users of chemicals. Giving them access to information on chemical hazards will at the same time empower them to ask for safer and better alternatives, a chance few of them have had in the past. The downstream users and retailers are the ones who put their brand name on a certain article and hence it is their company whose reputation will suffer if it is revealed to contain hazardous chemicals.

Unfortunately, owing to the lack of action from the authorities, NGOs were forced to file a lawsuit against the European Chemicals Agency (ECHA) in order to gain actual access to the information. The reason why it was necessary to bring ECHA to court was that they refused to disclose the names of chemical producers of very hazardous chemicals on the SIN List, and also the production volumes for these substances. ChemSec together with another NGO, Client Earth, found it necessary to file a formal complaint with the European Court of Justice. Following the lawsuit, ECHA has promised to disclose the names of producers and tonnage bands for most producers. However, even though they have promised to publish almost all of the requested information,

---

[†]The CSR is a key source of information provided to all users of a certain chemical in the whole supply chain from producer to consumer containing information on intrinsic properties and hazards, conditions of use, expected emissions/exposures and risks following such emission/exposure.

in court, they stand by their position that it should not be disclosed since it is considered to be CBI.

10.2.4   Business Initiatives

The ChemSec Business Group is a collaboration among companies working together to inspire concrete progress on toxic use reduction. It gathers market-leading companies, across a diversity of sectors, for the development of effective corporate practices in the substitution of hazardous substances. It also raises public awareness of companies' efforts to be drivers on this issue. The goal is to find common ground between ChemSec's pursuit of a world free from hazardous chemicals and individual companies' ambitions to be successful and more sustainable in their operations. Where regulation has been slow and inconsistent around the globe, companies are forced to develop their own strategies to comply with their own standards and consumer demands, resulting in increased costs.

The ChemSec Business Group consists of downstream enterprises, such as retailers, manufacturers of consumer goods and construction businesses. These progressive companies have expressed their support for stricter chemicals legislation and/or they are actively seeking to avoid hazardous substances in their production, going beyond current legislation, but rather in response to consumer demand, risk management and other responsible business priorities. Companies active within very different sectors of industry nonetheless often face similar challenges and can apply similar solutions.

The group is also used as a platform for public activities or partnership opportunities aimed at promoting the understanding of, and providing solutions to, emerging chemicals issues. The objective is to drive the public and regulatory debate in addition to increasing engagement in chemicals-related issues.

ChemSec has together with the group organised a number of activities to push for progressive chemicals policies, including seminars in the European Parliament, business-to-business conferences, and the production of a number of publications demonstrating the business sense of sustainable chemicals policies in industry. Apart from the close cooperation within the business group, we are also in dialogue and in *ad hoc* partnerships with a number of other companies. This cooperation has been especially fruitful in areas such as SUBSPORT, the SIN List and the joint efforts to reduce the use of hazardous materials in electronics. An example of such a partnership is described in detail below.

10.2.5   NGO–Business Cooperation – Electronics

Discarded electronic products are one of the fastest growing waste streams in the world, causing health and environmental problems far beyond where they are produced and used. Major electronics manufacturers have already initiated substitution away from problematic brominated flame retardants (BFRs) and

PVC and adopted a methodology for future limitation of such substances in electronic products. During the revision of the EU RoHS directive, NGOs and some of the world's largest electronics manufacturers had a common interest in making the RoHS directive more efficient and creating a level playing field with regard to environmental performance.[4] The result was that major companies in the industry, otherwise often seen as pulling backwards, joined forces with NGOs in order to strengthen the law, to give companies a better possibility in the future to phase out some of the most hazardous substances used in consumer articles. Partnerships such as this one are interesting for policy-makers since they not only identify an important problem, but also point towards alternative solutions that are readily available and in use.

10.2.6 Engaging Financial Investors

The investment sector has the potential to be an important stakeholder in promoting the production and use of safer chemicals. Especially among socially responsible investors (SRI), environmental aspects play a role alongside social and governance issues. From the environmental perspective, energy and climate change have been heavily prioritised lately and numerous theme funds are investing in renewable energy companies or financing activities to avoid CO_2 emissions. Water and waste management also attract attention, but unfortunately other environmental issues such as hazardous chemicals have so far played a subordinate role. However, with the implementation of REACH and future revisions of chemical regulations elsewhere, such as the Toxic Substances Control Act (TSCA) in the USA, this is starting to change. Regulations such as REACH will have a considerable impact on chemical manufacturers, and also on downstream companies and retailers. To avoid financial risks and lesser returns, hazardous chemicals are becoming an increasing issue on the investment horizon.

The so-called authorisation procedure within REACH states that substances subject to authorisation cannot be used in the EU after a specific date, called the sunset date, unless an authorisation has been granted for that specific use. The responsibility for seeking authorisation lies with the producer, importer or downstream user of a substance and can only be granted if the company proves that the risks of continued use are adequately controlled or that the societal benefits outweigh the risks to human health and the environment. Applications will be costly, €50 000 per substance and use, with no guarantee that an approval will be granted.

Producers and users of hazardous chemicals facing possible future restrictions, such as those listed on the SIN List, face the risk of increased costs associated with reformulating products and modifying processes, which can have significant implications for a company's financial performance. This implies elevated risks for companies with long production cycles. A product that is being formulated or designed today, but put on the market in a few years time, could require the use of a substance which by that time has been restricted through REACH.

Product recalls are also rising: in the UK there has been a 10% increase per annum since 2003 and they are very costly both financially and for a company's reputation.

At the same time, there are opportunities for winners to seize, in the form of both preparedness and innovation.[5] When substances are restricted, a need will arise for alternative chemicals. The producers and users of these alternatives can gain market shares and probably out-perform the sector.

Therefore, ChemSec has increasingly been in dialogue with the investment community about risks and opportunities with chemicals management. As investors need information about companies rather than substances, ChemSec has developed new tools for this group to use. These include the 'SIN producer list,' which names all companies that produce or import SIN List chemicals in Europe, and a comprehensive chemicals criteria catalogue, which can be used to evaluate chemical-producing companies.

10.3   The SIN List in Focus

The SIN (Substitute It Now) List has been developed to highlight the need for swift implementation of the REACH system for identifying and phasing out chemicals of high concern, creating a scientifically robust list of substances fulfilling the REACH criteria for Substances of Very High Concern.[6] This section describes how the SIN List has been developed and the methodology that has been used for identifying and evaluating substances for the SIN List.

The SIN List was developed by ChemSec in close collaboration with other leading NGOs. It also included toxicologists for assessment and review, and companies taking into consideration the business perspective as provided by the ChemSec business group. The SIN List is being used by a variety of stakeholders from companies, policy makers, interest organisations and financial investors.

10.3.1   REACH and Substances of Very High Concern

REACH aims to ensure that basic information on industrial chemicals used in the EU is provided and that the use of the most hazardous chemicals is limited or prohibited through either restriction or authorisation procedures. The success of REACH will depend on a prompt, effective process for identifying the most hazardous chemicals on the European market and replacing them with safer alternatives.

The most hazardous substances according to REACH should be identified on a case-by-case basis and be designated as Substances of Very High concern (SVHCs), which are subject to close scrutiny through a so-called 'authorisation process' (Figure 10.1). However, the mere fulfilment of the SVHC criteria does not mean that a substance is automatically placed on the candidate list, containing all identified SVHCs. In order for a substance to be listed on the candidate list, it must first be nominated by either an EU Member State or the European Chemicals Agency (ECHA) on behalf of the European Commission,

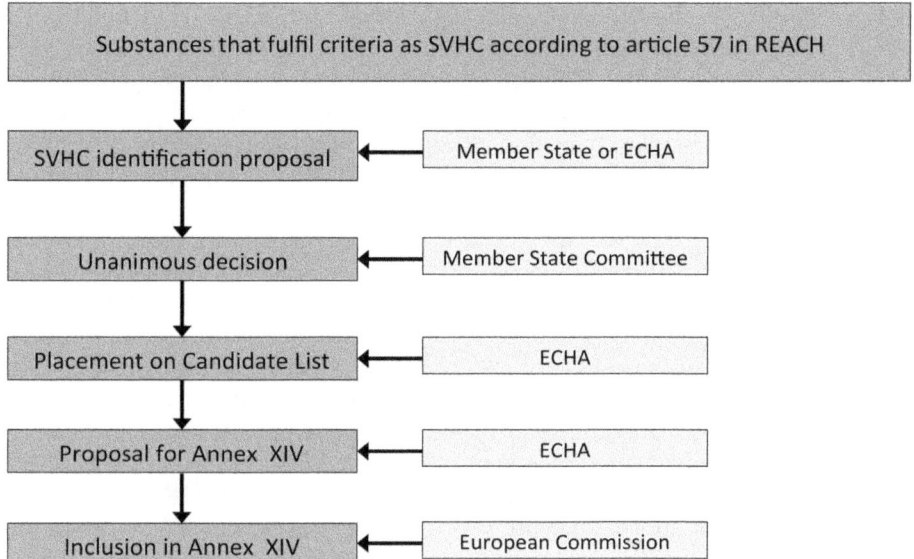

Figure 10.1   A schematic overview of the authorisation procedure in REACH and the responsible actors involved at different stages of the process.

a dossier needs to be prepared and subsequently all Member States must unanimously agree that it is indeed an SVHC. The candidate list is an important step in the authorisation process since placing of a substance on the candidate list triggers specific obligations for companies to inform downstream users and consumers about the presence of this substance in products lower down the supply chain.

The EU has started the process of populating this candidate list with these undesirable substances, but so far (February 2012) only 73 have been officially identified despite the fact that hundreds more are known to fulfil the set criteria via other EU-agreed processes. The current official list can be found on ECHA's webpage.[7] SVHCs are divided into six different categories (see below) based on the legal paragraph (§57) in REACH. However, a substance can fulfil more than one set of criteria. Depending on which criteria a substance fulfils for inclusion in the candidate list, different options and requirements apply in later stages of the process:

(a) carcinogenic [C]
(b) mutagenic [M]
(c) toxic to reproduction [R]
(d) persistent, bio-accumulative and toxic [PBT]
(e) very persistent and very bio-accumulative [vPvB]
(f) equivalent level of concern, such as endocrine disruptors [57(f)]

10.3.2 General Principles Used for the Compilation of the SIN List

All substances on the SIN List fulfil the criteria for SVHCs as defined in the REACH regulation and fall into at least one of the six REACH SVHC categories mentioned above. The first SIN List, 1.0, was released in September 2008 and has since proven to be valuable in providing advance guidance for both companies and regulators when identifying and phasing out hazardous substances. With the update of the SIN List 2.0 in May 2011, the spotlight was placed on endocrine-disrupting chemicals (EDCs) as SVHCs that need to be urgently addressed by the EU. All CMRs, PBTs or equivalent level of concern substances have been screened to identify primarily substances covered by the authorisation provisions in REACH. Substances exempt or otherwise not regulated by REACH, such as pesticides, unintentionally produced substances and substances already banned through global conventions, for example the Stockholm Convention, have accordingly been removed. However, a single substance can often have multiple uses and functions, including both included and exempt uses, as known to us. Such substances have hence been kept in the SIN List.

After the initial screening to find the relevant substances, CMR and PBT substances were easily added since use was made of their official classification and agreed criteria. Equivalent concern substances [Article 57(f)] added to the SIN List 1.0 or 2.0 went through a further literature research and case-by-case assessment phase to find the substances fulfilling REACH criteria for SVHCs. These additional steps on a case-by-case basis were needed since the available guidance document as provided by ECHA is rather general, lacking specific criteria and definitions. Technical guidance documents provide assistance on how to interpret the legal texts in REACH and have been developed by ECHA in cooperation with stakeholders and EU Member States. However, they are not legally binding documents.[8]

It should be clearly stated that the applied screening criteria and methodology do not capture all substances that could potentially fulfil the SVHC criteria. Further, the chosen starting points, the 25 black and grey lists for SIN List 1.0 and the European Commission database on EDCs for SIN List 2.0, limited the number of substances to those already suspected as substances of concern when the lists were drawn up. Therefore, the SIN List should not be considered as a final list, but merely a first step towards a more comprehensive list of SVHCs in need of regulation.

The SIN List should also be considered as a conservative list, which is exclusive rather than inclusive with regard to substances possibly meeting the criteria of SVHCs. Some substances not included, *i.e.*, not passing this conservative evaluation as SVHCs, might nonetheless still have strong indications of a number of concerning properties that should not be neglected and would thus need further evaluation.

10.3.3 How the SIN List Has Been Used and Received

Since the start, a large number of companies have become interested in the SIN List and its methodology, recognising the added value from a business perspective of being able to identify hazardous substances before any legal requirements and hence be ahead of the regulatory curve. Since the launch, many companies have incorporated the SIN List into their toxic use reduction programs in one way or another. Moreover, regulators have picked up an interest in the SIN List and started to look closer at the information contained therein for further regulatory action. The correspondence between the SIN List and the official candidate list is very good, confirming the relevance of the SIN List approach.

10.3.4 Detailed Methodology

10.3.4.1 Substances Officially Classified as Carcinogenic, Mutagenic or Toxic to Reproduction (CMRs). REACH automatically recognises substances that have been officially classified as CMR substances in the EU as by default meeting the criteria of SVHCs. However, even if CMR substances are fully recognised as SVHCs, they are not automatically included; therefore, the European Regulation on classification, labelling and packaging – the CLP regulation – was used.[9] The CLP regulation contains a register of all officially classified substances, including CMR substances category 1A or 1B. The CLP registry is updated regularly and successive updates after September 2008 have been subsequently incorporated into the SIN List to ensure a high concurrence between the SIN List and the CLP registry. All entries in the CLP registry concerning substances with CMR category 1A and 1B properties were used as the starting point. However, since not all substances included in the CLP registry are relevant from a REACH perspective, a further refinement was needed. First, entries lacking CAS numbers and EC numbers (a unique number under which a substance is registered in the EU) were removed since they did not identify a unique substance or a unique substance group. The registry also includes contaminants in otherwise less hazardous mixtures and such issues needed to be resolved. In particular, where a mixture included a substance classified as CMR, the mixture was excluded. However, the individual CMR substance contaminating the mixture was still included.

Substances exempted from the authorisation procedure were also removed, for example, complex hydrocarbon distillates occurring to a very large extent in product streams coming from refined or unrefined petroleum oil (fuels). These substances were easily identifiable because they have dedicated index number series in the CLP registry starting with 648- or 649- (648-xxx-xx-x and 649-xxx-xx-x).

Registered pesticides were also identified and all substances having a standardised pesticide name assigned by the International Organization for Standardization (ISO) were subsequently removed.

The above procedure resulted in the final selection of 317 CMR substances for the SIN List. Six of these substances are also PBT substances and identified below.

10.3.4.2 Substances Classified as PBT/vPvB. REACH puts a specific emphasis on targeting persistent and bio-accumulative substances through regulatory measures in order to phase them out of the market. In REACH, certain criteria have been established laying down how such substances should be identified and which criteria should be fulfilled. For persistence, depending on environmental compartment, a half-life from > 40 days in fresh water to > 180 days in marine sediment is considered to be a persistent chemical. To be recognised as a very persistent chemical, the half-life should be > 60 days in any water body and > 180 days in any sediment or soil. To be considered a bio-accumulating substance, a bio-concentration factor (BCF) of more than 2000 must be exceeded and over 5000 for very bio-accumulative substances. All BCFs are based on concentrations in water for aquatic species.

PBT/vPvB substances have no official classification, risk phrase or hazard statement connected with them and need to be evaluated on a case-by-case basis. However, a lot of work has previously been invested in the PBT Working Group, which was an official assembly of representatives from EU Member States and experts from the former European Chemicals Bureau (ECB). They reached a common agreement for a number of substances fulfilling the pre-REACH EU criteria for PBTs and vPvBs. These criteria are very similar to, although not identical with, those in REACH. Even if there is no direct link between an agreement reached in the PBT Working Group and REACH criteria for SVHCs, the differences between the criteria are small and all those substances had previously been agreed as being PBT/vPvB. Consequently, all 23 substances were initially included for screening.[‡,10] However, owing to the same exemptions as for CMR substances, REACH is not applicable to all of these substances and they were subsequently removed.

The remaining 17 substances fulfilling PBT and/or vPvB criteria were then added to the SIN List (six of these are also classified CMRs).

10.3.4.3 Substances Fulfilling REACH Criteria of Equivalent Level of Concern.
Whereas the selection of CMR and PBT substances was straightforward and based on official existing EU classifications and previous assessments and agreements, the selection for chemicals of equivalent concern required considerably more background research and a thorough case-by-case evaluation.

The sixth category of SVHCs mentioned above, equivalent level of concern, is a category that was introduced as a safety net in REACH in order to include those hazardous substances of equivalent level of concern to CMRs and/or PBTs where there is scientific evidence for probable serious effects (REACH article 57f). The aim is twofold. First, it is recognised that the other categories are rather limited in application since they require that the substance meet either the criteria as a CMR or that a substance will fit within the defined criteria as a PBT substance. However, for borderline cases that just miss both

---

‡REACH has an additional endpoint for Persistence (P) which is half-life in soil > 120 days (or > 180 days for very Persistent). The PBT Working Group has a wider Toxic criterion, including those which have been assigned the codes R45, R46, R60 and R61 or Xn – R62, R63 and R64 under the EU Classification and Labelling Directive 92/32/EEC1. The toxic criterion also includes long-term toxicity such as endocrine disruption on a case-by-case basis.

categories, the equivalent level of concern foresees that such substances could be just as dangerous for human health or the environment. Second, within the scientific community there is continuous development, improvement in knowledge and acceptance of new test methods. These future advances, which cannot be properly defined today, should not be excluded and prevent substances from being considered as SVHCs in the future. Substances with the ability to disrupt the endocrine system (EDCs) are mentioned as one specific example of a category causing such equivalent level of concern. In the future, nano-particles, neurotoxins and sensitisers might very well also be included in this category.

The common denominator for these substances is that there are no common classifications, as there are for CMR substances, nor common criteria as for PBT substances, for measuring any given property against. It will instead rely entirely on the availability of data and expert judgement to draw conclusions regarding their SVHC properties. There is a wide variety in intrinsic properties among these substances and the ECHA guidance calls for a substance-by-substance identification and assessment, which is also how the SIN list has been developed.

Worth noting is also that the entries of two SIN substances have been changed during the process. These two substances were initially included on the basis of the available scientific evidence as being of 'equivalent level of concern' but they have now received official recognition regarding their hazardous properties and have become officially classified as CMR substances.

10.3.4.4 Method for Inclusion of Equivalent Level of Concern Substances on the SIN List. In short, identifying equivalent level of concern substances as SVHCs and adding them to the SIN List has been a three-stage process that has been repeated twice independently (SIN List 1.0 and 2.0, respectively):

1. Screening phase – selection and filtering of substances relevant for REACH authorisation.
2. Literature research on selected substances.
3. Evaluation against REACH criteria for SVHCs.

We have gone through the same assessment process twice individually, focusing on two different categories of substances to cover as broad a spectrum as possible:

I The first time (SIN List 1.0), the aim was to identify substances in something that could be called a 'grey area' in regulatory terms, since these substances are known to be problematic but they have not yet been properly classified or regulated in a sufficient manner. In this first round, no specific kind of substance property or endpoint was favoured, even if, for instance, substances with potential endocrine-disrupting (ED) properties were specifically included. However, during the assessment, ED properties were assessed among all other endpoints such as persistence, carcinogenicity and mutagenicity.

II The second time (SIN List 2.0), the aim was specifically to identify EDCs of regulatory concern within the REACH context, since they have the ability to cause severe harm to both human health and the environment. Today they are only partly covered by official classification within the EU, for example, when acting on reprotoxic endpoints, but no EDC-specific classification exists. EDCs pose a threat to human health and the environment owing to their negative impacts on the hormone system that can lead to a variety of detrimental effects. Many of them are also readily available in consumer products, indicating a wide dispersive use and many of them are produced in large quantities. During the second assessment round, *only* chemicals with endocrine-disrupting properties have been considered and included in the assessment. This means that no other hazardous properties than the ED effect have been taken into consideration even if they exist to a higher or lesser degree for all substances.

10.3.5   The First Assessment Process (SIN List 1.0)

10.3.5.1   First Step – Screening Phase.   The screening phase had two aims. First, to cover as many 'grey zone' substances as possible, a first rough list was compiled of substances from many different records or lists of recognised hazardous chemicals. Examples of such lists are the OSPAR list of chemicals of possible concern and priority action, the EU Water Framework Directive, the Swedish Chemicals Inspectorate (KEMI) PRIO list, and lists compiled by the US and Canadian Environmental Protection Agencies.[11,12] Further, substances listed on collaborating companies' grey and black lists were included. This approach gave a very wide and inclusive selection of substances that in one way or another had been flagged owing to their hazardous properties in the past. The resulting rough list contained ∼4000 substances covering many different sorts of concern, both relevant and irrelevant. Throughout the compiling procedure, all risk phrases and classifications (official and unofficial) were kept for each substance to facilitate the subsequent screening process.

Second, the aim of screening was to find only the substances that would be relevant for REACH in general and consumers in particular with a proper dataset available. In order to have a positive and unique identification of each substance, any duplicate entries, references to substance groups and other substances not having a CAS number or EC number were removed. To identify the substances relevant to REACH and consumers, the Swedish Chemicals Agency (KEMI) was asked to search its 'Products Register' for the occurrence of these 4000 substances in chemical products and preparations available to consumers.[13,14] The response from KEMI was a refined list of ∼250 of the original 4000 substances. The chemical uses identified through the Products Register were used as proxy for identifying hazardous chemicals to which consumers might be exposed on an EU-wide basis. The actual uses, in Europe and globally, presumably go beyond the Swedish Product Register, hence the potential number of chemicals eligible for inclusion might be far greater. These 250 substances were cross-checked with information from the former European Chemicals Bureau to obtain information on only high production volume chemicals which normally are considerably better known, tested and have much more toxicological data available needed in later evaluation steps. This step reduced the list further to ~150 substances.

Substances whose hazardous properties were of a physical nature (corrosive, explosive, flammable, *etc.*) were removed, together with chemicals classified as CMRs (category 1A and 1B) already covered above, pesticides and other substances that are exempted from REACH as a whole or from the author-isation procedure.

Setting up the automatic and strict screening criteria as above was very useful in order to reduce the number of substances considerably and increase the possibility of providing available toxicity data. However, in order to reduce the numbers further to a manageable amount of substances, manual selection was needed. When manually selecting substances, priority was therefore given to substances whose (indicative) risk phrases suggested ED, CMR (category 2) or PBT (R51 and R50–53) properties, totalling 35 substances. However, these properties are based on the information from the original lists and the sub-stances are therefore not necessarily officially classified within the EU according to these risk phrases.

Owing to the strict automated screening criteria applied in the process, a number of well-known grey area substances and also EDCs fell out of the scope. These high-profile substances often found in human bio-monitoring studies or frequently mentioned in human health and environmental studies were manually included for evaluation. The presence of a man-made chemical in Nature or in human bodies often indicates persistence and possible bio-accumulation. The European Commission had started looking into the issue of EDCs and had commissioned different reports to identify EDCs of concern for regulators. Being an issue of increasing concern, EDCs assessed to be of high or medium concern in the European Commission report on EDCs (COM 2001/262) were also included.[15]

These substances added a further 25 substances to the list of substances to be assessed, making the final number of potential equivalent level of concern substance to be evaluated 60.

10.3.5.2 Second and Third Steps – Scientific Literature Research and Assess-ment. In order to make a proper assessment of each substance, external tox-icologists were assigned to first make an exhaustive literature search for each of the 60 substances of potential equivalent level of concern filtered out in the screening phase. After the collection of data, using publicly available records and published literature and also official risk assessments and sum-mary reports when available, they were instructed to conduct an in-depth assessment on each substance to determine whether these substances would qualify as SVHCs under REACH. As mentioned previously, there are still no established strict step-by-step criteria for determining if a substance is to be considered to be of equivalent level of concern or not. The toxicologists were therefore instructed to use the official REACH guidance document on how to identify 'equivalent level of concern' SVHCs and prepare an Annex XV dossier as stated in the *Guidance for the Preparation of an Annex XV Dossier on the Identification of Substances of Very High Concern* from June 2007.[16] This was then used as a basis for the SVHC assessment.

During the assessment phase, the toxicologists looked at the combined impact of the properties that had been confirmed during the literature survey. This resulted in a weighted approach where all known properties and gathered data were included. Bio-monitoring and environmental fate were also taken into consideration. The dataset included CMR and ED properties and also tendencies to persist in Nature and/or bio-accumulate and whether the substances had been detected in humans and biota. When assessed together, this combination of different hazards and properties, which individually might not have fulfilled the criteria of SVHC, built up a strong case for an equivalent level of concern substance.

After all assessments had been completed, the gathered background data and conclusions drawn by the toxicologists were subjected to further scrutiny by another group of external scientists in order to review and confirm the findings. The final outcome of the assessment process was that 30 substances of the assessed 60 were found to fulfil REACH criteria of SVHC as equivalent level of concern and were added to the SIN List 1.0. Among the remaining 30 substances not included in the SIN List, several were borderline cases but owing to the strict and conservative assessment applied they were consequently left out.

10.3.6   The Second Assessment Process (SIN List 2.0)

To ensure coherence in the SIN List, the same methodology used during the development of the SIN List 1.0 was used, but specifically adapted to identify and assess ED properties.

10.3.6.1   First Step – Screening Phase.   Since only the ED properties of the substances were to be addressed, a different starting point was needed. At the time work began on identifying these substances, there was only one official database listing potential EDCs in the EU, hence this became the natural starting point.[17] This was the same database that to a minor extent had been included during the first assessment process. Owing to the gap of 2 years between the first and second assessment processes, the database had now been supplemented and expanded with even more potential EDCs. The full database consists of 553 substances that have been grouped with regard to their ED potential. The substances are organised into three categories depending on the available information for each substance: category 1 for which there is at least one study *in vivo* indicating ED properties, category 2 for at least one *in vitro* study and category 3 for which there are no data indicating ED properties. This categorisation is not an official classification but a suggested system group for EDCs based on the available information. To focus on the most relevant ones and to narrow down the number of substances to assess further, it was decided to look only at the chemicals indicating ED properties, that is, category 1 and 2. This reduced the number of substances to 319. Based on the available information, substances were excluded from the evaluation list based on the same exclusion criteria as used for SIN List 1.0.

Further, an evaluation of possible uses for each substance was made. This evaluation was based on three sources. First, the assessments from the European Commission's database were used to identify uses as reported in the background documentation. Second, the Hazardous Substances Data Bank (HSDB) was used to obtain further information on potential uses.[18] Finally, for the substances for which no uses had been identified, an Internet search was carried out to check if there were any other probable uses that had not been addressed by the first two sources.

Substances having no known uses according to the above-mentioned sources were removed along with substances likely to be used only as intermediates or other uses not relevant to REACH, such as pharmaceuticals and registered pesticides.

To ensure consistency, the process and selection criteria were as similar as possible to those used for SIN List 1.0. After the application of these filters in the screening, a total of 41 substances were left to be assessed more closely by toxicologists with expertise in ED chemicals.

10.3.6.2 Second Step – Scientific Literature Research. The literature research was intended to give better understanding of the EDC properties associated with the selected 41 substances through verifying the existing data from the European Commission EDC database and also including the latest scientific research on these substances. The primary work was conducted by external experts from the Endocrine Disruption Exchange (TEDX).[19] The process included a literature search, initial screening, abstract review, selection of studies, data entry and verification, followed up by an internal peer review process.

First a comprehensive literature search in PubMed for each chemical was conducted.[20] Search terms were selected based on the research teams' long-term experience in reading endocrine-related literature. The general approach was to be inclusive, using wide terms such as endocrine, hormone and receptor, as well as terms for the many organs involved in endocrine activity. For a few chemicals, very little information was found in PubMed and additional searches were performed in Web of Science and ToxLine to give a more complete picture of the available literature.[21,22] The full list of search terms used during the literature search is shown in Figure 10.2.

Having used inclusive search terms, the literature search generated a long list of publications for each chemical that had to be checked to identify the most reliable and relevant ones. Therefore, the initial screening of these lists involved scanning abstracts to remove studies that were not published in peer-reviewed

endocrin* hormon* receptor* *estrogen*
androgen *testosterone* *thyr* adrenal
steroid *cortico* cortisol pituitary
hypothalam* hippocamp* pancrea* islet*
insulin testis testes testic* thymus thymic
prostate ovar* sertoli leydig follic* LH FSH
luteinizing *natal* gestat* lactat* sperm*
semen oocyte* oogene* retinoi* reprod*
fertil ahr "aryl hydrocarbon" *cyp*
cytochrome PPAR* "peroxisome proliferat*"
rat rats mouse mice hamster* "guinea pig*"
guinea-pig* murine rodent* human* woman
women men man child* adolescent* infant*
school day-care baby babies fetal foetal
fetus foetal worker* occup* epidemiol*

Figure 10.2   The search terms used when identifying studies with endocrine-
related effects. To be inclusive, '*' was used as a wildcard in these
searches. These terms were combined with individual chemical names/
CAS numbers.

journals, did not represent original primary research or were clearly irrelevant.
For example, studies of pest control, remediation, analytical methods and
toxico-kinetics were removed at this level of screening. Review articles and
other secondary research were used only to locate further primary research.
Most studies of human environmental exposure were removed at this level
primarily because they were based on retrospective self-report, failed to control
for simultaneous exposure to other chemicals and/or were unable to report any
measure of exposure dose.

Having removed the studies obviously not meeting a high scientific standard
following the initial review of abstracts, the remaining studies were downloaded
for review. The goal was to select the studies that provided the strongest evidence for endocrine effects. In addition to the estrogenic, anti-androgenic and
thyroid-based effects that tend to be the focus of regulatory attention, evidence
of hormonally based mechanisms of action in other organs, glands and systems
and at other levels of effects (*e.g.*, gene expression, signalling mechanisms) was
included. Every effort was made to select the most scientifically robust studies.
Studies that did not use appropriate control conditions or for which there were
inconsistencies in the text or tables were not selected. No studies in which null
findings directly contradicted significant findings from another study were
found.

High-dose studies measuring gross endpoints only (*e.g.*, organ weights) in
which the mortality rate was excessive were also not selected. Exceptions were
made for chemicals for which only high-dose studies were available and there
was evidence of an endocrine effect (not a toxic effect). Additionally, in some
cases, effects were found only at the lowest doses studied. Such studies were
evaluated carefully and were not rejected for this reason alone, as endocrine-
related effects are known to exhibit non-monotonic dose responses.

When all studies had been selected, the data for each substance were entered into a database and verified. Only statistically significant findings were reported, with the rare exception of particularly compelling results for which no statistical analyses were conducted (*e.g.*, gene arrays or changes in morphology). With regard to dose, it was not always practical to present the full range of doses used, as some studies used complex experimental designs and others only reported relative binding affinity.

To ensure the accuracy of database and conclusions drawn from them, an internal peer review process was conducted. The final analysis was achieved via a collaborative effort within the entire research team. The researchers reviewed the chemicals one by one, evaluating each study in the database. The test methods employed were discussed in addition to the assays used, whether the effects were truly endocrine related and how the authors interpreted their results. The principles of endocrinology state that endocrine effects encompass not only direct effects on traditional endocrine glands, their hormones and receptors, but also entire signalling cascades. Such cascades affect reproductive function and foetal development and also the nervous system, behaviour, immune system, liver, bone and many other organs and glands. These broad principles of the endocrine system were consequently included in the assessment process.

10.3.6.3   Third Step – Assessment Towards REACH Criteria for SVHCs.   The assessment process to determine whether each substance fulfils the REACH criteria for SVHCs was led by ChemSec with support from an external group of scientists and toxicologists based on the information found in the database in the previous step.

Again, the official REACH guidance document on how to identify equivalent level of concern SVHCs from June 2007 was used as a basis for the SVHC assessment.[23] As stated before, the guidance does not give clear criteria on how to assess EDCs other than that it should be applied on a case-by-case basis. The guidance mentions a few mechanisms and factors to be considered and acknowledges that substances displaying endocrine-active properties can result in changes in growth, development, reproduction or behaviour in the organism or in future generations. The guidance document and the definitions developed for the European Commission database, and also from external EDC experts, were used as the basis for assessment.

To obtain sufficiently robust data, all eligible substances needed to have test data primarily from *in vivo* tests obtained through studies of documented endocrine disruption in intact animals. This information was then supplemented with *in vitro* data from experiments performed in test-tubes and individual cell (lines), as supporting evidence. To establish a reliable and robust dataset, at least three studies were considered necessary with a minimum of two *in vivo* studies, to qualify for in-depth evaluation.

Following this approach and the subsequent evaluation, 22 substances were identified as having strong enough evidence to be considered SVHCs with regard to their ED properties. Again, it is important to stress that for almost all the 19 other substances that in this assessment did not meet the strict assessment criteria established by the research team and were consequently not included in the SIN List, strong evidence for their ED properties was found, but at present not strong enough to categorise them as SVHCs and include them in the SIN List.

For the SIN List, it is important to stress that even though we tried to a large extent to limit non-REACH relevant substances, we are aware that not all uses of the substances included in the SIN List will *always* fall under REACH or authorisation procedures. Specific uses might still be exempted, such as substances used as intermediates, in fuels or as pesticides. Also, some of the substances removed in the screening phase may indeed be classified as SVHCs for specific uses covered by REACH, but we do not have knowledge of such uses at present. Substances removed during the screening phase might therefore potentially be considered as SVHCs under REACH in the future.

10.4 Finding Safer Alternatives Through SUBSPORT

10.4.1 Background

Substitution has often been considered as something difficult and unachievable for most companies, both small enterprises and large multinational corporations. However, this view does not necessarily reflect the actual situation. Substitution may be a complex issue that can be time consuming and require financial resources, since it requires knowledge of both the toxicological and technical aspects of chemicals and production processes, and also detailed supply chain communication. However, it can also be relatively quick and easy, if you are inspired by what someone else has already done.

Most companies are actively working on substitution, but for a different reason than replacing hazardous substances with more benign alternatives: simply for product development. The core of the problem is the same, however: it deals with the replacement of one product with a better one that still fulfils the customers' needs and expectations. To assist with this process of substituting hazardous substances and in order to meet future demands regarding knowledge on substitution as required by regulations, the SUBSPORT project offers an Internet portal that constitutes a state-of-the-art resource for safer alternatives to hazardous chemicals.

The free-of-charge multilingual portal is supporting companies in fulfilling substitution requirements posed by EU regulations, such as those specified under the REACH authorisation procedure, the Water Framework Directive or the Chemical Agents Directive. These regulations have specific provisions to minimise and restrict the use of hazardous substances. Specifically, REACH has a very strong legal text that in fact will restrict all uses of substances that have been identified as SVHCs and placed on a specific Annex in REACH, Annex XIV. Such substances may not be used unless a specific authorisation has been granted by the European Chemicals Agency. Such an authorisation may be granted only if all possible routes of substitution have been explored and deemed not to be possible. The routes that need to be explored include an ambitious research and development plan and input from authorities, competitors and third parties, and will also include non-chemical alternatives and technical solutions. It is therefore clear that SUBSPORT has an important role to fill, not only with regard to the implementation of chemical regulations such as REACH, but also for companies that want to find better and safer solutions even if there are no regulatory requirements yet in place. Furthermore, other stakeholders such as authorities, environmental and consumer organisations, trade unions ands scientific institutions can benefit from the information found in SUBSPORT.

SUBSPORT includes a number of different pieces of the substitution puzzle, covering different parts of the process and sharing the knowledge of practical substitution examples in an easily accessible database. As an added feature, the project will also provide training sessions in the art of substitution.

10.4.1.1 Partners. SUBSPORT has been developed by four partners. In addition to ChemSec, the German consultancy Kooperationsstelle Hamburg, the Spanish technical trade union foundation ISTAS and the Danish consultancy Grontmij have all been involved in the work. All partners have a different background, which has helped to give the project a very broad network in addition to incorporating different points of view and perspectives. The common denominator is the experience in building databases on hazardous chemicals or safer alternatives and how to handle hazardous substances in regulation and practice.

Kooperationsstelle Hamburg, based in Germany, is a consultancy firm that specialises in matters involving occupational safety and health and protection of the environment. Kooperationsstelle Hamburg has developed special service activities, which range from the measurement of hazardous substances in ambient air to the development of multilingual and interactive tools for better process design in industry, namely metal surface cleaning.[24]

ISTAS (Instituto Sindical de Trabajo Ambiente y Salud) is a self-managed trade union technical foundation based in Spain and is supported by the Spanish Trade Union Confederation (CCOO) to promote the improvement of working conditions, occupational health and safety and environmental protection in Spain. ISTAS has extensive experience in the substitution of hazardous chemicals for many different sectors and products. ISTAS has developed an online tool for the assessment of alternatives and maintains an alternatives database.[25,26]

Grontmij A/S is a Danish consultancy company, which offers services within the spheres of building, construction, water, environment, occupational health, energy and industry. The consultancy has a special focus on small- and medium-sized enterprises concerning tools and information activities. Grontmij carries out occupational health consultancy in the area of chemical risks and substitution and holds a database for substitution, CATSUB.[27]

10.4.1.2 Network. In addition, the project has created a network of experts and stakeholders who are active in substitution. The network assists in content development and promotion of the portal as well as ensuring sustainable updates and maintenance. This has been an important contribution to the project's goal of raising awareness and promoting safer alternatives. Furthermore, training on substitution methodology and alternatives' assessment is being provided.

10.4.2   The Web Portal

10.4.2.1   Key Pieces of the Substitution Puzzle.   The web portal is the node for accessing all information and constitutes a state-of-the-art resource on safer alternatives to the use of hazardous chemicals. It is a source not just of information on alternative substances and technologies, but also of tools and guidance for substance evaluation and substitution management. From the portal, all resources can be easily accessed and, when applicable, also searched directly.

10.4.2.2   Identification of Hazardous Substances.   Even if there were no immediate legal demand to substitute a particular substance that might affect a specific company, it is always beneficial to identify which substances could be regulated in the future or are of concern for other reasons, such as public concern or pressure by consumer organisations. To assist in this process, a number of lists of individual substances of known concern from different sources have been assembled, making it possible to search and identify potential hazardous substances in use today based on CAS number, EC number or chemical name.[28] The sources of these lists are diverse to reflect the different approaches and needs covered by different stakeholders, including international agreements, authorities, companies and NGOs. Any list, even if very inclusive, will, however, never be able to cover all known hazardous substances and to compensate for this alternative identification criteria for hazardous substances can also be found in the portal.[29] The properties include endpoints and properties such as endocrine disruption, persistence, carcinogenicity and toxicity.

10.4.3   The Substitution Case Story Database

10.4.3.1   Finding the Better Alternatives.   One of the unique features of SUBSPORT is the case story database and substitution inspiration found in the database. The database provides substitution examples and also information on alternative substances and technologies from enterprises, published reports and other sources. The case stories can serve as inspiration and offer concrete help to companies or organisations searching for substitutes for hazardous chemicals and may also prove useful in procurement and in legislative processes, for example. The substances mentioned in the case stories have all been pre-evaluated regarding potential hazards according to the SUBSPORT methodology, which includes a check of the substance database according to SUBSPORT screening criteria. By the end of 2012, there will be at least 500 case stories published in the public database. These will all be pre-assessed with regard to their properties in comparison with the hazardous alternative they are substituting.

Most of these alternatives are actual cases that have already been implemented in practice in different industries. These have been submitted by cooperating companies that are willing to share their knowledge on substitution of hazardous substances and describing how they made the change, but sometimes also indicating what could be improved in future substitutions. Furthermore, there will also be considerably more detailed alternative assessments available for a number of identified substances. Ten selected substances and substance groups of high concern are being assessed in detail, looking in depth at selected uses, functions and alternatives. The substances selected for this closer scrutiny are chloroalkanes, chromium(VI) compounds, bisphenol A, dialkyl phthalates, lead and its inorganic compounds, alkylphenols and their ethoxylates, tri- and tetrachloroethylene, formaldehyde, brominated flame retardants and parabens.

10.4.3.2 Substitution Steps. If a hazardous substance has been identified and an effort to phase it out of a product or a process has been initiated, SUBSPORT can provide guidance, inspiration and practical tips on how to proceed with the process of finding a better alternative. For simplicity, the process has been divided into a stepwise approach in six phases.

1. First, it is important to identify the problem, which properties are desired to move away from. Describe both hazards and useful properties of the substance to be substituted. Ask suppliers and use *reliable sources* to verify chemical hazards. Describe the function of the substance and the working conditions needed to make it work at the desired performance level. Operational parameters such as pH, temperature, quantity and existing equipment: substitution might require changing some of these also. When choosing among different substances to prioritise, consider applicable legislation, company policies and customers' requests.
2. Set clear criteria to eliminate alternatives that are not safer or not safe enough. When establishing elimination criteria, check which substances and substance criteria can be found on *priority lists* of legal bodies or companies to identify potential bad substitution examples that might have been identified by others. The elimination criteria as supported by SUBSPORT are described above.
3. Carry out thorough research on potential alternatives, aiming at finding similar cases that could be used as starting point for a more in-depth assessment. Natural starting points are SUBSPORT, the Internet and asking authorities, professional associations, NGOs and trade unions for advice. Look for *alternatives* already elaborated and implemented: this may lower the substitution cost and associated risks. It is also important

to expand the view beyond a mere drop-in substitution, carefully investigating technical solutions and potential process changes that could fulfil
the function performed by the substance to be substituted. For downstream users of chemicals, one way could also be to ask suppliers to
formulate a safer alternative directly or to provide a technical solution to
solve the identified problem. Time spent on this step could tentatively
save large resources and effort if successful.

4. If several potential alternatives have been identified, they need to be
   assessed in a comparable way together with the existing solution using the
   same method and criteria considering the elimination criteria previously
   established. Analyse costs and benefits for both the shorter term and the
   longer term. Consider that not all benefits might be clearly visible, such
   as reducing days for sick leave, a better working environment and less
   staff turnover. Even if a general risk reduction solution is the ideal
   alternative, most solutions are better suited for some of the risks than for
   others. This makes the adoption of a final solution a balancing act
   matching safety, feasibility, costs and benefits in both the long term and
   short term.

5. To facilitate implementation, a smaller, pilot scale is recommended in
   order to eliminate potential problems before a full-scale operation is
   initialised. Also, careful planning of possible technological and organi-
   sational changes that might be needed should be done at this stage. Assess
   substitution as regards functional performance, impact on workers,
   environment or consumers. Special attention should be paid to possible
   shifts of risks and necessary control measures since new risks might have
   appeared or moved up or down the process.

6. To establish full implementation, consideration should also be given
   to what other measures would be needed when the substitution
   has reached full capacity. The supply chain, which ideally will have
   been involved from the start, and also customers and downstream
   users, should be informed of the change. If the substitution was the
   first to solve a specific problem or included a new and improved
   technology, potentially there could be a new market for knowledge and
   expertise that could be explored. The final step in this phase is to collect
   extended feedback from staff, clients and suppliers to identify issues to
   improve.

10.4.3.3 Making Use of SUBSPORT in Practice. Being a multifaceted
tool, there are many different ways in which SUBSPORT could be used by
different stakeholders. The primary target groups for SUBSPORT are companies active in substitution and also regulators and enforcers that make sure
substitution is actually taking place when mandated by law. From a EU regulatory perspective, SUBSPORT could be used to verify company compliance when applying for a permit to continue to use a substance requiring
authorisation permission. It will be easy to check if there are any possible
alternative avenues found in SUBSPORT that have not been properly

explored by an applicant. Further, it will be possible for regulators to see that some companies have managed to phase out a specific substance that might not need a permit for its use, but still is of high regulatory concern, and if viable alternatives are on the market, the regulatory barrier might be lowered to restrict the use of that substance.

Companies, on the other hand, can make similar use of SUBSPORT to prepare for their authorisation requests. If an alternative for a specific use has already been established, they could potentially save time and money on further investigating new alternatives for other uses not yet covered instead. Responsible companies can also proactively identify good substitution possibilities ahead of regulation in order to pre-empt regulatory demands or customers' requests. When using the hazard criteria compilation and aggregated list of identified hazardous substances, they can identify potentially hazardous substances and avoid future problems and potentially bad substitutions.

10.4.3.4 Training Sessions. SUBSPORT also provides concepts and materials for substitution training programmes specially aimed at small- and medium-sized enterprises, trade unions and other parties working on substitution. These training sessions follow a methodology that has been developed within the project and adopted based on feedback received from participants. The overall goal of the alternatives identification and assessment training is to provide basic concepts and tools to help participants to get started on substitution processes and understand the different points of view of stakeholders involved and their specific interests. Areas covered during these training sessions are the identification of chemicals of high concern, how to identify alternatives and get new ideas, introduction to alternatives assessment and basic cost assessments.

10.5 Conclusion

Even with a proactive chemicals regulation such as REACH in place, it still needs to be properly used and enforced to have the desired effect – protection of human health and the environment. As shown, NGOs can play an important role in the application of the regulation, not only as watchdogs, but also as active stakeholders providing input and solutions for better implementation and compliance. The two tools discussed in this chapter, the SIN List and SUBSPORT, are good examples of concrete instruments for a better chemicals management useful for both companies and regulators in their strive for better control of hazardous substances. We have also described other important NGO activities, including promoting transparency surrounding the production of hazardous materials, increasing dialogue and cooperation between NGOs and companies, demonstrating financial risks for investors and providing concrete alternative solutions. All these actions are small pieces of a greater puzzle creating a better tomorrow.

References

1. SAICM – Strategic Approach to International Chemicals Management, http://www.saicm.org (last accessed May 2012).
2. Stockholm Convention – Protecting human health and the environment from persistant organic pollutants, http://chm.pops.int/ (last accessed May 2012).
3. European Commission – the REACH regulation, http://ec.europa.eu/environment/chemicals/reach/reach_intro.htm (last accessed May 2012).
4. European Commission – RoHS, http://ec.europa.eu/environment/waste/rohs_eee/index_en.htm (last accessed May 2012).
5. Henderson Global Investors, Chemical Safety in Consumer Products Industries, SRI Briefing Paper, Henderson Global Investors, London, 2009.
6. ChemSec – The SIN List, http://www.sinlist.org (last accessed May 2012).
7. European Chemicals Agency - REACH Candidate List, http://echa.europa.eu/web/guest/candidate-list-table (last accessed May 2012).
8. European Chemicals Agency – Guidance for the preparation of an Annex XV dossier on the identification of substances of very high concern, http://echa.europa.eu/documents/10162/13638/svhc_en.pdf (last accessed May 2012).
9. European Chemicals Agency – CLP Regulation, http://echa.europa.eu/web/guest/information-on-chemicals/cl-inventory (last accessed May 2012).
10. European Commission Joint Research Centre – PBT Information System, http://esis.jrc.ec.europa.eu/index.php?PGM = pbt (last accessed May 2012).
11. OSPAR Commission – OSPAR, http://www.ospar.org/ (last accessed May 2012).
12. Swedish Chemicals Agency – PRIO a tool for risk reduction of chemicals, http://www2.kemi.se/templates/PRIOEngframes____4144.aspx (last accessed May 2012).
13. KEMI – Swedish Chemicals Agency, http://www.kemi.se/en/ (last accessed May 2012).
14. Swedish Chemicals Agency – About the products register, http://www.kemi.se/en/Content/Products-Register/About-the-Products-Register/ (last accessed May 2012).
15. European Commission – Endocrine disruptors website, http://ec.europa.eu/environment/endocrine/strategy/substances_en.htm (last accessed May 2012).
16. European Chemicals Agency – Guidance for the preparation of an Annex XV dossier on the identification of substances of very high concern, http://echa.europa.eu/documents/10162/13638/svhc_en.pdf (last accessed May 2012).
17. European Commission – EDC database, http://ec.europa.eu/environment/endocrine/strategy/substances_en.htm (last accessed May 2012).
18. US National Institutes of Health – Hazardous Substances Data Bank, http://toxnet.nlm.nih.gov/cgi-bin/sis/htmlgen?HSDB (last accessed May 2012).

19. TEDX – The Endocrine Disruption Exchange; http://endocrinedisruption. com/home.php (last accessed May 2012).
20. US National Institutes of Health – Pubmed, http://www.ncbi.nlm.nih.gov/ pubmed (last accessed May 2012).
21. Thomson Reuters – Web of Science, http://thomsonreuters.com/ products_services/science/science_products/a-z/web_of_science/ (last accessed May 2012).
22. US National Institutes of Health – Toxline, http://toxnet.nlm.nih.gov/ cgi-bin/sis/htmlgen?TOXLINE (last accessed May 2012).
23. European Chemicals Agency – Guidance for the preparation of an Annex XV dossier on the identification of substances of very high concern, http:// echa.europa.eu/documents/10162/13638/svhc_en.pdf (last accessed May 2012).
24. Cooperation Centre Hamburg – Cleantool, http://www.cleantool.org/ (last accessed May 2012).
25. ISTAS – Risctox, Evalúa lo que usas, http://www.istas.net/risctox/evalua (last accessed May 2012).
26. ISTAS – Alternativas, http://www.istas.net/risctox/alternativas (last accessed May 2012).
27. Catsub – Catalogue of substitution, http://www.catsub.eu (last accessed May 2012).
28. SUBSPORT – Restricted and Priority Substances Database, http:// www.subsport.eu/listoflists (last accessed May 2012).
29. SUBSPORT – Identifying substances of concern, http://www.subsport.eu/ identifying-substances-of-concern/compilation-of-criteria (last accessed May 2012).

Alternatives Assessment in Regulatory Policy: History and Future Directions

J. A. TICKNER,* K. GEISER, C. RUDISILL AND J. N. SCHIFANO[†]

ABSTRACT

As the use of alternatives assessment as a tool to support the adoption of safer chemicals continues to evolve, the role of government in forming successful policies and initiatives needs thoughtful analysis. Although many chemical policies restricting or phasing out chemicals exist, very few mandate alternatives assessment or have adequate frameworks or tools to support successful implementation while avoiding regrettable substitutions. This chapter explores the justification and rationale for requiring alternatives assessment in the formation of chemical policies that support 'informed substitution.' A historical and current overview of chemical restriction and alternatives assessment policies is then provided and also a typology of current policies and initiatives that require or incentivize alternatives assessments. Five case studies highlighting regulatory policies requiring alternatives assessments, along with other examples, are used to support several lessons learned for future government policies that support informed substitution. Taken together, these policies illustrate that (1) there is a need for policies that require alternatives assessment, (2) carefully designed incentives and disincentives can encourage adoption safer alternatives, (3) alternatives assessment requirements should be tied to initiatives that incentivize adoption of safer alternatives and (4) there is a need for clear, yet flexible guidance and criteria for alternatives

*Corresponding author

[†]The views expressed in this chapter are the personal views of the authors and do not purport to reflect the official views or positions of the Occupational Safety and Health Administration (OSHA) or the US Department of Labor.

Issues in Environmental Science and Technology, 36
Chemical Alternatives Assessments
Edited by R.E. Hester and R.M. Harrison
© The Royal Society of Chemistry 2013
Published by the Royal Society of Chemistry, www.rsc.org

assessment processes. Although alternatives assessment requirements are necessary, it is critical that they not become overly prescriptive, burdensome or scientized in a way that inhibits their ability to achieve the goal of safer chemistry.

11.1   Introduction

Alternatives assessment is a critical tool in advancing the informed substitution of hazardous chemicals. However, alternatives assessment is rarely required or incentivized in policy. The purpose of this chapter is to characterize government policies that incentivize or mandate alternatives assessment as part of requirements to reduce, phase out or substitute chemicals of concern. Whereas other chapters in this volume have focused on particular alternatives assessment frameworks, tools and case studies, this chapter evaluates the role of policy in stimulating the informed quest for safer chemistries. For the purpose of this chapter, we define alternatives assessment as a process for identifying and comparing potential chemical and non-chemical alternatives that could replace chemicals or technologies of concern on the basis of their hazards, performance and economic viability. The goal of alternatives assessment processes is to support informed substitution, or 'a considered transition from a chemical of particular concern to safer chemical or non-chemical alternatives.'[1]

Government efforts to drive chemical substitution (including mandated chemical phase outs, reductions and restrictions that lead to substitutions) are not new, but with the adoption of the European Union (EU)'s Registration, Evaluation, Authorization and Restriction of Chemicals (REACH) program and several European and US state initiatives, there is greater government attention to policies that support the informed transition to safer chemicals. In this chapter, we argue that to achieve informed substitution, there is a need for policies that not only provide incentives to substitute, but also require an adequately flexible evaluation of alternatives and provide structures to support the informed transition to safer chemicals. We explore the rationale for integrating alternatives assessment in policies that drive substitution. We then provide a historical overview of government policies that support chemicals restrictions and alternatives assessment and characterize the range of current policies and programs that include alternatives assessment elements, mainly in North America and Europe. Following this overview, we outline five case examples of regulatory policies that require assessment of alternatives and lessons learned from them.

11.2   Rationale for Informed Substitution

Substitution requirements can play an important role in promoting a solutions-oriented approach to toxic chemical problems. Rather than focusing on establishing 'acceptable levels of exposure,' as many risk-based policies do, chemical reduction, restriction and substitution requirements focus attention on risk prevention and opportunities to identify and adopt more sustainable chemistries.[2,3]

Efforts to reduce or eliminate problematic chemical use through chemical substitution or process redesign can lead to changes to production processes (along a whole supply chain), products, or work practices and use patterns that can enhance ecosystems and human health. However, reducing or eliminating the use of chemicals – even relatively dangerous ones – can also result in unintended adverse consequences, or trade-offs. Examples abound where well-intentioned efforts (government or otherwise) to restrict a chemical of concern without a clear plan, information, processes, or requirements to evaluate substitutes or the implications of substitution processes has resulted in risk trade-offs. Some notable examples include the following:

- *Flame retardants*. Global concern has been raised about the class of flame-retardant chemicals polybrominated diphenyl ethers (PBDEs). These chemicals are now subject to government restrictions throughout the world. Most of these policies have directed little attention toward potential substitutes. However, manufacturers are still required to comply with flame retardancy (performance) requirements for materials. Some manufacturers have switched to substitutes that are functionally similar and technically and economically feasible, such as brominated phthalates, without undertaking broad evaluations of alternative materials. This has resulted in increased levels of these substitutes in the environment due to the persistence and/or bio-accumulation of some of the alternatives or their breakdown products.[4,5] Some evidence indicates that these alternatives may also be toxic.[6]
- *Bisphenol A*. Consumer pressure on bisphenol A (BPA) in baby bottles and water containers led manufacturers and retailers to replace polycarbonate plastics containing BPA. Although policies restricting BPA use in bottles have now emerged in many places, none has required a review of the substitutes. Research indicates that a number of plastic materials that could serve as replacements for BPA may also be toxic and exhibit estrogenic activity, a health endpoint of concern for BPA.[7,8]
- *Solvents*. Air quality regulations in various jurisdictions have forced the substitution of hazardous air pollutants. In some cases, this has led to substitution with chemicals that are more hazardous to workers, as such hazards were not considered in the substitution decision process, nor were changes in work practices that might occur from a chemical substitution. For example, restrictions on the use of perchloroethylene led to a number of problematic substitutions, including *n*-hexane in vehicle repair applications and *n*-propyl bromide in dry-cleaning and degreasing. Although these chemicals were relatively easy to implement as 'drop-in' substitutes (important for small businesses without technical resources to redesign production processes), both pose neurotoxicological risks to workers.[9,10]

These examples of 'uninformed substitution' illustrate important short-comings of chemical substitution efforts that do not adequately consider safer alternatives. These include:

- Policies to restrict a chemical of concern to a particular population or media may result in substitutions that shift the risks to another population or media, including ones at different phases of a chemical or product life-cycle. This occurs in part because chemical restrictions are often implemented by government agencies or divisions representing a particular population, life-cycle stage, or medium. Assessments of substitutes, if undertaken at all, may fail to consider other populations or media.
- Chemical restriction policies generally fail to consider how chemical changes may change process chemistries, work practices or exposure patterns.
- Efforts focused only on eliminating chemicals of concern fail to consider the 'functional use' of the chemical, instead focusing on the chemical and not the service it provides. By focusing on functional use, an agency or manufacturer may be able to identify non-chemical or process options to fulfill the function or determine that the function is not even necessary.
- Chemical restriction policies often focus on the creation of lists of 'chemicals of concern.' Given the large percentage of chemicals lacking hazard data and the detailed processes often required to establish 'authoritative' lists for some chemical endpoints (such as carcinogens), these lists may miss many chemicals of concern and lead to manufacturers or consumers assuming that chemicals *not* on these lists (or that are not regulated) are therefore 'safer.'

Ultimately, unintended consequences of chemical restrictions can undermine efforts to transition to safer chemistries, diverting attention away from encouraging solutions. If the goal of chemical restriction policies is to promote the identification and adoption of safer chemicals, there is a clear need to ensure that these types of chemicals management efforts are integrally tied to policies, guidance and support that ensure a thoughtful evaluation of and transition to safer alternatives.

11.3 Evolution of Alternatives Assessment Elements in Government Chemicals Reduction Policies

Government policies to restrict or limit chemicals of concern are not new. Nor are policies that require proponents of potentially harmful activities to consider a range of alternatives to reduce harm from those activities. What is relatively new is the integration of these two concepts in different policy contexts.

11.3.1 Chemical Restriction and Phase-out Policy Development

Government policies in the USA and Europe that restrict or require reduction or substitution of chemicals of concern date back to the 1950s, such as the

Delaney Clause of the US Federal Food, Drug and Cosmetic Act prohibiting the inclusion of carcinogenic additives in processed foods. The philosophy behind such policies is that the most effective way to address chemical risks is not through exposure controls but elimination of the chemical. Restrictive policies give clear signals to the marketplace that can also push innovation towards safer substitutes.[11,12] They are also seemingly easy to implement in practice. However, these policies have remained largely silent on the issue of what will replace the chemical of concern or how alternatives should be evaluated. For example, government initiatives in the 1970s called for phase outs of discharges of chemicals of concern to critical aquatic ecosystems. The 1977 Great Lakes Water Quality Agreement called for 'the virtual elimination,' the reduction of the concentration of the substance to below detectable levels, of discharges of persistent and bio-accumulative chemicals in the Great Lakes basin.[13] The US–Canada International Joint Commission, a bi-national body dedicated to protecting boundary waters, issued numerous reports calling for precautionary policies that lead to substitution of chemicals of concern in the region but mentioning little about evaluating those alternatives.[14]

Similarly, regional agreements in Europe for the protection of the Baltic and North Seas also included goals for substitution of chemicals of concern in these ecosystems with little guidance as to the nature of the substitutes. In fact, Greenpeace commissioned a report which noted that to implement these precautionary goals, a focus on achieving safer alternatives through clean production methods was necessary.[15]

US and European governments have undertaken chemical restriction efforts on specific chemicals of concern, such as lead, polychlorinated biphenyls (PCBs) and organochlorine pesticides, since the 1970s. Much of the focus of restrictions was on chemicals that ended up in various media – air, water, waste – from manufacturing processes or at the end of life of products. For example, the European Union Limitations Directive, passed in 1976 (now part of the EU's REACH regulation), authorized the EU to restrict or ban chemicals of concern across Member States. More than 900 chemicals have been restricted under this authority; many of these are petroleum products restricted as carcinogens, mutagens and reproductive toxicants in consumer-available preparations.[16]

Similar types of chemical restrictions increased in the late 1990s and early 2000s as concerns about the links between chemical exposures and human health impacts grew.[17,18] European and US governments initiated policies to restrict individual chemicals of concern (often called bans or phase outs), such as mercury, PBDEs, phthalates, BPA, short-chain chlorinated paraffins, and formaldehyde. For example, 32 US States have some type of policy on mercury in products.[19] In the 2009–2010 legislative session, 19 US States proposed restrictions on BPA, 12 proposed restrictions on PBDEs and 24 proposed restrictions on chemicals of concern (phthalates, BPA, lead, *etc.*) in children's products.[20]

Various European countries (Sweden, Denmark, Norway) and US States (Washington, Maine, Minnesota) also developed lists of 'chemicals of concern' to provide clear signals to the marketplace of the types of chemicals that were of concern to regulators, providing a driver for substitution. An early example of this is the California Safe Drinking Water and Toxic Enforcement Act of 1986 (Prop 65), which establishes a list of chemicals known to cause cancer, reproductive toxicity, or developmental toxicity. While the Swedish Observation List and the Danish List of Undesirable Substances listed only several hundred chemicals of concern, the more recent US State lists, built from authoritative lists of chemicals of concern for specific endpoints (carcinogens, persistent chemicals, reproductive toxicants, etc.), contain more than 1000 chemicals.[16,21]

These chemicals policies have had an impact in moving manufacturers away from chemicals of concern and reducing quantities of the chemicals in the environment.[22,23] However, the question remains as to whether they have stimulated the adoption of safer alternatives, as evidenced in the case of PBDE replacements. Nonetheless, more recent policies, such as some State-level chemical restrictions in the USA (PBDEs in Washington and Minnesota) and some European restrictions (flame retardants in Sweden and phthalates in Denmark) have explicitly linked restrictive policies to research on alternatives.

11.3.2   Alternatives Assessment Policy Development

The evolution of alternatives assessment in government policy parallels that of chemical restrictions. An early example of a policy that requires alternatives assessment is the Environmental Impact Statement (EIS) process under the 1970 National Environmental Policy Act (NEPA) and similar state programs. Under NEPA, major federal actions affecting the quality of the human environment must undergo an EIS process. The goal of NEPA is to foster better decisions and 'excellent action' through the identification of reasonable alternatives that will avoid or minimize adverse impacts.[24] NEPA regulations specify an environmental impact assessment process that proponents must follow before initiating an activity. Through an interdisciplinary approach, proponents must (1) comprehensively identify and examine environmental effects and values, (2) rigorously study, develop and describe appropriate (reasonable) alternatives in comparative form, including not moving ahead with an activity, and (3) recommended courses of action. Proponents are instructed to undergo a 'scoping process' to broadly define potential impacts and to examine them in detail including direct and indirect impacts, cumulative effects, effects on historical and cultural resources, impacts of alternatives and options to mitigate potential impacts.

In the chemicals area, the 1987 Montreal Protocol on Substances that Deplete the Ozone Layer represents the first global effort to restrict and evaluate substitutes for a group of chemicals of concern. The Montreal Protocol Chemical Technical Options Committees review alternatives for ozone depleting substances and some national programs, such as the US Environmental Protection Agency (EPA)'s Significant New Alternatives Policy (SNAP) Program, require evaluation of alternatives to ozone-depleting substances.

The principle of informed substitution for dangerous chemicals was codified in EU Member State policies in the 1980s and 1990s (most notably in Sweden). For example, the amended 1990 Swedish Act on Chemical Products states that:[25]

> *Anyone handling or importing a chemical product shall take such steps and otherwise observe such precautions as are necessary to prevent or minimize harm to human beings or to the environment. This includes avoiding chemical products for which less hazardous substitutes are available.*

In applying the substitution principle, the Swedish government called for a progressive elimination of chemicals of concern, a process called 'sunsetting,' whereby the government would provide generic criteria for undesirable chemicals and substitutes and also ambitious but long-term targets and lead times to develop new processes and products.[26]

The passage of State-level pollution prevention requirements in the USA and cleaner production initiatives and the Integrated Pollution Prevention and Control Directive in Europe in the 1990s sparked the development of facility planning processes to support waste, chemical and emissions reduction policies. Facility planning (also known as cleaner production, pollution prevention, or source reduction planning) involves characterizing and understanding why and how chemicals are used in a particular production process and evaluating options for reducing use, waste, or emissions. Geiser characterized such facility plans as documents 'describing the means and timing by which corporations will reduce the risks of toxic chemicals in production.'[27] The goal of the plan is 'to serve as a guide for raising the level of attention about toxic chemicals, increasing motivation to change, presenting alternatives, guiding decision-making, advocating for resources and providing information to evaluate the consequences of change.'[27] The passage of pollution prevention and cleaner production policies in the USA and Europe sparked the development of dozens of new tools for chemical ranking and scoring to support reduction and sub-stitution efforts,[28] case studies of safer alternatives, such as those developed by the Massachusetts Toxics Use Reduction Program's Office of Technical Assistance, voluntary demonstration projects and challenges, such as the US EPA's Project XL and European Initiatives such as the PRISMA Project and frameworks for alternatives assessment, such as the US EPA's Cleaner Technologies Substitutes Assessments and Use Cluster Scoring System.[29,30]

Beginning in the late 1990s, initiatives in EU Member States, such as those in Germany, The Netherlands, Sweden and Denmark, and discussions leading up to the EU's REACH proposal reinvigorated the development of policies and tools for chemical alternatives assessment. For example, the Swedish government developed the PRIO chemical assessment tool to provide chemical users with an ability to evaluate chemicals of concern and safer alternatives. The Danish government developed chemical action plans for several chemicals of concern (such as phthalates) that outline problems with the substance, goals for reducing hazards, evaluation of alternatives and costs of implementation.

As part of the Dutch government's Strategy on Management of Substances, the Ministry of Housing, Spatial Planning and the Environment developed the Quick Scan rapid hazard assessment method to provide decision guidance to support voluntary chemical substitution efforts by firms.[16] The European Commission developed several reports and a guidance document outlining decision frameworks and tools for advancing informed substitution of chemicals.[31] The European Commission and German government funded the development of the SUBSPORT project, a database of tools, guidance and case studies of safer alternatives.[32]

Similarly, occupational policies in Europe addressing chemicals in the workplace, such as the Chemical Agents Directive, led to the development of detailed alternatives assessment regulations for substitution in Germany. The Spanish Trade Union Confederation Institute for Health, Work and Environment has worked with Spanish authorities to establish its RiscTox database, designed to provide workers with tools and information to evaluate chemical hazards and substitutes.

In the USA, the European Restrictions on Hazardous Substances (RoHS) in Electronic and Electrical Products Directive and some State chemical restrictions led to the establishment of a variety of State and federal initiatives around alternatives assessment for chemicals of concern in products. New policies, such as Maine's Toxic Chemicals in Children's Products Law, require the evaluation of alternatives to chemicals of concern to children. In 2005, the Massachusetts legislature requested that the Toxics Use Reduction Institute (a State-funded research institute based at the University of Massachusetts Lowell) undertake an alternatives assessment of five chemicals of concern used in products: lead, formaldehyde, diethylhexyl phthalate, perchloroethylene and chromium(VI). The Institute developed the Five Chemicals Alternatives Assessment methodology for the assessment (adapted from its pollution prevention planning guidance), engaging stakeholders in reviewing potential alternatives for priority uses of those chemicals in the state on the basis of health and safety, economic and technical feasibility considerations.[33] The US EPA's Design for the Environment (DfE) program established its Alternatives Assessment and Safer Product Labeling programs to evaluate alternatives for chemicals of concern and provide recognition for product formulations containing safer chemical ingredients. The EPA developed guidance on comparative chemical hazard assessment, a process for comparing chemicals on the basis of their intrinsic hazards in 2011,[34] and also a list of chemicals that meet the DfE criteria for safer chemistry for use in chemical cleaning formulations in 2012.[35] In 2008, the California legislature passed its Safer Consumer Products legislation, requiring the evaluation of alternatives to chemicals of concern in products.

11.3.3 Convergence of Chemical Restriction and Alternatives Assessment Policies

Both policies that require the phase out of chemicals of concern and those that require evaluation of alternatives initially focused on chemicals used in manufacturing. As information on the dispersive nature of chemicals in products and chemical hazards, particularly concerns about exposures to children, has increased, the direction of these two policy types has evolved to emphasize substitution of chemicals in products. Similarly, there has been increased government and industrial attention to 'integrated product policy,' with policies and tools designed to address product life-cycles, including energy, materials use and chemical toxicity. With the focus on products, there is an increasing convergence of chemicals restrictions linked to alternatives assessment requirements. There are various reasons for this convergence, including (1) greater awareness of the negative impacts caused by uninformed substitutions, due to notable failures of substitution, (2) a greater understanding of and support for the application of 'green chemistry,' the design of less hazardous chemicals throughout their life-cycles, (3) the increased attention to evaluating alternatives to reduce life-cycle impacts, particularly energy and material use, and (4) increasing marketplace pressure for safer chemicals and products.[36]

An interesting result of this convergence is the establishment of new structures for government agencies to collaborate and share information on chemicals and alternatives and develop consistent approaches. For example, 11 US States have formed the Interstate Chemicals Clearinghouse (IC2) to share information on chemical hazards and priorities, chemical use in products and safer alternatives, and also to develop consistent frameworks for alternatives assessment. Similarly, EU Member States and other stakeholders have collaborated to develop guidance for alternatives assessment under REACH and to implement substitution requirements in occupational health directives. The Organization for Economic Co-operation and Development (OECD) established an *ad hoc* working group on chemicals alternatives assessment to identify common tools and frameworks for alternatives assessment.[37]

Despite this convergence, as previously noted, most chemical restriction policies still do not include alternatives assessment elements. A review of the Interstate Chemicals Clearinghouse state chemicals policy database (a database of over 1200 proposed and enacted State chemicals policies) shows that from 1997 to 2012 more than 400 individual chemical restrictions have been proposed but less than 100 of those proposed policies contained alternatives assessment elements.[38] The establishment of alternatives assessment in restriction policies is still in its infancy, with varied application and emerging decision-frameworks and tools.

11.4 Alternatives Assessment in Government Chemicals Policies: a Categorization

When they have formed part of government chemicals reduction policies, alternatives assessment elements have been integrated in a variety of different ways. It is instructive to understand the various types of alternatives assessment policies and how these, on their own or together, can support informed substitution of chemicals of concern. A summary of European, North American and global government regulatory and non-regulatory policies and initiatives that support informed substitution is provided in Table 11.1.

The table is not meant to be exhaustive but rather to provide an instructive sample of some of the most prominent policies and to characterize the range and types of efforts that have been initiated. The table outlines the focus and key elements of the policies and identifies what tools or guidance may have been developed to support informed substitution in their application. The policies or initiatives are summarized below and can be categorized using the following typology. In some cases, policies or initiatives can fit multiple sub-categories:

- *Non-regulatory* – Voluntary policies that support the evaluation and adoption of safer alternatives without requirements on regulated entities. There are two main types of non-regulatory informed substitution policies:
  - **Evaluative.** These policies and efforts are designed to review or designate safer alternatives to chemicals of concern or provide tools for companies to evaluate substitutes. Such efforts include those (1) where governments conduct evaluations of alternatives to support informed substitution, such as the EPA DfE's alternatives assessment on PBDEs in furniture foam; (2) where governments identify safer substitutes for chemicals in specific applications, such as the EPA's Safer Consumer Labeling Program or alternatives assessments for ozone-depleting substances under the Montreal Protocol Technical Options Committees; (3) where governments categorize chemicals in commerce and potential alternatives by their hazards, such as the Danish government's effort to categorize its existing substance list by hazard classification using quantitative structure–activity relationship analysis; (4) where governments provide criteria for evaluating safer materials, such as the German Federal Environment Agency's Sustainable Chemistry Guidance, which provides a framework for developing safer chemicals; or (5) where governments provide tools for chemical users to evaluate and categorize alternatives, such as PRIO or the Quick Scan.
  - **Supportive.** These policies and efforts support decision-making processes and adoption of safer chemicals. They include: (1) development of guidance documents, such as the REACH authorization substitution guidance, developed by the European Chemicals Agency or the guidance documents being developed by the Washington Department of Ecology, which provide detailed guidance – tailored to smaller and larger enterprises – for the various steps of alternatives assessment; (2) technical support such as pollution prevention research and engineering support provided by the New York Pollution Prevention Institute; (3) case studies and demonstration projects of safer alternatives, such as the wire and cable partnership of the Massachusetts Toxics Use Reduction

Table 11.1 Policies and initiatives supporting informed substitution.

Jurisdiction	Name	Date	AA/substitution Elements	Guidance/tools Developed
Non-regulatory – Evaluative				
United Nations	Montreal Protocol on Substances that Deplete the Ozone (Article 9)[67]	1987 (signed)	Supports international research and information exchange on alternatives to ozone-depleting substances. Ratification by countries may include regulatory elements	No guidance, but Technology and Economic Assessment Panel addresses issues concerning alternative technologies[68]
Denmark	Chemicals Action Plan 2010–2013	2010	Includes an initiative focusing on the continued development of (Q)SARs to prioritize chemicals and evaluate substitutes.[69] Also, includes chemical action plans for substitution of chemicals of concern	Oversees (Q)SAR database that can be used by industry to identify substitutes. Also hosts the Catsub database, which includes case studies for successful substitutions of hazardous chemicals (mostly in the occupational health arena)[70]
United States	US EPA – Design for the Environment (DfE)[71]	Ongoing	Alternatives assessments are conducted for Action Plan chemicals for either broad or specific end uses.[72] Assessment focuses on comparative hazard screening. Identifies safer substitutes for chemicals in specific applications in its Safer Product Labeling Program[73]	Developed hazard assessment criteria incorporating persistence, bioaccumulation, ecotoxicity and human health endpoints[74]
Germany	German Federal Environment Agency – Sustainable Chemistry Activities	2011	Promotes the systematic identification and adoption of sustainable chemicals by businesses	Published *Guide on Sustainable Chemicals*, which provides a framework and criteria to evaluate and compare chemicals[75]

Non-regulatory – Supportive

United Nations	Stockholm Convention on Persistent Organic Pollutants (POPs) (Articles 5, 9 and 11)[76]	2001 (signed)	Aimed at global reduction of priority POPs. Intersessional workgroups develop alternative assessment and substitution policy. Created POPs Free initiative to assist substitution efforts in developing nations.[77] Ratification by countries may include regulatory elements	General guidance for review of alternatives to potential POPs[78]
The Netherlands	Strategy on Management of Substances (SOMS)[79]	2002–2003	Main purpose of this program was not AAs. It included nine pilot case studies in various industrial sectors aimed at identifying, prioritizing and managing hazardous substances. It was discontinued after the adoption of REACH	QuickScan was used as a screening prioritization tool that takes into account the potential risks and hazards at each life-cycle stage. The analysis is based mostly on EU risk phrases[80]
Sweden	Environmental Quality Objectives[81]	1999	One of the 16 environmental quality objectives includes a 'non-toxic environment.' Includes interim targets that may involve regulatory measures	Hosts a Restricted Substances database.[82] Published a 2007 report on the Substitution Principle, providing broad overview of the elements of the principle.[83] Developed PRIO, a web-based tool for chemical risk reduction that is available to the public[84]
United Kingdom	UK Chemicals Stakeholder Forum (UKCSF)[85]	Ongoing	Composed of representatives from government, industry, trade organizations and NGOs. It is an effort that aims to advise the UK government on how industry should reduce risks from hazardous chemicals	Published a general guide to substitution in 2010[86]
United States		2009–2011	Multi-stakeholder discussion forum recommending	No guidance

Table 11.1 Continued.

Jurisdiction	Name	Date	AA/substitution Elements	Guidance/tools Developed
	Centers for Disease Control (CDC) – National Conversation on Public Health and Chemical Exposures		governments and organizations work towards substituting hazardous chemicals with safer alternatives. Also, calls for enhancement of research and development in this area[87]	
United States – Washington	Reducing Toxic Threats Initiative[88]	2011	Focuses on preventing the use of toxic substances by encouraging safer alternatives and promoting green chemistry	AA guidance document is currently under development. Endorses GreenScreen and QCAT for conducting hazard assessment[89]
Regulatory – Classification-based Substitution Requirements				
European Union	Chemical Agents Directive (98/24/EC)[90] and Carcinogens and Mutagens Directive (2004/37/EC)[91]	1998 and 2004	98/24/EC is an industrial hygiene hierarchy that emphasizes substitution first. 2004/37/EC substitution of carcinogenic and mutagenic substances in the workplace is first priority, where feasible	Impact of 98/24/EC is monitored by project Chemical Agents Directive Implementation (CADimple).[92] Developed a substitution guidance document for small- and medium-sized enterprises
European Union	Cosmetics Directive (EC 1223/2009), Toy Safety Directive (2009/48/EC)[93,94]	2009	Requires documentation of alternatives analysis for inclusion of CMR chemicals in products	No guidance
France	Labor Code Article R4412		Requires employers to replace CMR substances, to the extent technically feasible, with a substance, preparation or process which is less hazardous to health[95]	Substitution-CMR database containing an inventory of CMR substances, alternatives and case studies

| Germany | Hazardous Substances Ordinance (GefStoff V) | 2008 | Requires substitution analysis for hazardous substances in the workplace[96] | TRGS 600 Guidance covers the full scope of substitution from determination of potential substitutes through technical suitability. Developed the Column Model, a tool for simple comparison of hazards and risks of identified alternatives |
| United Kingdom | Control of Substances Hazardous to Health (COSHH) Regulations[97] | 2002 | Requires that, where reasonably practical, employers prevent the exposure of employees to substances hazardous to health through substitution | Developed COSHH Essentials web tool designed to help companies implement COSHH regulations, although it focuses mostly on chemical management activities[98] |

Regulatory – Pollution Prevention Planning Requirements

United States – Massachusetts	Toxic Use Reduction Act	1989	Calls for a 50% reduction in the use of hazardous byproducts within the State. Companies must create a reduction plan and are encouraged to substitute with less hazardous substances. Created the Toxic Use Reduction Institute to provide resources and education	Conducted the Five Chemicals Study in 2006, which analyzed alternatives for five hazardous chemicals.[99] P2OASyS is a comparative hazard assessment tool developed by TURI.[100] Supports the development of demonstration projects, case studies and research efforts on safer alternatives
Canada – Ontario	Toxics Reduction Act (TRA), Ontario Regulation 455/09[101]	2009	Requires regulated facilities to track, quantify and develop plans to reduce the use of toxic chemicals	Published AA guidance which covers topics such as hazard assessment, life-cycle, technical feasibility and cost–benefit[102]
China	Promotion of Clean Production[103]	2003 (enacted)	Calls for substitution of toxic and hazardous materials during technical upgrades	Legislation authorized the development of guidance. Current status is unknown

Table 11.1 Continued.

Jurisdiction	Name	Date	AA/substitution Elements	Guidance/tools Developed
Regulatory – Alternatives Assessment to Support Regulatory Decision-making				
European Union	REACH – Annex XIV Substances of Very High Concern and Annex XV Restricted Use Substances (EC 1907/2006)[104]	2006	Companies must include an alternatives assessment in their application for authorization of any Annex XIV or Annex XV chemicals	General guidance for both Annex XIV[105] and Annex XV[106]
European Union	Biocidal products (EU 528/2012, replaces 98/8/EC)	2012	Prohibits CMRs, sensitizers and bio-accumulative chemicals from use as active ingredients in biocidal products. Requires evaluation of alternatives for active ingredients under review for approval[107]	No guidance
United States – California	Safer Consumer Products Regulations [Draft][108]	2012	Companies are required to conduct alternatives analysis of any chemical of concern in priority products	Guidance is currently under development
United States – Minnesota	Toxic Free Kids Act	2009	Requires Agencies to evaluate alternatives for chemicals of high concern used in children's products. Published a report in 2010 concerning options for reducing and phasing out chemicals[109]	No guidance
Regulatory – Single or Multiple Chemical Restrictions with Alternatives Assessment Requirements				
European Union	VOC Directive (1999/13/EC), Article 7[110,111]	1999	Calls for substitution of VOCs at specific installations that operate above consumption thresholds	Provides guidance documents and resources for substitutes in various VOC-containing products

European Union	Restrictions of Hazardous Substances in Electrical and Electronic Equipment (RoHS) (2002/95/EC)	2002	Eliminates cadmium, lead, mercury, PBDEs and PBBs in electronic products and encourages substitution with safer substance where technically feasible[112]	No guidance
United States	Consumer Product Safety Commission (CPSC) – Consumer Product Safety Improvement Act (CPSIA) (S.108)[113]	2008	CPSC is required to assess health effects of all phthalate alternatives used in toys and other childcare products. Activities are conducted through the Chronic Hazard Advisory Panel (CHAP)[114]	No guidance
China	Administration on the Control of Pollution Caused by Electronic Information Products (Joint Ministerial Decree No. 39) (China RoHS)[115]	2007	Calls for producers of electronics to substitute safer alternatives	Unknown
United States – Minnesota	Products Containing Polybrominated Diphenyl Ethers[116]	2007	Requires the State to conduct assessment for deca-BDE to identify safer alternatives	No guidance
Regulatory – Requirements for use of Acceptable Substitutes				
United States	US EPA – Significant New Alternatives Policy (SNAP) Program (Clean Air Act §612(c))[117]	1990	SNAP identifies chemical substitutes that reduce risk compared with Class I and Class II ODS for multiple industries. Rules and regulations are promulgated based on results of the substitution analysis	An instruction manual on analyzing potential alternatives is available, which includes criteria for assessing hazards, exposure and atmospheric effects[118]
Regulatory – Requirements for use of Safer Alternatives in Procurement				
United States – New York	Establishing a State Green Procurement and Agency Sustainability Program (Executive Order No. 4)[119]	2008	Establishes processes for identifying preferred products, and also a list of chemicals to avoid in purchasing	Established criteria for environmentally sensitive cleaning products[120]

Institute designed to support firms in substituting lead in electronics applications; (4) databases of alternatives such as the Danish Catsub database providing information on safer chemicals in specific applications or the French Ministry of Environment's Substitution CMR database of alternatives to chemicals that may cause cancer or reproductive impacts in the workplace; and (5) stakeholder dialogues designed to identify challenges to safer chemicals and opportunities for advancing adoption, such as the UK Stakeholder Forum and the US Centers for Disease Control National Conversation on Chemical Exposures.

- *Regulatory* – Policies that require chemical substitution and alternatives assessment for chemicals used in manufacturing processes or products. There are several types of identified regulatory informed substitution policies:

  ○ **Classification-based substitution requirements.** Several European Directives on chemicals (such as the Chemical Agents Directive, the Cosmetics Directive and the Toys Directive) derive from European and now Globally Harmonized System of Classification and Labeling (GHS) classifications, particularly for carcinogens, mutagens and reproductive toxicants. These policies require evaluation of alternatives for continued use of such substances.

  ○ **Pollution prevention planning requirements.** These policies require manufacturing firms to characterize chemical use in processes and evaluate alternatives to reduce or eliminate toxics use and waste. The Massachusetts Toxics Use Reduction Act and Ontario Toxics Reduction Act both require manufacturers to characterize chemical use and evaluate alternatives to chemicals of concern in manufacturing processes.

  ○ **Alternatives assessment to support regulatory decision-making.** Under these policies, alternatives assessments are required prior to regulatory decisions either to permit continued use of a chemical of concern or to undertake regulatory restrictions. For example, REACH requires that firms seeking authorization undertake an alternatives assessment to demonstrate the need for authorization and their processes to adopt safer alternatives.

  ○ **Single or multiple chemical restrictions, with alternatives assessment requirement.** These are policies that restrict a particular chemical or class of chemicals, but also require either government agencies or regulated companies to evaluate alternatives to demonstrate availability or lack of availability of alternatives or to avoid regrettable substitutions. For example, the Minnesota Act on Products Containing Polybrominated Diphenyl Ethers (and other State policies restricting PBDEs, phthalates and lead), requires the state Pollution Control Agency to undertake an evaluation of alternatives prior to finalizing the restrictions process.

  ○ **Requirements for use of acceptable substitutes.** These represent policies where manufacturers are required to use safer (or approved) substitutes to a chemical of concern with mandated requirements to list a substance as acceptable. The EPA SNAP Program requires companies to seek approval for substitution of ozone-depleting substances.

  ○ **Requirements for use of safer alternatives in procurement.** Many jurisdictions have enacted policies that require government agencies to 'lead by example' to choose the least toxic alternatives for particular chemical or product classes. New York Executive Order No. 4, Establishing a State Green Procurement and Agency Sustainability Program, establishes processes for identifying preferred products, for example, cleaning products, and also a list of chemicals to avoid in purchasing.

Table 11.1 demonstrates that there have been a wide range of voluntary and regulatory initiatives to support informed substitution to date. Voluntary government initiatives, often in response to regulatory programs, have focused for the most part on government-initiated alternatives assessments for chemicals of concern to either inform the marketplace or support restrictive policies and action plans; development of substitution guidance documents, tools and criteria for evaluating alternatives; and stakeholder dialogues on substitution methods, examples and policies, engaging a range of chemical users, manufacturers, environmental groups and government agencies. Regulatory policies around informed substitution have generally centered on single chemical restrictions with alternatives assessment requirements prior to substitution and classification-based substitution requirements. There is an increasing focus on requiring alternatives assessment to support the regulation of chemicals of concern, for example, under REACH and the proposed California Safer Consumer Products regulations, which would require chemical manufacturers and users to conduct alternatives assessments for chemicals of concern in specific products of concern prior to government regulation to restrict or control that chemical. Interestingly, these requirements for alternatives assessment, primarily for chemicals in products, build off of the facility planning steps inherent in policies such as the Massachusetts Toxics Use Reduction Act.

11.5 Examples of Regulatory Informed Substitution Policies

In this section, we provide five case examples of government regulatory policies that support informed substitution, covering several of the policy categories outlined above. They provide details of how informed substitution requirements are being implemented in different regions and in different contexts. These, along with the evaluation of policies above, provide important lessons as to gaps in current policies and their implementation, and also the strengths and weaknesses of particular policy types.

11.5.1   Pollution Prevention Planning Requirements –
Massachusetts Toxics Use Reduction Act

The Massachusetts Toxics Use Reduction Act, passed in 1989, represents a successful example of required alternatives assessment in the context of facility planning. The Act requires that manufacturers producing, processing or using some 1000 chemicals over threshold amounts conduct a materials accounting every year to understand chemical throughputs and undertake a detailed planning process every 2 years to identify options for toxics reduction. About 600 firms in Massachusetts are required to comply with the program. These firms pay a relatively small fee on chemicals that funds both the regulatory program but also a voluntary technical assistance program for firms (the Office of Technical Assistance for Toxics Use Reduction) and a research and education center [the Massachusetts Toxics Use Reduction Institute (TURI)].

The results of the program have been impressive. From 2000 to 2009, Massachusetts companies reduced toxics use by 21% (adjusted for production), toxic byproducts by 38% and emissions by 56%. Since its inception, Massachusetts firms have reduced toxics use by more than 50%, with even greater reductions in some uses, such as trichloroethylene use in surface cleaning, which has been reduced by about 95%.[39]

The success of the toxics use reduction/facility planning model is due in part to two main elements: recommended materials accounting and required planning supplemented by technical and research support. The materials accounting process forces firms to understand how chemicals are being used in various production processes and also costs and inefficiencies. The planning process requires that firms carefully identify a wide range of potential alternatives. As noted in the 2011 Program Report, 'TURA does not require toxics users to stop using a chemical, but instead requires them to examine how they use it and what their alternatives might be. When toxics users are required to evaluate how they use chemicals and to identify alternatives, they often find ways to improve manufacturing and develop safer products and more efficient operations, boosting the competitiveness of Massachusetts firms.'[39] Educational programs, tools development and required certification by trained planners ensure that manufacturers' planning processes are as comprehensive and of as high quality as possible. The planning requirements force firms to think about the 'functional use' of a chemical in a production process or product, whether that function is indeed necessary and whether safer chemical or process or product design alternatives can technically and economically fulfill that function.

Technical, educational and research support supplements the planning requirements, providing an incentive to innovate. A large challenge to firms, particularly small and medium sized ones, is capacity to innovate. Government partners provide services, demonstration projects and evaluation support to help firms overcome barriers to adoption of safer chemicals. For example, TURI's Surface Solutions Laboratory tests alternative cleaning agents (including water-based systems) to ensure their performance and safety, lowering the technological risk to adoption. Supply chain partnerships, to support substitution of lead in electronics, allow firms to share experiences and resources in solving pre-competitive challenges.

Nonetheless, toxics use reduction has focused primarily on chemicals in manufacturing processes and not on products. For example, there is no planning requirement for chemicals in products manufactured outside the State. Also, firms are not required to adopt alternatives, even if they are technically and economically viable. However, market demands and regulatory requirements outside Massachusetts, such as the RoHS Directive in Europe, are forcing Massachusetts manufacturers to have to evaluate alternatives to chemicals of concern in products they manufacture. As such, the Toxics Use Reduction Institute has adapted its guidance and outreach for alternatives assessment and informed substitution in facility planning to also include product level alternatives assessment.

11.5.2   Alternatives Assessment as a Precursor to Regulation – California Safer Consumer Product Regulations

Through the passage of Assembly Bill (AB) 1879 in 2008, the State of California began an effort to regulate chemicals of concern in products based fundamentally on the thorough evaluation of safer alternatives.

AB 1879 has resulted in extensive discussions, conferences, legal, scientific and technical opinions and the development of multiple iterations of regulations to implement the mandate. The law requires the Department of Toxic Substances Control (DTSC) to develop processes for: (1) evaluating and prioritizing chemicals (Chemicals of Concern) and products of concern (Priority Products) in the State; (2) determining who is required to undertake alternatives assessments; (3) undertaking comprehensive alternatives assessments that consider life-cycle impacts of alternatives and critical exposure pathways; and (4) establishing which regulatory responses to reduce exposure and risks are warranted based on the results of alternatives assessments.

The requirements of the law are fairly simple, yet comprehensive, representing the first regulatory policy explicitly to require consideration of life-cycle impacts in the context of chemical substitution processes. Any manufacturer, distributor or retailer that sells products in California would be subject to the requirements. The goal of the law is to encourage the marketplace to substitute chemicals of concern in products of high concern while avoiding regrettable substitutions that may shift risks through the product life-cycle.

The California DTSC developed proposed regulations to implement the law in 2012. Based on a Priority Product designation (which may focus on a particular component or material in a product), the draft regulations would require manufacturers of the chemical/product to undertake a multi-stage, detailed alternatives assessment.[40] DTSC has indicated that only a few Chemical of Concern–Priority Product combinations would be selected at first to 'test' the program and its implementation. DTSC will develop guidance for the alternatives assessment processes, but expects some level of flexibility provided that the regulatory requirements for evaluation are met. Companies would be allowed to submit previously conducted alternatives assessments or to use different approaches to evaluate alternatives if these meet the minimum requirements of the regulations. As currently proposed, the alternatives assessment process would be overseen by a certified alternatives assessment planner, a trained professional, who would ensure the quality and detail in the plan, much like the planners in Massachusetts. The proposed alternatives assessment process consists of the following two steps:

a. *First stage*: This is an initial alternatives assessment to identify whether options exist that could lead to immediate substitution of the chemical of concern. In this first stage of the alternatives assessment, the 'responsible entity' (which may be a consortium of companies), must undertake the following steps:
   i. Identification of Product Requirements and Function of Chemical(s) of Concern. If the function is not necessary, the company may choose to eliminate the chemical altogether.
   ii. Identification of a range of possible alternatives.
   iii. Initial chemical hazard screening, comparing the chemical of concern to possible alternatives.
   iv. Identification of next steps in the full alternatives evaluation. Following completion of a first-stage alternatives assessment, the responsible entity submits the report to DTSC for review.
b. *Second stage*. This is the more detailed alternatives assessment that considers economic and technical feasibility along with life-cycle impacts and potential exposures. In the second-stage alternatives assessment, the responsible entity must identify the life-cycle stages and hazards for comparing alternatives, particularly those where differences might exist between alternatives. Based on the identification of these life-cycle stages, alternatives are compared on their ecological and human health impacts. The comparison of alternatives must also include an evaluation of potential exposures and exposure pathways to the alternatives, including an evaluation of quantities of the alternative chemical used. Finally, the analysis must review economic impacts (jobs, market, costs of manufacturing) of alternatives in addition to technical feasibility. The second-stage alternatives assessment report must identify location of key suppliers of the chemical in the product, a justification of the alternative chosen, and proposed regulatory options based on the results of the alternatives assessment. Based on the assessment reports, DTSC then determines appropriate regulatory responses for the chemical–product combination.

Although the Safer Consumer Products regulations are not finalized, it is clear that the California proposal will require more comprehensive alternatives assessments than any other policy to date.

11.5.3 Alternatives Assessment as a Precursor to Regulation – REACH

The EU's REACH Regulation[41] provides an example of a policy that utilizes alternatives assessment to support regulatory decision-making. REACH requires manufacturers, importers and downstream users applying for authorization (*i.e.*, seeking permission to continue to use a chemical on the authorization list) to include 'an analysis of alternatives considering their risks and the technical and economic feasibility of substitution and including, if appropriate information about any relevant research and development activities by the applicant' [REACH, Article 62(4)(e)]. Where this analysis demonstrates that suitable alternatives exist, the applicant must also develop a 'substitution plan including a time table for proposed actions' [REACH, Article 62(4)(f)], prior to being given authorization for continued use of the chemical. REACH also requires the European Chemicals Agency (ECHA) or EU Member States to present an analysis of alternatives in dossiers for identifying candidate chemicals for authorization and in dossiers for restriction proposals.

The information developed on alternatives is used to make and support various decisions in the authorization and restriction processes. Based on the availability of suitable alternatives, government authorities decide whether an authorization should be granted and, if so, the length of time for which the continued use of a chemical is permitted. For example, where suitable alternatives are available and the substance cannot be adequately controlled, an application for authorization is unlikely to be granted. Where suitable alternatives are available and the substance can be adequately controlled, the decision to grant authorization and the conditions of the authorization are shaped by the required substitution plan. If no suitable alternatives are available and the substance can be adequately controlled, authorization may be granted.

Subsequent information about alternatives plays an important role in the renewal of authorizations or, in some cases, the withdrawal of existing authorizations.[42]

With regard to the restriction process, information on alternatives is required in order to evaluate the effectiveness and practicality of the proposed restriction, identify certain uses that ought to be excluded from restriction, identify other risk management options that ought to be considered and ensure the time for compliance is appropriate.[43]

In order to support its regulatory decision-making with alternatives analyses that are consistent and comprehensive, the ECHA has developed detailed guidance documents outlining the information that must be incorporated into such analyses. For example, according to ECHA's guidance for developing applications for authorization, an analysis of alternatives involves: identifying possible alternatives for each use; assessing the technical feasibility of possible alternatives identified; assessing possible alternatives for their potential risks to the environment and human health; assessing the economic feasibility of possible alternatives identified; identifying relevant research and development; assessing the suitability and availability of possible alternatives; and determining the actions and time scales that may be required to make possible alternatives suitable and available for the applicant. A substitution plan includes a description of proposed actions and justifications as to why those actions are required, who will conduct the proposed actions, a timetable for proposed actions that will lead to the transferal to the substitute and justifications as to why the action requires the time allocated and what the uncertainties are in achieving the actions within the time scale and what possible mitigation is to be considered.[42]

Since the first applications for authorization are not due until 2013 and no dossiers for restriction have been developed to date, the extent to which information on alternatives will be included in these submissions and how the information will be utilized in the decision-making process remain to be seen. While some information on alternatives is included in dossiers prepared by the ECHA and EU Member States for identifying candidate chemicals for authorization, the consistency and comprehensiveness of this information varies widely.

11.5.4 Requirements for Use of Acceptable Substitutes – US EPA SNAP Program

Section 612(c) of the Clean Air Act provides an example of a policy that requires the use of acceptable substitutes where certain chemicals of concern are phased out. Implementation of such a policy involves the development of criteria for acceptability, identification of possible alternatives for particular functional uses, the evaluation of those alternatives based on the criteria developed and the determination that an identified substitute is either 'acceptable' or 'unacceptable.' This type of regulatory model provides a mechanism for minimizing risk trade-offs and driving the use of safer alternatives to restricted chemicals of concern.

Specifically, Section 612(c) makes it unlawful to replace certain ozone-depleting chemicals with any substitute that the US EPA determines may present adverse effects to human health or the environment when an alternative has been identified that (1) reduces the overall risks to human health and the environment and (2) is currently or potentially available. In order to achieve this mandate, Section 612(c) requires that EPA publish a list of acceptable and unacceptable substitutes for specific uses, and also maintain a public clearinghouse of alternative chemicals, product substitutes and alternative manufacturing processes available for products and manufacturing processes that use ozone-depleting chemicals.[44]

The EPA established the Significant New Alternatives Policy (SNAP) program to implement these legal requirements. The purpose of the program is to allow a quick, orderly transition away from ozone-depleting compounds by identifying substitutes, evaluating the acceptability of substitutes, promoting the use of those substitutes believed to present lower overall risks to human health and the environment, relative to compounds being replaced, and to other substitutes for the same end use and prohibiting the use of substitutes found to increase overall risk.[45]

Manufacturers, formulators and end users introducing a substance, process, or product for sale, import, export, or use in industrial sectors that historically have used ozone-depleting substances, including refrigeration and air conditioning, foam blowing, solvent cleaning, fire suppression and explosion protection, aerosols, sterilants, tobacco expansion and adhesives, coatings and inks, may be required to submit information to the SNAP program.[46] Applicants must submit data on atmospheric effects, exposure assessments, toxicity, flammability and other environmental impacts (*e.g.*, ecotoxicity, local air impacts and impacts on aquatic life).

The EPA evaluates identified substitutes within a comparative risk framework, comparing alternative compounds with those of ozone-depleting compounds and the available alternatives. The environmental risk factors that are considered include ozone depletion potential, flammability, toxicity, occupational health and safety, and contributions to global warming and other environmental factors. Risk factors associated with quality of information, uncertainty of data and economic factors, including feasibility and availability, are also taken into account. Substitutes are also evaluated by use, as environmental and human health exposures can vary significantly depending on the particular application of a substitute. The EPA has developed specific risk screen methodologies for each industrial sector to analyze the information and determine the acceptability of a proposed replacement. When making decisions regarding the acceptability of substitutes, the EPA does not require that substitutes be risk free to be found acceptable, restricts only those substitutes that are significantly worse and defers to other environmental regulations when warranted.[47] To date, the EPA has identified hundreds of acceptable substitutes and tens of unacceptable substitutes for 37 end uses in industrial sectors using ozone-depleting substances.

In some cases, the EPA acknowledges that substances identified as acceptable alternatives under the SNAP program framework, while reducing the potential for ozone depletion, can still pose risks to human health and the environment. For example, the EPA approved methylene chloride as an acceptable alternative for its use as a foam blowing agent, a solvent in electronics cleaning, metals cleaning and precision cleaning, a solvent in adhesives, coatings and inks and an aerosol solvent, despite the fact that its 'higher toxicity' posed a 'potential risk to workers and residents in nearby communities.' This decision was justified by a desire to 'immediately transition from certain ozone-depleting substances as well as the belief that the risks to human health and the environment could be controlled by adhering to existing regulatory standards' (*i.e.*, State and local restrictions, workplace permissible exposure limits, RCRA waste disposal requirements, future regulation as a hazardous air pollutant under the Clean Air Act).[47]

11.5.5 Classification-based Substitution Requirements – European Union Regulation of Chemicals in the Workplace

The EU Chemical Agents Directive[48] and the Carcinogens and Mutagens Directive[49] provide examples of policies that require substitution for certain classes of chemicals. These regulations place a general duty on employers to replace the use of hazardous chemicals in the workplace with safer substitutes, whenever it is technically possible. As the Member States are responsible for implementing and enforcing these regulations, they have developed in different ways throughout the EU. This type of regulatory model provides a mechanism for the systematic assessment and identification of chemical hazards in the workplace and the continuous transition to less hazardous substances.

The Chemical Agents Directive requires employers to determine whether any hazardous chemical agents are present at the workplace, assess any risk to the safety and health of workers arising from the presence of those chemical agents and take necessary preventive measures where hazardous chemical agents are identified. Specifically, Article 6(2) details that 'substitution shall by preference be undertaken, whereby the employer shall avoid the use of a hazardous chemical agent by replacing it with a chemical agent or process which, under its condition of use, is not hazardous or less hazardous to workers' safety and health.' Similarly, Article 4(1) of the Carcinogens and Mutagens Directive requires employers to 'reduce the use of a carcinogen or mutagen at the place of work, in particular by replacing it, in so far as is technically possible, by a substance, preparation or process which, under its conditions of use, is not dangerous or is less dangerous to workers' health or safety.' Both directives utilize EU/GHS classifications to identify hazardous chemicals for which substitution is required.

The UK and Germany have developed their own regulatory frameworks, guidance documents and tools to support the employer's duty, as mandated in the EU Directives, to evaluate substitutes to hazardous substances and implement substitution where less hazardous alternatives are identified. For example, the UK has implemented the general duty to substitute under its Control of Substances Hazardous to Health Regulations (COSHH). COSHH requires that, where reasonably practical, employers prevent the exposure of employees to substances hazardous to health through substitution. In order to help employers meet these requirements, the UK Health and Safety Executive produced guidance on the topic of substitution, which describes a process to follow when considering alternatives to hazardous chemicals.[50,51]

In Germany, a regulatory framework, along with detailed guidance and tools to assist small- and medium-sized enterprises, has been developed. To implement the employer's duty to evaluate substitutes to hazardous substances and implement substitution where less hazardous alternatives are identified, as mandated in the German Hazardous Substances Ordinance,[52] the German Federal Institute for Occupational Safety and Health established guidance, TRGS 600,[53] which includes a framework for identifying and evaluating substitutes and establishes criteria and decision rules for assessing and comparing the health risks, physico-chemical risks and the technical suitability of identified alternatives. In order to facilitate the application of the TRGS 600 guidance in small and medium enterprises, the Institute for Occupational Safety and Health of the German Federation of Institutions for Statutory Accident Insurance and Prevention developed the Column Model tool.[54] The tool provides a scheme for evaluating and comparing chemicals based on six hazard categories using information obtained from chemical safety data sheets.

Despite the existence of these policies, employers are often not well equipped to compare chemicals and select suitable alternatives. As a result, the extent to which employers are able to implement effectively their duty to substitute and transition their workplaces to safer chemicals is unclear.[55]

11.6 Lessons Learned from Alternatives Assessment in Regulatory Policy

The five case studies and policies outlined above provide some important lessons to guide the development of government alternatives assessment policies in the future. A core lesson is that informed substitution cannot be achieved by a single policy; rather, a mix of policies that require evaluation of alternatives, regulatory incentives and disincentives (such as restrictions) and support structures that facilitate evaluation and adoption of safer alternatives is needed. Nonetheless, two of the policies outlined, the EU's REACH authorization process and the California Safer Consumer Products regulations, are relatively new and it is yet to be established whether their implementation will or will not drive the adoption of safer alternatives. The REACH authorization process, however, forms part of a broader chemicals management policy overhaul that includes requirements to classify chemicals, characterize uses through the supply chain and develop toxicological data. Some of the main lessons from our analysis of policies and case examples include the following:

- *To drive informed substitution, there is a need for policies that require evaluation of alternatives.* While there are many regulatory policies requiring the elimination of chemicals of concern, policies that require alternatives assessment are far less common. As noted, alternatives assessment processes have tended to be discretionary in nature, involving research, partnership projects, case studies, guidance development and voluntary technical support databases on chemicals. Such efforts are important and critical to supporting informed substitution but are not sufficient in and of themselves to encourage it. Voluntary initiatives in and of themselves have varied participation which tends to be dominated by leading edge firms, those already with a culture, willingness and interest in sustainability and transitioning to safer products and that see a benefit in participation.[11,56] While chemical restrictions alone may force laggards to remove a chemical of concern from a production process or product (particularly with drop-in substitutes), they are unlikely to force the investment in systems and processes to ensure informed substitution. Numerous analysts have documented the importance of regulatory requirements in encouraging innovation in a particular firm or sector. For example, Porter and van der Linde, showed through case studies that stringent but well-designed regulations are a critical component of innovation.[57] Regulation can provide signals about inefficiencies, can increase awareness within the firm, reduce uncertainty and stimulate progress.[12] Geiser suggested that laws requiring alternatives assessment are needed as there is little incentive for firms to plan or reduce toxics otherwise.[58] O'Rourke and Lee reported that facility planning requirements in Massachusetts have not only been instrumental in the success of the Massachusetts program in reducing chemical use and waste, but also led companies to establish structures and procedures that support future innovation, what they call 'command and innovate policy.'[59] While the incentives to substitute chemicals of concern are clearly changing, with greater market pressures on manufacturers, government-required substitution planning can help facilitate informed transitions to safer chemicals and improve the knowledge base on which substitution occurs.

- *Carefully designed policy incentives and disincentives can encourage the adoption of safer alternatives.* Although safer alternatives may be available for a particular chemical, economic viability, technical challenges or lack of markets may inhibit the willingness of some firms to adopt the alternative. As such, alternatives assessment requirements alone may not be sufficient to encourage firms to adopt safer substitutes (and they may not encourage firms to think broadly enough about the range of alternatives available). In the context of Environmental Impact Assessment, Steinemann noted that despite NEPA requirements for broad analysis of impacts and alternatives, such broad evaluations of alternatives do not tend to occur in practice.[60] This may in part be because there is no requirement to adopt the safest alternative for a particular project and the exercise may simply be an effort to justify an already made decision. In an analysis of implementation of the European Chemical Agents Directive, Lissner and Zayzon found that despite a policy mandate to substitute certain chemicals where feasible, such substitution does not occur regularly in practice due to a lack of incentives, guidance and tools for implementation.[55] O'Rourke and Lee found that despite the requirements to plan under the Toxics Use Reduction Act, many firms choose not to implement alternatives identified in their plans, particularly if the payback period is not sufficiently short.[59] This may also be the case where the alternatives are harder to implement and can open the firm to criticism for not acting fast enough. As such, tying alternatives assessment requirements with requirements for substitution of problem chemicals (or other incentives) may facilitate adoption. For example, many firms had undertaken assessment processes to evaluate alternatives to lead in electronics applications but had not adopted alternatives due to technical challenges. The adoption of the EU's Restrictions on Hazardous Substances, which prohibited the use of lead, provided the impetus to solve those challenges quickly and effectively in a pre-competitive sector-wide manner.[61] Similarly, governments can recognize or list safer alternatives as an incentive to stimulate the marketplace for safer materials, as demonstrated by the U.S. EPA's SNAP program and its DfE Safer Product Labeling program. In the latter case, the EPA provides the criteria for defining safer chemicals, an approach that can be used to stimulate the marketplace for alternatives for multiple functional uses of chemicals.

- *To support informed substitution, alternatives assessment requirements should be tied to government initiatives that support the adoption of safer alternatives.* Ashford[12] and O'Rourke and Lee[59] noted that technological innovation in safer chemicals requires three elements: motivation, capacity (or facilitation) and opportunities for change. Motivation and opportunity can be addressed through government policies that (1) provide strong incentives and disincentives, (2) establish clear goals and metrics for measuring progress and encouraging accountability and (3) establish mandated yet flexible procedures that require regulated entities to understand the impacts of technology changes. However, the fact that safer alternatives may be available is meaningless if these cannot be adopted in practice by a wide range of firms. While many firms may wish to adopt a safer technology, technical or institutional barriers may inhibit adoption. Capacity can be addressed through government initiatives that provide research, guidance, information, technical assistance, databases and networking of firms that support the adoption of safer chemicals. Massachusetts' experience with toxics use reduction indicates that carefully designed policies that require understanding chemical use and evaluation of alternatives are more successful when linked to technical support structures to facilitate adoption. Implementation of the European Chemical Agents Directive has been supported at the EU and Member State level with the development of tools, guidance and support for firms to identify safer alternatives to chemicals of concern. Where such support is available, substitution is more successful.[55]

- *There is a need for clear and consistent, yet flexible, definitions, guidance and criteria for alternatives assessment processes.* Despite the development of numerous policies that require substitution or evaluation of safer alternatives, there has been little consistency in how this process should occur. While the steps of alternatives assessment processes have been fairly consistent (from chemical and functional use prioritization, to product or process characterization to comparative evaluation to implementation), there is much less consistency in how alternatives should be compared. Lack of consistency may make it difficult for alternatives assessments produced in one region to be applicable in other regions. Hence having clear guidance and criteria that define a 'thorough' alternatives assessment is important. A key barrier to consistency in approaches is the lack of data on chemical use through supply chains (what chemicals are in what products, particularly for complex supply chains) and also data on chemical toxicity and hazards. Chemical toxicity data are important to understanding toxicity trade-offs between options. Life-cycle inventory data are important to understanding energy and materials use through product life-cycles. Consistency in the types of data required for alternatives assessment and how data gaps are addressed is important in ensuring compatibility between alternatives assessments conducted from one region to another.

The case examples and policies reviewed suggest some important differences in alternatives assessment processes, including:

- *How higher/lower concern is defined to determine a safer alternative.* Some policies, such as REACH and the SNAP program define higher or lower concern based on chemical risk or elimination of a particular risk (without clarity of what level of risk is acceptable for an alternative) whereas others, such as the Massachusetts Toxics Use Reduction program, PRIO and the EPA alternatives assessment program, focus on comparison of key hazard criteria, such as persistence and carcinogenicity. Even these comparative chemical assessment schemes can vary in the criteria used to compare chemicals.[28] The adoption of the Globally Harmonized System of Classification and Labeling may support greater harmonization of processes for comparing chemical alternatives.
- *How economic and technical feasibility should be evaluated.* Alternatives are not viable if they are too costly or do not adequately perform their functions. While many policies require regulated parties to consider economic and technical feasibility in the context of alternatives assessment processes, how these are addressed vary widely. Guidance developed under REACH, German Hazardous Substances Ordinance and the Massachusetts Toxics Use Reduction Act provide details as to how economic and technical feasibility should be addressed, whereas others do not.
- *How exposures should be considered.* Requirements to consider exposures (to consumers, workers, or the environment) vary between policies as to how these should be evaluated. The EPA alternatives assessment approach is primarily 'hazard based' (with persistence, bio-accumulation and physical chemistry characteristics serving as a surrogate for exposure), whereas others require a comparison of risks. Some include detailed exposure measures, whereas others evaluate surrogates or indicators for exposure – persistence and bio-accumulation, vapor pressure, use scenario, *etc.*
- *How impacts along a chemical/product life-cycle should be considered.* The draft California Safer Consumer Products regulation is the first substitution policy to require explicitly consideration of life-cycle impacts. Whereas the EPA SNAP program requires consideration of life-cycle trade-offs, through indicators such as ozone depletion and greenhouse gas emissions, the draft California regulations require the regulated party to evaluate explicitly life-cycle segments that might differ between

alternatives (for example, if one increases worker risks in manufacturing). Although there are extensive life-cycle evaluation methods available, many include limited evaluations of chemical hazards. An EU project, OMNIITOX, was initiated to support consideration of life-cycle impacts in the context of REACH implementation and the State of Washington is working on guidance for integration of life-cycle considerations in alternatives assessment processes.[62]

While it is important to develop policies that require alternatives assessment and ensure consistency in definitions and approaches, it is equally important that these alternatives assessment processes do not become overly prescriptive or burdensome with more scientific detail than is necessary to accomplish the goal of informing substitution. The history of existing decision-support processes, such as risk assessment and life-cycle assessment, offer some important warnings about the possible pitfalls that may lie ahead in the application of alternatives assessment requirements in policy. Ultimately, although careful consideration of alternatives to chemicals of concern is an important policy goal, alternatives assessment itself is simply a decision-support method or process not unlike risk assessment and life-cycle assessment and how it is used in policy can either promote or limit its promise in making more effective decisions.

We see ahead two potential missteps in the development and use of alternatives assessment in policy that are worth some cautious consideration. These involve the degree to which the methods and uses of alternatives assessment become too heavily weighed down by science or too formally constrained by regulation:

1. The promise of risk assessment particularly in chemical assessment applications has been compromised by the determination of some users and those who contest the decision outcomes to set overly high standards for the scientific evidence that can be used, to rely primarily on variables for which there is extensive research (while largely ignoring the many variables where there is little scientific study) and to discount large amounts of emerging science, novel test methods and thoughtful professional judgment. The calls for 'sound science' and 'rigorous science' are too often code terms for a narrow, reductionistic approach to using science to constrain decision-making.[63] By limiting focus to those variables where there is substantial research and then setting a very high bar for scientific evidence, some risk assessment initiatives have turned a laudable decision-assisting tool into a slow, costly and cumbersome drag on effective decision-making. The recent EPA federal risk assessments on trichloroethylene and dioxin provide telling illustrations of the costs of encumbering such a decision-assisting tool.[63] It is important that alternatives assessment policies address the use and misuse of science in conducting alternatives assessments and that agencies and other promoters develop tools that are flexible yet provide sufficient information to make informed decisions. Indeed, the amount of information needed

to compare two alternatives is different to the amount of information needed to determine whether a chemical is definitely safe or dangerous.

2. Overly detailed requirements in alternatives assessment regulations can inhibit the ability to complete such assessments and overburden both the regulated community and regulators themselves. The statutory requirements in California's Safer Consumer Product Act (AB 1879) requires the use of a life-cycle analysis that includes consideration of 13 specified variables in alternative assessments that will meet regulatory standards. As the regulations mature in California, it is important that they do not become too formulaic, complex and burdensome. Such a 'locked down' specification of the design and conduct of alternatives assessments can not only reduce the creative construction of a decision-assisting tool and render it inappropriate to the many and varied conditions that it may need to address, but also tend to convert the tool into a formulistic, compliance-oriented device that encourages a routinized exercise that is neither substantive, innovation generating, nor respected. Overly specifying the logic and applications that must, by mandate, be followed either by statute, government regulation, court dictate or private standard-setting body can drive out the creative, appropriate and accurate uses of the tool. For example, experience with Toxics Use Reduction planning in Massachusetts indicates that if procedures are too narrowly defined and inflexible, firms may complete the required process without necessarily engaging in thoughtful alternatives assessment.[64] The NEPA EIS process provides one model procedure for alternatives assessment, which lays out the components and minimum requirements but does not provide such specificity.

Promoters of alternatives assessment should be wary of efforts to narrow, formalize or routinize the tool too quickly in such a way that limits its flexibility and versatility. Such detailed requirements can also limit the ability of policies and the agencies that implement them to address more than a small number of the thousands of chemicals of concern in commerce. These potential misdevelopments could be advanced with the best of intentions by well-meaning promoters. Scientists might press for tighter scientific foundations and lawyers might press for clear and formulistic standards. If alternatives assessment is to become a truly reliable and effective decision-assistance tool in regulatory policies that supports the transition to safer chemistries, its promoters must work not only to improve its methods and applications, but also to resist these potentially damaging pressures.

11.7 Conclusion

While chemical restrictions and phase outs can lead to the adoption of safer alternatives, they can also lead to regrettable substitutions. As such, carefully designed chemicals policies that require the thorough evaluation of alternatives can help minimize the potential for such trade-offs and also increase the potential benefits of restrictive policies in supporting innovation in safer materials. If agencies' actions require or promote the phase out of a specific chemical, this creates a responsibility to oversee and support the transition to safer chemicals.

Alternatives assessment provides an important tool for promoting the informed transition to safer and more sustainable chemicals, materials and products. In its generic form it is nothing more than a logical, transparent and replicable set of methods for evaluating the desirability of one or several alternatives over a currently used chemical, material or product of concern. Alternatives assessment requirements encourage firms to understand the functional uses of chemicals of concern, their role in manufacturing processes and products, a wide range of options to reduce or eliminate a chemical of concern, the pros and cons of reasonable options and the opportunities and challenges to implementation. Combined with policies that provide incentives, disincentives, tools, support and information, these alternatives assessment requirements can move firms from simply examining options to their ultimate adoption.

Nonetheless, in some cases, technologically viable safer alternatives may not be available, may be difficult to implement in a particular process or may present risk trade-offs. In these cases, it may be useful to link alternatives evaluation with policies that support green chemistry research and development. Green chemistry is the design of inherently safer chemicals throughout their life-cycle.[65] For example, the California Safer Consumer Products draft regulations require companies to engage in green chemistry research when safer alternatives are not available. In Massachusetts, the Toxics Use Reduction Institute provides seed research support to academic researchers to investigate alternatives to chemicals of concern across sectors. The German Environmental Protection Agency's guidance on Sustainable Chemistry supports companies in the development of new molecules.

Although chemical restriction policies have not typically included alternatives assessment requirements, this seems to be changing. This is evident in the policies analyzed – which include various types of alternatives assessment requirements – and the increasing trend towards intergovernmental and industrial collaboration and cooperation in sharing information on alternatives and developing common approaches to alternatives assessment. It is likely that more collaboration will be needed to ensure that alternatives assessment requirements serve the goal of advancing the development, evaluation and adoption of safer chemicals and products without creating overburdensome requirements that inhibit the ability of such policies to be innovation and solutions stimulating. Given the relatively recent history of mandatory alternatives assessment requirements in the context of government chemicals restriction policies, there is a need for metrics in policies and post-implementation evaluation to understand what the most effective policy structures are to ensure that such policies support informed substitution and ultimately live up to their stated goals. Such analyses, which have been done for the

Massachusetts Toxics Use Reduction program[66] and the European Chemical Agents Directive,[55] provide the opportunity for authorities to take stock and evaluate the most effective means of integrating alternatives assessment requirements in policy as well as to adapt to new knowledge or changing circumstances.

References

1. US EPA, *Design for the Environment Safer Product Recognition Program Elements: a Discriminating and Protective Approach to Product Review and Recognition*, US EPA, Washington, DC, 2009.
2. J. Tickner and K. Geiser, *Environ. Impact Assess. Rev.*, 2004, **24**, 801–824.
3. M. O'Brien, *Making Better Environmental Decisions: an Alternative to Risk Assessment*, MIT Press, Cambridge, MA, 2000.
4. V. Brown, *Environ. Health Perspect.*, 2012, **120**, A280–A283.
5. H. M. Stapleton, J. G. Allen, S. M. Kelly, A. Konstantinov, S. Klosterhaus, D. Watkins, M. D. McClean and T. F. Webster, *Environ. Sci. Technol.*, 2008, **42**, 6910–6916.
6. California Office of Environmental Health Hazard Assessment, *Brominated and Chlorinated Organic Chemical Compounds Used as Flame Retardants. Materials for the December 4–5, 2008 Meeting of the California Environmental Contaminant Biomonitoring Program (CECBP) Scientific Guidance Panel (SGP)*, California Office of Environmental Health Hazard Assessment, Sacramento, CA, 2008.
7. C. Z. Yang, S. I. Yaniger, V. C. Jordan, D. J. Klein and G. D. Bittner, *Environ. Health Perspect.*, 2011, **119**, 989–996.
8. J. Tickner, *J. Epidemiol. Community Health*, 2011, **65**, 649–650.
9. D. Kriebel, M. Jacobs, J. Tickner and P. Markinnen, *Lessons Learned: Solutions for Workplace Safety and Health*, University of Massachusetts Lowell, Lowell, MA, 2011.
10. M. Wilson, S. K. Hammond, A. Hubbard and M. Nicas, *J. Occup. Environ. Hyg.*, 2007, **4**, 301–310.
11. N. Ashford and C. Caldart, *Technology, Law and the Working Environment*, Island Press, Washington, DC, 1996.
12. N. Ashford, in *Innovation-oriented Environmental Regulation: Theoretical Approach and Empirical Analysis*, ed. J. Hemmelskamp, K. Rennings and F. Leone (Proceedings of the International Conference of the European Commission Joint Research Centre, Potsdam, 27–29 May 1999), Springer, Berlin, 2000, pp. 67–107.
13. *Canadian Environmental Protection Act of 1999*, 1999.
14. US–Canada International Joint Commission, *Seventh Biennial Report on Great Lakes Water Quality*, 1994.
15. L. Baas, H. Hofman and H. Huising, *Protection of the North Sea: Time for Clean Production*, Erasmus Centre for Environmental Studies, Erasmus University, Rotterdam, 1990.

16. J. Tickner and K. Geiser, *New Directions in European Chemicals Policies: Drivers, Scope and Status*, Lowell Center for Sustainable Production, University of Massachusetts Lowell, Lowell, MA, 2003.

17. T. Colborn, J. P. Myers and D. Dumanoski, *Our Stolen Future*, Dutton Books, New York, 1996.

18. National Research Council, Committee on Pesticides in the Diets of Infants and Children, *Pesticides in the Diets of Infants and Children*, National Academies Press, Washington, DC, 1993.

19. J. Schifano, J. Tickner and Y. Torrie, *State Leadership in Formulating and Reforming Chemicals Policy: Actions Taken and Lessons Learned*, Lowell Center for Sustainable Production, University of Massachusetts Lowell, Lowell, MA, 2009.

20. *2009–2010 State Legislative Sessions Summary*, Lowell Center for Sustainable Production, University of Massachusetts Lowell, Lowell, MA, 2010.

21. A. Stone and D. Delistraty, *Environ. Impact Assess. Rev.*, 2010, **30**, 380–387.

22. Y. Lind, P. O. Darnerud, S. Atuma, M. Aune, W. Becker, R. Bjerselius, S. Cnattingius and A. Glynn, *Environ. Res.*, 2003, **93**, 186–194.

23. S. D. Grosse, T. D. Matte, J. Schwartz and R. J. Jackson, *Environ. Health Perspect.*, 2002, **110**, 563–569.

24. Council on Environmental Quality, *Regulations for Implementing the Procedural Provisions of the National Environmental Policy Act: 40 CFR 1500–1508*, Council on Environmental Quality, Washington, DC, 1992.

25. B. Wahlstrom, in *Protecting Public Health and the Environment: Implementing the Precautionary Principle*, ed. C. Raffensperger and J. Tickner, Island Press, Washington, DC, 1999, pp. 51–70.

26. B. Wahlstrom, *Nature*, 1989, **341**, 276.

27. K. Geiser, *Source Reduction Plans: a Proposal*, Tufts Center for Environmental Management, Medford, MA, 1985.

28. G. Davis, M. Swanson and S. Jones, *Comparative Evaluation of Chemical Ranking and Scoring Methodologies*, Center for Clean Technologies and Clean Products, University of Tennessee, Knoxville, TN, 1994.

29. L. Kincaid, J. Meline and G. Davis, *Cleaner Technologies Substitutes Assessments: a Methodology and Resource Guide*, EPA 744-R-95-002, US Environmental Protection Agency, Washington, DC, 1996.

30. L. W. Baas, *Cleaner Production and Industrial Ecology: Dynamic Aspects of the Introduction and Dissemination of New Concepts in Industrial Practice*, Eburon Academic Publishers, Delft, 2005.

31. J. Lohse, M. Wirts, A. Aherns, K. Heitmann, S. Lundie, L. Lissner and A. Wagner, *Substitution of Hazardous Chemicals in Products and Processes*, Report compiled for the Directorate General Environment, Nuclear Safety and Civil Protection of the Commission of the European Communities, Hamburg, 2003.

32. *SUBSPORT – Substitution Support Portal*, http://www.subsport.eu/.

33. P. Eliason and G. Morose, *J. Cleaner Prod.*, 2011, **19**, 517–526.

34. US EPA Design for the Environment Program, *Design for the Environ-ment Program Alternatives Assessment Criteria for Hazard Evaluation*, US EPA, Washington, DC, 2011.
35. US EPA Design for the Environment Program, *DfE's Safer Chemical Ingredient List*, US EPA, Washington, DC, 2012.
36. J. Tickner and M. Coffin, in *The Business of Sustainability: Trends, Policies, Practices and Stories of Success, I, II, III*, ed. S.G. McNall, J.C. Hershauer and G. Basile, Praeger Press, Westport, CT, 2011, pp. 123–143.
37. Organization for Economic Cooperation and Development, Environment Directorate, *The OECD Environment, Health and Safety Programme: Achievements, Strengths and Opportunities, 40 Years of Chemical Safety at OECD Planning for the Next Decade*, ENV/JM(2011)17, OECD, Paris, 2011.
38. *State Chemicals Policy Database*, Interstate Chemicals Clearinghouse, Boston, MA, 2012.
39. Massachusetts Executive Office of Energy and Environmental Affairs, *Toxics Use Reduction in Massachusetts: a Progress Report for FY 2011*, EOEEA, Boston, MA, 2011.
40. California Department of Toxic Substances Control, *Proposed Regulation for Safer Consumer Products [Draft]*, DTSC, Sacramento, CA, 2012.
41. European Commission, *EC Regulation on Registration, Evaluation, Authorization and Restriction of Chemicals (REACH)*, No. 1907/06, European Commission, Brussels, 2006.
42. European Chemicals Agency (ECHA), *Guidance on the Preparation of an Application for Authorization*, ECHA, Helsinki, 2011.
43. European Chemicals Agency (ECHA), *Guidance for the Preparation of an Annex XV Dossier for Restrictions*, ECHA, Helsinki, 2007.
44. *Safe Alternatives Policy*, 42 USC §7671k.
45. *Significant New Alternatives Policy Program*, 40 CFR §82.170.
46. US EPA, *Instructions for the Significant New Alternatives Policy (SNAP) Program Information Notice and TSCA/SNAP Addendum*, US EPA, Washington, DC, 2011.
47. *Protection of Stratospheric Ozone – Final Rule*, 59 FR 13044, 1994.
48. European Commission, *Directive on the Protection of the Health and Safety of Workers from the Risks Related to Chemical Agents at Work*, 98/24/EC, European Commission, Brussels, 1998.
49. European Commission, *Directive on the Protection of Workers from the Risks Related to Exposure to Carcinogens or Mutagens at Work*, 2004/37/EC, European Commission, Brussels, 2004.
50. UK Legislation, *The Control of Substances Hazardous to Health Regula-tions 2002*, No. 2677, 2002.
51. UK Health and Safety Executive, *COSHH and Substitution*, OC 273/17, HSE, Bootle, 2009.
52. Federal Institute for Occupational Safety and Health (BAuA), *Hazardous Substances Ordinance*, Germany, 2010.
53. German Federal Institute for Occupational Safety and Health, *Technical Rules for Hazardous Substances – Substitution*, TRGS 600, BAuA, Dortmund, 2008.

54. German Federation of Institutions for Statutory Accident Insurance and Prevention, *Column Model*, HVBG, Sankt Augustin, 2011.
55. L. Lissner and R. Zayzon, *Gefahrstoffe Reinhalt. Luft*, 2011, **71**, 247–254.
56. S. Arora, *J. Environ. Econ. Manage.*, 1995, **28**, 271–286.
57. M. E. Porter and C. van der Linde, *J. Econ. Perspect.*, 1995, **9**, 97–118.
58. K. Geiser, *Source Reduction Plans: a Proposal*, Tufts Center for Environmental Management, Medford, MA, 1985.
59. D. O'Rourke and E. Lee, *J. Environ. Planning Manage.*, 2004, **47**, 181–200.
60. A. Steinemann, *Environ. Impact Assess. Rev.*, 2001, **21**, 3–21.
61. *Clean Tech: an Agenda for a Health Economy. Preliminary Report*, Lowell Center for Sustainable Production, University of Massachusetts Lowell, Lowell, MA, 2007.
62. F. Christensen and S. Olsen, *J. Life-cycle Assess.*, 2004, **9**, 327–332.
63. J. Tickner, in *Risk Assessment in Public Health*, ed. W. Toscano and M. Robson, Jossey-Bass, San Francisco, 2006, 423–462.
64. R. Campbell, *Planning Without a Public: Legitimacy and Action in Toxics Use Reduction Planning*, University of Massachusetts Lowell, Lowell, MA, 1999.
65. P. T. Anastas and J. C. Warner, *Green Chemistry: Theory and Practice*, Oxford University Press, Oxford, 2000.
66. Massachusetts Toxics Use Reduction Institute, *Toxics Use Reduction Program Assessment*, TURI, University of Massachusetts Lowell, Lowell, MA, 2009.
67. SUBSPORT – Substitution Support Portal, Substitution in Legislation, 2010, http://www.subsport.eu/substitution-in-legislation/montreal-protocol-on-ozone-depleating-substances-ods (last accessed 13 August 2012).
68. UNEP, Ozone Secretariat, Technology and Economic Assessment Panel (TEAP), http://ozone.unep.org/new_site/en/assessment_panels_bodies.php?committee_id = 6 (last accessed 31 August 2012).
69. Danish EPA, Chemicals Action Plan 2010–2013, http://www.mst.dk/NR/rdonlyres/3B097825-39FB-4E30-94DB-04CAA2BC5A1C/0/chemicals_action_plan_20102013.pdf (last accessed 27 August 2012).
70. Catsub: Catalogue of Substitutions, http://www.catsub.dk/ (last accessed 27 August 2012).
71. US EPA, Design for the Environment (DfE), Alternatives Assessment, 2012, http://www.epa.gov/dfe/alternative_assessments.html (last accessed 14 August 2012).
72. US EPA, Existing Chemical Action Plans, 2012, http://www.epa.gov/oppt/existingchemicals/pubs/ecactionpln.html (last accessed 28 August 2012).
73. US EPA, Design for the Environment (DfE), Safer Product Labeling Program, 2012, http://www.epa.gov/dfe/pubs/projects/formulat/saferproductlabeling.htm (last accessed 27 September 2012).
74. US EPA, Design for the Environment (DfE), Design for the Environment Program Alternatives Assessment Criteria for Hazard Evaluation,

Version 2.0, 2011, http://www.epa.gov/dfe/alternatives_assessment_criteria_for_hazard_eval.pdf (last accessed 28 August 2012).

75. German Federal Environment Agency, Guide on Sustainable Chemicals, 2011, http://www.umweltdaten.de/publikationen/fpdf-l/4169.pdf (last accessed 27 September 2012).

76. SUBSPORT – Substitution Support Portal, Substitution in Legislation, 2011, http://www.subsport.eu/substitution-in-legislation/stockholm-convention-on-persistent-organic-pollutants-pops (last accessed 13 August 2012).

77. UN, POPs Free Initiative, http://chm.pops.int/Programmes/POPsfreeinitiative/tabid/2194/language/en-US/Default.aspx, 2008 (last accessed 13 August 2012).

78. UN, Fifth Meeting of the Persistent Organic Pollutants Review Committee (POPRC5), Meeting Documents, General Guidance on Considerations Related to Alternatives and Substitutes for Listed Persistent Organic Pollutants and Candidate Chemicals, http://chm.pops.int/Convention/POPsReviewCommittee/POPRCMeetings/POPRC5/POPRC5-Documents/tabid/592/Default.aspx, 2009 (last accessed 13 August 2012).

79. European Agency for Safety and Health at Work, How to Convey OSH Information Effectively – The Case of Dangerous Substances, 2003 http://osha.europa.eu/en/publications/reports/312 (27 August 2012).

80. SUBSPORT – Substitution Support Portal, Substitution Tools, Quick Scan, http://www.subsport.eu/substitution-tools/quick-scan (last accessed 27 August 2012).

81. Swedish Environment Ministry, Sweden's Environmental Quality Objectives, http://www.sweden.gov.se/sb/d/5775 (last accessed 30 August 2012).

82. Swedish Chemicals Agency – KEMI, Restricted Substances Database, 2012, http://apps.kemi.se/begransningsdatabas/index.aspx?sprak = en (last accessed 27 August 2012).

83. Swedish Chemicals Agency – KEMI, The Substitution Principle, Report No 8/07, 2007, http://www.kemi.se/Documents/Publikationer/Trycksaker/Rapporter/Report8_07_The_Substitution_Principle.pdf (last accessed 27 August 2012).

84. Swedish Chemicals Agency – KEMI, Prio – A Tool for Risk Reduction of Chemicals, http://www2.kemi.se/templates/PRIOEngframes____4144.aspx (last accessed 27 August 2012).

85. UK Department for Environment Food and Rural Affairs (DEFRA), UK Chemicals Stakeholder Forum, 2012, http://www.defra.gov.uk/chemicals-forum/ (last accessed 27 August 2012).

86. UK Chemicals Stakeholder Forum (UKCSF), A Guide to Substitution: An Information Note form the UK Chemicals Stakeholder Forum, 2010, http://archive.defra.gov.uk/environment/quality/chemicals/csf/documents/forum-guide-substitution.pdf (last accessed 27 August 2012).

87. Consumer Product Safety Improvement Act of 2008, http://www.cpsc.gov/cpsia.pdf (last accessed 28 August 2012).

88. Washington Department of Ecology, Reducing Toxic Threats, http://www.ecy.wa.gov/toxics/index.htm (last accessed 27 September 2012).

89. Washington Department of Ecology, Pollution Prevention, Assessing the Safety of Chemical Alternatives, http://www.ecy.wa.gov/programs/hwtr/ChemAlternatives/index.html (last accessed 14 August 2012).

90. European Agency for Safety and Health at Work, Directive 98/24/EC – Risks Related to Chemical Agents at Work, http://osha.europa.eu/en/legislation/directives/exposure-to-chemical-agents-and-chemical-safety/osh-directives/75 (last accessed 27 August 2012).

91. European Agency for Safety and Health at Work, Directive 2004/37/EC – Carcinogens or Mutagens at Work, http://osha.europa.eu/en/legislation/directives/exposure-to-chemical-agents-and-chemical-safety/osh-directives/directive-2004-37-ec-indicative-occupational-exposure-limit-values (last accessed 27 August 2012).

92. Cardiff Work Environment Research Centre (CWERC), Chemical Agents Directive Implementation (CADImple), http://www.cf.ac.uk/cwerc/research/cadimple/index.html (last accessed 13 August 2012).

93. *Official Journal of the European Union*, Regulation (EC) No 1223/2009 of the European Parliament and of the Council of 30 November 2009 on Cosmetic Products, http://eur-lex.europa.eu/LexUriServ/LexUriServ.do?uri = OJ:L:2009:342:0059:0209:en:PDF (last accessed 27 August 2012).

94. *Official Journal of the European Union*, Directive 2009/48/EC of the European Parliament and of the Council of 18 June 2009 on the Safety of Toys, http://eur-lex.europa.eu/LexUriServ/LexUriServ.do?uri = OJ:L:2009:170:0001:0037:en:PDF (last accessed 27 August 2012).

95. www.subsitution-cmr.fr (last accessed 27 August 2012) (website in French, translated using Google).

96. Committee on Hazardous Substances, AGS Management, BAuA, Technical Rules for Hazardous Substances, Substitution, TGRS 600, 2008, http://www.baua.de/en/Topics-from-A-to-Z/Hazardous-Substances/TRGS/pdf/TRGS-600.pdf;jsessionid = E90D80C2234B2C564F24E85C-CD6753EB.1_cid253?__blob = publicationFile&v = 3 (last accessed 27 August 2012).

97. UK Health and Safety Executive, COSHH Basics, Substance Substitution, http://www.hse.gov.uk/coshh/basics/substitution.htm (last accessed 1 September 2012).

98. UK Health and Safety Executive, COSHH Essentials, http://www.hse.gov.uk/coshh/essentials/index.htm (last accessed 1 September 2012).

99. Toxic Use Reduction Institute (TURI), Five Chemicals Alternatives Assessment Study, 2006, http://www.turi.org/About/Library/TURI_Publications/2006_Five_Chemicals_Alternatives_Assessment_Study (last accessed 3 September 2012).

100. Toxic Use Reduction Institute (TURI), P2OASys Tool to Compare Materials, http://www.turi.org/About/Library/TURI_Publications/P2OASys_Tool_to_Compare_Materials (last accessed 3 September 2012).

101. Ontario Regulation 455/09 of the Toxics Reduction Act 2009, http://www.e-laws.gov.on.ca/html/regs/english/elaws_regs_090455_e.htm#BK25 (last accessed 27 August 2012).

102. Ontario Toxics Reduction Program, Reference Tool for Assessing Safer Chemical Alternatives, 2012, http://www.ene.gov.on.ca/stdprodconsume/groups/lr/@ene/@resources/documents/resource/stdprod_095226.pdf (last accessed 27 August 2012).

103. Subsport – Substitution Support Portal, Substitution in Legislation, China, Chinese Law on Promotion of Clean Production, http://www.subsport.eu/substitution-in-legislation/chinese-law-on-promotion-of-clean-production (last accessed 14 August 2012).

104. European Commission, REACH, 2012, http://ec.europa.eu/environment/chemicals/reach/reach_intro.htm (last accessed 13 August 2012).

105. ECHA, Guidance on the Preparation of an Application for Authorisation (p. 45), http://echa.europa.eu/documents/10162/13637/authorisation_application_en.pdf, 2011 (last accessed 13 August 2012).

106. ECHA, Guidance for the Preparation of an Annex XV Dossier for Restrictions (pp. 68–75), http://echa.europa.eu/documents/10162/13641/restriction_en.pdf, 2007 (last accessed 13 August 2012).

107. *Official Journal of the European Union*, Regulation (EU) No 528/2012 of the European Parliament and of the Council of 22 May 2012 Concerning the Making Available on the Market and Use of Biocidal Products http://eur-lex.europa.eu/LexUriServ/LexUriServ.do?uri = OJ:L:2012:167:0001:0123:EN:PDF (last accessed 27 August 2012).

108. California Department of Toxic Substances Controls (DTSC), Safer Consumer Products Regulations, 2012, http://www.dtsc.ca.gov/SCPRegulations.cfm (last accessed 14 August 2012).

109. Minnesota Department of Health, Toxic Free Kids Act, Reports, December 2010, http://www.health.state.mn.us/divs/eh/hazardous/topics/toxfreekids/reports.html (last accessed 3 September 2012).

110. SUBSPORT – Substitution Support Portal, Substitution in Legislation, 2010, http://www.subsport.eu/substitution-in-legislation/voc-solvents-directive (last accessed 13 August 2012).

111. European Commission, VOC Solvents Directive, 2012, http://ec.europa.eu/environment/air/pollutants/stationary/solvents/guidance_en.htm (last accessed 13 August 2012).

112. *Official Journal of the European Union*, Directive 2002/95/EC of the European Parliament and of the Council of 27 January 2003 On the Restriction of the Use of Certain Hazardous Substances in Electrical and Electronic Equipment, http://eur-lex.europa.eu/LexUriServ/LexUriServ.do?uri = OJ:L:2003:037:0019:0023:en:PDF (last accessed 27 August 2012).

113. Consumer Product Safety Improvement Act of 2008, http://www.cpsc.gov/cpsia.pdf (last accessed 28 August 2012).

114. US Consumer Product Safety Commission (CPSA), Chronic Hazard Advisory Panel (CHAP) on Phthalates, http://www.cpsc.gov/about/cpsia/chapmain.html (last accessed 15 August 2012).

115. SUBSPORT – Substitution Support Portal, Substitution in Legislation, China, Administration on the Control of Pollution Caused by Electronic Information Products (refers to Joint Ministerial Decree No. 39), http://www.subsport.eu/substitution-in-legislation/chinese-rohs (last accessed 14 August 2012).
116. Lowell Center for Sustainable Production, Chemicals Policy and Science Initiative, US State chemicals Policy, http://www.chemicalspolicy.org/chemicalspolicy.us.state.database.php (last accessed 28 August 2012).
117. US EPA, Significant New Alternatives Policy (SNAP) Program, 2012, http://www.epa.gov/ozone/snap/about.html#q4 (last accessed 14 August 2012).
118. US EPA, Instructions for the Significant New Alternatives Policy (SNAP) Program Information Notice and TSCA/SNAP Addendum, September 2011, http://www.epa.gov/ozone/snap/submit/appguide.pdf (last accessed 1 September 2012).
119. Executive Order No. 4, Establishing a State Green Procurement and Agency Sustainability Program, 2008, http://www.ogs.state.ny.us/purchase/spg/pdfdocs/EO4.pdf (last accessed 27 September 2012).
120. New York State Office of General Services, Green Cleaning Program, https://greencleaning.ny.gov/Policies.asp (last accessed 27 September 2012).

Subject Index